# Antioxidants and Oxidative Stability in Fats and Oils

# Antioxidants and Oxidative Stability in Fats and Oils

Guest Editor

**Joaquin Velasco**

MDPI

Basel • Beijing • Wuhan • Barcelona • Belgrade • Novi Sad • Cluj • Manchester

*Guest Editor*
Joaquin Velasco
Lipid Characterization and
Quality
Instituto de la Grasa (CSIC)
Seville
Spain

*Editorial Office*
MDPI AG
Grosspeteranlage 5
4052 Basel, Switzerland

This is a reprint of the Special Issue, published open access by the journal *Antioxidants* (ISSN 2076-3921), freely accessible at: www.mdpi.com/journal/antioxidants/special_issues/antioxidant_fats_oils.

For citation purposes, cite each article independently as indicated on the article page online and using the guide below:

Lastname, A.A.; Lastname, B.B. Article Title. *Journal Name* **Year**, *Volume Number*, Page Range.

**ISBN 978-3-7258-4198-1 (Hbk)**
**ISBN 978-3-7258-4197-4 (PDF)**
**https://doi.org/10.3390/books978-3-7258-4197-4**

# Contents

# About the Editor

**Joaquin Velasco**

Joaquín Velasco is a Senior Researcher at the Instituto de la Grasa–CSIC (Spain) and currently heads the Department of Lipid Characterization and Quality. His research focuses on the chemical, technological, and analytical aspects of lipids, with the primary objective of ensuring food quality and extending shelf life. He leads research efforts aimed at elucidating the physicochemical mechanisms and factors involved in the oxidative degradation of fats, with particular emphasis on the role of antioxidants in preventing food rancidity. His recent work explores the role of spontaneously formed colloidal structures in fats and oils and their influence on antioxidant mechanisms. He primarily employs Electron Paramagnetic Resonance (EPR) spectroscopy to assess lipid oxidation and antioxidant efficacy. One of his main goals is to improve the quality of confectionery and baked goods through the development of effective antioxidant strategies.

# Preface

Oxidative degradation remains a major challenge in food production, processing, and storage. As saturated fats are increasingly replaced by vegetable oils and polyunsaturated lipids are incorporated into foods, the need to develop and understand effective antioxidant systems has become urgent. This reprint offers an exploration of the chemistry, functionality, and recent innovations related to phenolic antioxidants and other lipid stabilization strategies.

We begin by examining the paradoxical behavior of phenolic compounds with varying polarities, such as $\alpha$-tocopherol and Trolox, in lipid systems containing phosphatidylcholine reverse micelles. These nuanced studies reveal that antioxidant behavior in real oil systems may deviate from simplified model predictions, emphasizing the need for context-specific insights.

Building on this foundation, this reprint presents a kinetic and mechanistic analysis of phenolic antioxidants. Structural evaluations provide crucial information on radical stability, chain propagation, and antioxidant efficacy, factors essential for designing the next generation of protective agents.

This volume also illustrates an integrated approach to antioxidant strategies in diverse edible oils, linking mechanisms to practical applications. Studies on cold-pressed black cumin oils highlight variability in composition and oxidative stability due to processing. Investigations into thymoquinone and thymohydroquinone show how molecular structure shapes antioxidant performance. Flavoring virgin olive oil with herbs like rosemary enhances both stability and aroma, while rosemary extracts, and particularly carnosic acid, effectively protect the novel Acer truncatum seed oil.

Beyond bulk oils, this reprint examines emulsions, nanoemulsions, and encapsulated lipid systems. A theoretical framework is reviewed for understanding antioxidant partitioning across phases, a key determinant of efficacy. In $\beta$-lactoglobulin-stabilized emulsions, phenolic acid derivatives improve oxidative stability depending on pH and molecular structure. Another study explores potato protein hydrolysates as multifunctional agents, showing improved emulsion physicochemical stability. In nanoemulsions of red raspberry seed oil, surfactant selection and the use of natural antioxidants like oregano and oak extracts offer sustainable alternatives to synthetic compounds.

This volume concludes with the development of biopolymer films enhanced with green carbon dots derived from apple pomace and rosemary. These films illustrate a multifunctional, biodegradable approach to oxidative protection.

This reprint is intended for researchers, technologists, and graduate students in food science, chemistry, biotechnology, and materials science, as well as industry professionals aiming to improve product stability and shelf life through efficient antioxidant systems.

**Joaquin Velasco**
*Guest Editor*

*antioxidants*

MDPI

*Article*

# New Insights into the Loss of Antioxidant Effectiveness of Phenolic Compounds in Vegetable Oils in the Presence of Phosphatidylcholine

**Joaquín Velasco [1,*], María-Jesús Gil [1], Yun-Qi Wen [2], Aída García-González [1] and María-Victoria Ruiz-Méndez [1]**

[1] Instituto de la Grasa, Consejo Superior de Investigaciones Científicas (CSIC), Ctra. de Utrera, km 1, 41013 Sevilla, Spain; mjgil@ig.csic.es (M.-J.G.); aida.garcia@csic.es (A.G.-G.); mvruiz@ig.csic.es (M.-V.R.-M.)

[2] College of Food Science and Technology, Ocean University of China, No. 5 Yu Shan Road, Qingdao 266003, China; wenyq0715@163.com

* Correspondence: jvelasco@ig.csic.es

**Abstract:** It has been proposed that lipid oxidation reactions in edible oils primarily occur in reverse micelles (RM) of amphiphilic components. While the prooxidative effect of RM has been demonstrated, the mechanism involved is not fully understood. Both reductions and enhancements in the antioxidant efficacy (AE) of α-tocopherol and Trolox have been observed in different studies when phosphatidylcholine (PC) was added and PC RM were formed. However, most of these investigations employed lipid systems consisting of stripped vegetable oil diluted in saturated medium-chain triacylglycerols (MCT) and utilized antioxidant concentrations well below those found in edible oils. These two specific factors were investigated in the present study. The effect of RM of purified egg yolk PC on the AE of 1.16 mmol $kg^{-1}$ α-tocopherol or Trolox in stripped sunflower oil (SSO) was studied by the Rancimat (100 °C) and oven (50 °C) tests. Increasing PC concentrations (50–1000 ppm) had no significant impact on α-tocopherol, but substantial reductions in AE were observed for Trolox. This phenomenon may be attributed to the partitioning of Trolox into the pre-existing PC micelles, suggesting that primary oxidation reactions occurred in the continuous lipid phase. In addition, the effectiveness of both antioxidants decreased significantly in the presence of PC when a low antioxidant concentration (0.06 mmol $kg^{-1}$) was assayed in SSO:MCT (1:3, *w*/*w*).

**Keywords:** reverse micelles; lipid oxidation; antioxidant; phosphatidylcholine; α-tocopherol; Trolox; sunflower oil; Rancimat

## 1. Introduction

Edible oils comprise triacylglycerols in around 98 wt% and a variety of minor components, included sterols, hydrocarbons, free fatty acids, diacylglycerols, monoacylglycerols, tocopherols, and phospholipids. Many of these are amphiphilic and they can associate with water trace, whose contents can oscillate between 200 and 2000 mg $kg^{-1}$ in refined oils [1]. In turn, they self-aggregate to form reverse micelles, in which the surface-active molecules orientate with their polar heads towards the aqueous core of the micelles and the nonpolar tails towards the apolar continuous phase [1]. These nanoscale colloidal units change the physical and chemical properties of oils and have an impact on lipid oxidation [2,3]. Through complex reactions, unsaturated fatty acyl chains react with oxygen and give rise to a number of compounds that impart off-flavors and make the oil unacceptable to consumers [4].

It has been demonstrated that reverse micelles in oils play a catalytic role in lipid oxidation [2]. It was suggested that hydroperoxides accumulate at the interfacial region of these colloids due to their interfacial activity and decompose into free radicals by the action of transition metals concentrated at the interface. However, the prooxidative mechanism involved is not fully understood. Different results have been obtained in a number of reports. Dioleylphosphatidylcholine (DOPC) association colloids promoted both primary and secondary lipid oxidation in purified or stripped soybean oil (SSbO), and mixtures of SSbO or stripped corn oil (SCO) with saturated medium-chain triacylglycerols (MCT) [5–9]. In addition, a mixture of surface-active components commonly present in oils, such as DOPC, dioleylphosphatidylethanolamine (DOPE), diolein, oleic acid, and stigmasterol, also formed reverse micelles in blends of SCO and MCT and promoted lipid oxidation at concentrations above the critical micelle concentration (CMC) [10]. In contrast, no effects were found for similar mixed reverse micelles in stripped high oleic or high linoleic safflower oils [11].

It has been stated that the greater antioxidative efficacy of polar antioxidants in bulk oils, compared to lipophilic antioxidants, may be attributed to their greater affinity for the water–oil interfaces created by reverse micelles of oil minor components, where oxidative reactions appear to predominate [1,2].

A comprehensive understanding of the physicochemical mechanisms of antioxidants in bulk oils is of utmost importance to enhance their effectiveness and achieve better control of lipid oxidation. Interactions between antioxidants and association colloids, and particularly their placement in the micellar structure, can significantly impact their antioxidant efficacy [2]. Koga and Terao postulated that phospholipids enhanced the accessibility of α-tocopherol to the free radicals generated by a water-soluble radical initiator in a bulk oil model system, improving its antioxidant activity [12]. They affirmed that α-tocopherol was able to scavenge the chain-initiating free radicals generated in an aqueous microenvironment and the chain-propagating peroxyl radicals in the continuous lipid phase.

Studies on the effect of reverse micelles on the efficacy of antioxidants, such as α-tocopherol and its synthetic polar homologue Trolox, have shown varying results depending on the antioxidant concentration. DOPC reverse micelles improved the activity of low α-tocopherol or Trolox concentrations (10 μM) in a blend of SSbO and MCT (1:3, *w/w*) [6], but decreased the activity of high antioxidant concentrations (100 μM) [6,13,14]. A synthetic phospholipid unable to form reverse micelles, such as 1,2-dibutyryl-*sn*-glycero-3-phosphocholine ($DC_4PC$), also decreased the antioxidant activity of both α-tocopherol or Trolox in SSbO:MCT (1:3, *w/w*) at low (10 μM) and high (100 μM) concentrations [6].

For convenience, a common practice in these studies is the use of a stripped vegetable oil diluted in MCT oil as a lipid substrate (e.g., [6,13,14]). This approach allows one to prepare a large volume of samples in a short time because MCT oil does not require any tedious purification step to remove naturally occurring antioxidants and other amphiphilic minor oil components. In addition to the purification advantage, the use of MCT allows testing antioxidants at very low concentrations. The degree of unsaturation is considerably reduced with the addition of MCT, which comprises saturated medium-chain fatty acids, resulting in increased oxidative stability. To shorten the time scale of oxidation experiments, antioxidants are used at very low concentrations. This model system may not accurately represent common vegetable oils, which typically contain much higher concentrations of natural antioxidants, such as tocopherols. For example, the concentration of α-tocopherol in refined sunflower oil ranges between 400 and 900 mg $kg^{-1}$ [15] or between 854 μM and 1922 μM, considering that the oil density is 0.92 kg $L^{-1}$ [15]. More importantly, the interactions between antioxidants and reverse micelles may be overstated due to the extremely high differences in concentrations between reverse micelle components and antioxidants.

The main goal of the present work was to study in more detail the effect of reverse micelles on the antioxidant efficacy of phenolic antioxidants in bulk oil using a model system which was closer to refined vegetable oils. Neat stripped sunflower oil (SSO) was chosen as lipid substrate, and purified egg yolk phosphatidylcholine (PC) was selected as the surfactant. PC was chosen because it plays a major role as a surface-active component in the formation of reverse micelles in oils [10]. The effect of increasing concentrations of PC on the effectiveness of α-tocopherol at concentrations normally found in sunflower oil was studied. For comparative purposes, Trolox, the polar homologue of α-tocopherol, was also investigated. The Rancimat test at 100 °C was the method applied to evaluate the effectiveness of antioxidants and results were validated by oxidation experiments in an oven at 50–60 °C (oven test). For comparative purposes, the effect of PC on the effectiveness of the antioxidants was also evaluated at conditions that were similar to those in reported studies [6,12–14]. In this regard, a very low concentration of the antioxidants (0.06 mmol $kg^{-1}$) was assayed in SSO:MCT (1:3, *w/w*).

## 2. Materials and Methods

### 2.1. Materials

High linoleic sunflower oil was acquired from a local supermarket. Pure C8 medium chain triacylglycerol (MCT) oil (100% coconut sourced) was purchased from Ketosource B.V. (Amsterdam, The Netherlands). (±)-α-Tocopherol (>97%), (±)-6-hydroxy-2,5,7,8-tetramethylchormane-2-carboxylic acid (Trolox) (97%), gallic acid, propyl gallate (≥98%), dodecyl gallate, L-α-phosphatidylcholine from egg yolk (~60%), 7,7,8,8-tetracyanoquinodimethane (TCNQ) (98%), and hydranal® Coulomat oil were acquired from Sigma-Aldrich (St. Louis, MO, USA). Aluminum oxide 90 active neutral was purchased from Merck KGaA (Daemstadt, Germany). n-Hexane (95%) (HPLC grade), chloroform stabilized with 0.5–0.8% ethanol (HPLC grade), methanol for UV, IR, HPLC, ACS, n-heptane (HPLC grade), and anhydrous methanol (max. 0.005% water) were obtained from Panreac Química S.L.U. (Barcelona, Spain), diethyl ether stabilized with 1% (*v/v*) ethanol (Super purity solvent, HPLC grade) was from Romil, LTD (Cambridge, UK), 2-propanol (gradient grade for HPLC) from Merck KGaA (Daemstadt, Germany), and dichloromethane (≥99.8%, HPLC grade) from Honeywell GmbH (Seelze, Germany). Except for the L-α-phosphatidylcholine, the chemicals and solvents were used as received.

### 2.2. Oil Stripping Procedure

In order to remove the minor components naturally present in the oil, the sunflower oil was purified by adsorption chromatography using a glass column packed with aluminum oxide according to Fuster et al. [16]. Briefly, the oil was dissolved in an equal volume of n-hexane and passed through the column containing alumina activated at 200 °C during 20 h and conditioned with a volume of n-hexane equal to its mass. Nitrogen was bubbled through the eluate over the process. The solvent was removed in a Büchi R-210 rotary evaporator (Büchi Ibérica, Barcelona, Spain) at 35 °C and finally by bubbling the oil with nitrogen at room temperature. The oil was protected from light with aluminum foil during all the procedure and it was preserved under nitrogen at −20 °C until analyses. According to Fuster et al. [16], the purification method gives rise to an oil containing undetectable amounts of tocopherols, peroxides, and iron and copper levels below 0.01 and 0.001 ppm, respectively. The absence of tocopherols [17], hydroperoxides [18], free fatty acids [19], and mono- and di-acylglycerols [19] were verified in the present study by HPLC.

### 2.3. Phosphatidylcholine Purification

L-α-phosphatidylcholine (PC) from egg yolk was selected owing to its high level of PC (~60%) and low price. It was purified by adsorption chromatography using a glass column packed with alumina 90 according to Singleton et al. [20]. The alumina was not activated and used as received. The elution solvent was chloroform:methanol 9:1 (*v/v*). This was also used for packing and conditioning the alumina. A fraction collector was used and the progress of the fractionation was followed by thin layer chromatography (TLC), employing 50% sulfuric acid to reveal the spots. The solvent was evaporated in the rotary evaporator at 35 °C and finally with a nitrogen stream at room temperature until dryness. The PC so obtained was chromatographically pure. This was preserved under nitrogen at −20 °C until analysis. The fatty acid composition of PC was determined by standard IUPAC methods 2.301 and 2.302, applying the special preparation of the methyl esters of fatty acids with potassium hydroxide in methanol at room temperature [21]. The fatty acid composition of PC did not change with the purification and their main fatty acids were 0.14% C14:0, 30.49% C16:0, 0.26% C16:1 n-9, 1.26% C16:1 n-7, 0.16% C17:0, 12.29% C18:0, 28.70% C18:1, 17.10% C18:2, 0.32% C18:3, 0.27% C20:3 n-6, 4.36% C20:4 n-6, 0.27% C22:4 n-6, 1.51% C22:5 n-6, and 1.26% C22:6 n-3.

### 2.4. Sample Preparation

Stripped sunflower oil (SSO) or a blend of SSO and MCT oil (SSO-MCT) in proportions 1:1 (*w*/*w*) or 1:3 (*w*/*w*) were the lipid substrates used in this study. The MCT oil was not purified, but it was dried at 80 °C and 100 mbar in a Büchi R-210 rotary evaporator (Büchi Ibérica, Barcelona, Spain) due to its elevated moisture content (ca. 1000 mg $kg^{-1}$). The moisture content was reduced to 50 mg $kg^{-1}$, which was virtually the same as that measured for the SSO.

The antioxidants were added to the oils using solutions in organic solvents. The antioxidant concentrations of the solutions were adjusted to add only 1 mL solution to an amount of oil that ranged between 40 and 240 g. α-Tocopherol was dissolved in n-heptane, whereas the rest of the antioxidants tested were dissolved in anhydrous methanol. In order to avoid unnecessary additional variables, 1 mL of anhydrous methanol was added to the oils stabilized with α-tocopherol and 1 mL of n-heptane to the rest of the samples when antioxidants were compared. After the antioxidant addition, the oils were stirred for 10 min under nitrogen and the solvent was removed at 35 °C and vacuum conditions (100 mbar) in the rotary evaporator.

The PC was dissolved in the oils using a dichloromethane solution of the phospholipid (40 mg $mL^{-1}$). After addition of the PC, vigorous stirring was applied under nitrogen for 10 min. The solvent was removed at 35 °C and vacuum conditions (100 mbar) in the rotary evaporator. From concentrations above 1000 mg $kg^{-1}$, the PC was not completely soluble in the SSO and the samples presented noticeable turbidity and solid particles in suspension. PC concentrations above the solubility limit were only assayed in the Rancimat test.

In the last step, the moisture content in the oils was adjusted to 200 ppm mg $kg^{-1}$. Deionized Milli-Q water was added and vigorous stirring was applied under nitrogen for 5 h. The moisture content was measured using a C-20 Coulometric Karl-Fischer Titrator from Mettler Toledo S.A.E. (Barcelona, Spain). Hydranal®—Coulomat oil from Sigma-Aldrich (St. Louis, MO, USA) was the reactive employed.

### 2.5. Determination of the Critical Micelle Concentration (CMC)

The CMC was measured by applying the spectrophotometric assay proposed by Kanamoto, Wada, Miyajima, and Kito [22]. It is based on charge transfer interactions between TCNQ and phospholipids. TCNQ has a low solubility in oils, but this increases in the presence of reverse micelles. The absorbance of TCNQ at 480 nm increases drastically when phospholipids reverse micelles form. The CMC was taken as the break in the curve between the absorbance and the logarithm of phospholipid concentration.

### 2.6. Rancimat Test

The oxidative stability index (OSI) was determined at 100 °C or 70 °C (SSO with no antioxidants) in the Rancimat test using a 679 Rancimat device from Metrohm (Herisau, Switzerland), according to AOCS Cd 12b-92 method [23]. An amount of 2.5 g oil and 20 L $h^{-1}$ air were used.

### 2.7. Oven Test

The oil samples (15 g) were placed into 250-mL amber glass bottles and these were tightly closed with a lid and screw cap. The specific area of the oil samples was 1 $cm^{-1}$. Each sample was oxidized in triplicate. The bottles were placed in an oven at 50 °C or 60 °C (±0.1) and sample aliquots were taken periodically. The oxidation was monitored by the direct analysis of triglycerides containing linoleate hydroperoxides or hydroperoxy-dienes by HPLC-DAD following the method proposed by Velasco et al. [18] as a measure of primary oxidation. The HPLC chromatograph, the chromatographic conditions, and hydroperoxydiene quantitation can be found in detail in [18]. In addition, triglyceride polymers were also analyzed by High-Performance Size-Exclusion Chromatography (HPSEC) as a measure of advanced oxidation, applying ISO 16931:2009 standard method [19]. A 1200 AT HPLC chromatograph from Agilent Technologies (Agilent Technologies, Palo Alto, CA, USA) equipped with two Agilent HPSEC columns containing respectively 100- and 500-Å particles of highly cross-linked styrene-divinylbenzene copolymers, connected in series, and a refractive index detector was used. The separation was performed at 35 °C. Tetrahydrofuran at 1 mL min$^{-1}$ was the elution solvent. The HPSEC chromatograms were also employed to confirm the absence of di- and mono-acylglycerols, and free fatty acids in the SSO. In fact, oil compounds with lower molecular weight than triacylglycerols were not detected in the analysis.

### 2.8. Statistical Analyses

All experiments were run in triplicate. Results were expressed as means and standard deviations ($n = 3$). One-factor analysis of variance was applied. Levene's test based on the mean value was used to verify that variances were equal across groups. For multiple mean comparisons, Duncan's test was used. Comparisons between two samples were made by Student's *t*-test. Significance was defined at $p < 0.05$. The analyses were performed using 29.0 IBM SPSS Statistics program (IBM Corp., Chicago, IL, USA).

## 3. Results

### 3.1. Effect of PC on the Antioxidant Effectiveness of α-Tocopherol and Trolox in SSO

α-Tocopherol is by far the major and most important chain-breaking antioxidant in refined sunflower seed oil, with a typical concentration ranging between 400 and 900 mg kg$^{-1}$ [15]. During the refining process of vegetable oils, phospholipids are removed to levels below 10 mg kg$^{-1}$ of phosphorus. This limit corresponds to 230 mg kg$^{-1}$ of phospholipids when applying the conversion factor of 23.0 proposed by Carelli et al. [24] to transform phosphorus into phospholipid content in degummed sunflower oils. This estimate is coherent with the content of 254 mg kg$^{-1}$ obtained when phosphorus is expressed as dioleylphosphatidylcholine, whose molecular weight is 786.1. Therefore, levels higher than 300 mg kg$^{-1}$ of phospholipids are not expected in refined oils.

The antioxidant efficacy (AE) of α-tocopherol in stripped sunflower oil (SSO) containing varying concentrations of pure egg yolk phosphatidylcholine (PC) was evaluated. The concentration of the antioxidant tested was that normally found in refined sunflower oil (500 mg kg$^{-1}$), whereas that of the phospholipid ranged between 50 to 10,000 mg kg$^{-1}$, i.e., from levels that can be found in commercial oils to those much higher than expected. For comparative purposes, Trolox, the hydrophilic counterpart of α-tocopherol, was also tested at the same experimental conditions, using the same molar concentration of 1.16 mmol kg$^{-1}$ as α-tocopherol.

The Rancimat test run at 100 °C was one of the methods selected in this study to evaluate the AE, because it allows one to make a number of experiments in a relatively short time. The results showed that the oxidative stability index (OSI) of the oil containing α-tocopherol did not significantly change in the presence of PC in a wide range of phospholipid concentrations (50–5000 mg kg$^{-1}$) (Figure 1). A slight increase was only observed at the highest PC concentration tested, when a great amount of the phospholipid was not dissolved and the sample presented solid particles in suspension. The solubility limit of PC in the oil was found to be 1000 mg kg$^{-1}$. Conversely, the OSI of the oil containing Trolox substantially decreased in the presence of PC in a concentration-dependent way (Figure 1). Nevertheless, Trolox was more effective than α-tocopherol when the PC was completely dissolved in the oil. The reductions in effectiveness observed for Trolox were consistent with most studies in this field, which reported the prooxidative activity of PC at levels above the critical micelle concentration (CMC) [3,5,6]. In contrast, the results obtained for α-tocopherol at a concentration normally found in commercial sunflower oil differ from those by other authors, who assayed the antioxidant at much lower concentrations [6,13,14].

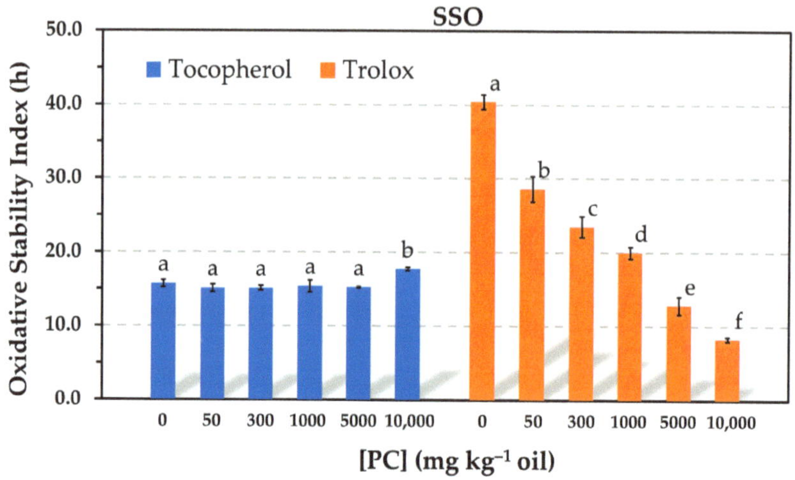


**Figure 1.** Influence of the phosphatidylcholine (PC) concentration on the oxidative stability index determined in the Rancimat test at 100 °C in stripped sunflower oil containing 1.16 mmol kg$^{-1}$ α-tocopherol (500 mg kg$^{-1}$) or Trolox (290 mg kg$^{-1}$). Data represent means and standard deviations ($n$ = 3). For a given antioxidant, different letters indicate significant differences according to Duncan's test ($p < 0.05$).

The Schaal oven test was the other method used to evaluate the AE. This method is commonly used in studies addressed to investigate the role of reverse micelles of amphiphilic oil components in lipid oxidation. Oxidation experiments are normally run at 55 °C, and the levels of hydroperoxides and hexanal are measured to evaluate the extent of oxidation [6,13,14]. The Rancimat test applies higher temperatures to accelerate lipid oxidation, normally 100 °C for sunflower oils and other common vegetable oils [25]. In addition, the oil is oxidized by continuous forced aeration, which is achieved by bubbling air through it. To ensure non-limiting oxygen conditions, a considerable flow of air (15–20 L h$^{-1}$) is applied, which makes the oil oxidize in a rough continuous movement. These particular dynamic conditions differ from the ones in the oven test, where the oil is oxidized at static conditions. Consequently, owing to the different experimental conditions applied in the Rancimat and oven tests, the AE of α-tocopherol and Trolox in the presence of PC was also evaluated at 50 °C by heating selected samples in an oven. The levels of hydroperoxides were measured to determine the oxidation extent. The results obtained at 50 °C were consistent with those by the Rancimat test. The levels of hydroperoxides in the sample containing α-tocopherol were not affected by the addition of PC at concentrations of 300 and 1000 mg kg$^{-1}$, whereas hydroperoxides increased more rapidly with the addition of PC in the sample containing Trolox, showing a clear reduction in AE that was dependent on the PC concentration (Figure 2). Therefore, these results supported those obtained by the Rancimat test.

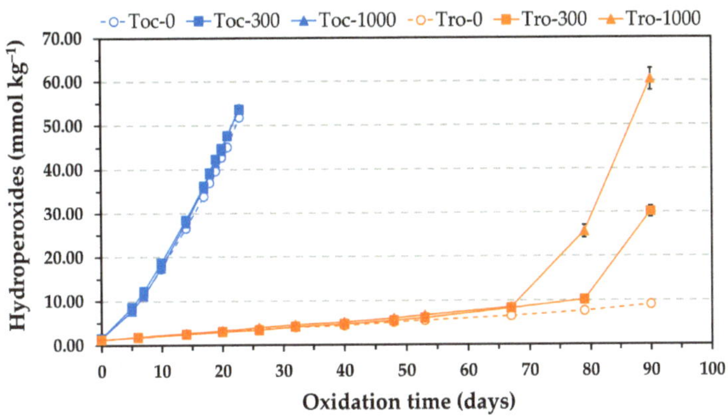


**Figure 2.** Influence of the phosphatidylcholine (PC) concentration on the formation of hydroperoxides during the oxidation at 50 °C of stripped sunflower oil containing 1.16 mmol kg$^{-1}$ α-tocopherol (500 mg kg$^{-1}$) or Trolox (290 mg kg$^{-1}$). Toc and Tro in the sample codes represent α-tocopherol and Trolox, respectively, and the number after the hyphen indicates the PC concentration expressed in mg kg$^{-1}$. Data represent means and standard deviations (*n* = 3). Some of the error bars are within data points.

In order to evaluate whether the antioxidants had an impact on the PC reverse micelle formation at such a relatively high concentration (1.16 mmol kg$^{-1}$), so far undocumented, the CMC was determined in SSO at the temperatures used in the Rancimat and oven tests. The TCNQ method applied presented a few issues. In the absence of antioxidants, the samples rapidly faded and no clear changes in color were noticed over a wide range of PC concentrations (0–1000 mg kg$^{-1}$), neither at 50 °C nor at 100 °C. While a sharp rise in the absorbance was found in the presence of Trolox when the PC concentration was increased, the absorbance only increased moderately, and also unevenly, in the presence of α-tocopherol, which made it difficult to determine the CMC (Figure S1). The CMC did not show changes with temperature, and it was found to be 60–80 mg kg$^{-1}$ PC when Trolox was tested. The absorbance increments also appeared to occur in the same range of PC concentrations in the presence of α-tocopherol. Even though the reactant was more soluble in the presence of Trolox when the PC micelles were formed, as higher absorbances were found compared to α-tocopherol (Figure S1), these results suggest that the type of antioxidant did not affect the CMC. Therefore, the CMC determined using the TCNQ method was unaffected by both the type of antioxidant and the temperature (50 °C compared to 100 °C).

The CMC found in this study is coherent with the result reported by Laguerre et al. [26] for dioleylphosphatidylcholine (DOPC) in stripped corn oil (SCO), which was found to be 65 μM at 55 °C, or 56 mg kg$^{-1}$ considering that the density of SCO is 0.92 kg L$^{-1}$. By small-angle X-ray scattering (SAXS) measurements, Chen et al. [6] found that the presence of α-tocopherol or Trolox, assayed at much lower concentrations (100 μM) than that of the present study (1.16 mmol kg$^{-1}$ or 1067 μM considering that the density of SSO is 0.92 kg L$^{-1}$), did not have any impact on the structure of DOPC reverse micelles in stripped soybean oil.

Despite the limitations of the TCNQ method to measure accurately the CMC, which is based upon the intersection of two linear regression curves that depend on the number and selection of data points, the results of this study exhibited that the antioxidant activity of Trolox was significantly reduced even when the PC was tested at a concentration below the CMC and that the antioxidant activity of α-tocopherol was not affected even in the presence of PC reverse micelles. Compared to studies reported by other authors, the main difference of the present study is that the antioxidants were tested at much higher concentrations to study a model that is closer to a commercial oil. It is evident that interactions between the antioxidants and phospholipids depend on their relative concentrations. In order to study the impact of PC on the antioxidant activity of α-tocopherol in more detail, similar experiments were run in the Rancimat test using decreased α-tocopherol concentrations. When the antioxidant concentration was reduced ten-fold, the PC had an effect on the oil stability, but this was only marginal. Very slight but significant losses of stability were detected in the presence of PC. The same was true when the antioxidant concentration was halved (Figure 3).

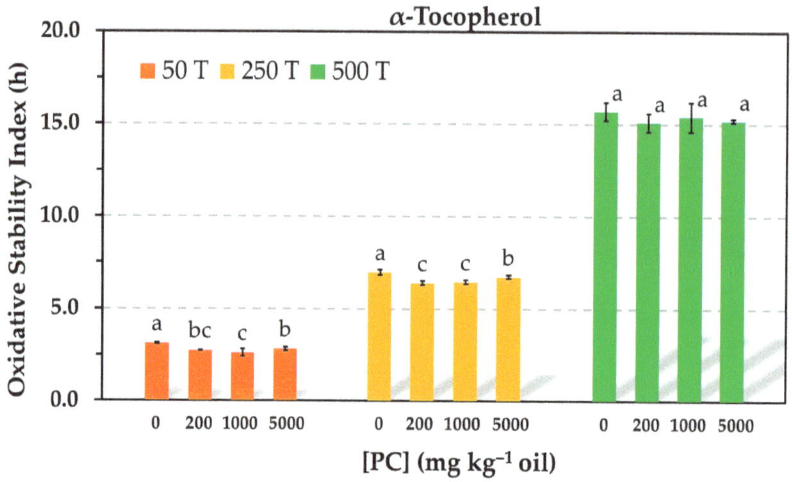


**Figure 3.** Influence of the phosphatidylcholine (PC) concentration on the oxidative stability index determined in the Rancimat test at 100 °C in stripped sunflower oil containing 50 (50 T), 250 (250 T), or 500 (500 T) mg kg$^{-1}$ α-tocopherol. Data represent means and standard deviations ($n$ = 3). For a given antioxidant concentration, different letters indicate significant differences according to Duncan's test ($p < 0.05$).

### 3.2. *Effect of PC on the Antioxidant Effectiveness of α-Tocopherol and Trolox at Reduced Concentration in a Blend of SSO and MCT (1:3, w/w)*

Concentrations of α-tocopherol in SSO below 50 mg kg$^{-1}$ were ruled out in the Rancimat test owing to the expected low OSI values. According to the AOCS official method [23], time periods less than 4 h in the Rancimat test result in a wider variation of the end point determination. To test a lower concentration of the antioxidant in a fast way and so confirm the loss of effectiveness of α-tocopherol in the presence of PC, the SSO was diluted in saturated medium-chain triacylglycerols (MCT) in a 1:3 (*w*/*w*) proportion as reported elsewhere [6,10,13]. The MCT oil used mainly comprised caprylic acid (C8), which is a saturated fatty acid. The dilution of SSO in such a saturated oil considerably increases the oxidative stability due to the resulting reduction in the degree of unsaturation. The antioxidant concentration assayed was 0.06 mmol kg$^{-1}$, which is an intermediate concentration compared to those used by other authors [6,13,14]. For comparative purposes, Trolox was also tested at the same experimental conditions. The AE was evaluated by the Rancimat test at 100 °C and by the oven test, this time at 60 °C to shorten the time of the experiments. Hydroperoxides were measured to determine the oxidation extent. In addition, the levels of polymers were also analyzed because substantial polymerization is a marker of the onset of advanced or uninhibited oxidation, i.e., when the antioxidants are exhausted [27,28], in a similar way to the measure upon which is based the Rancimat test. Results showed that, similar to Trolox, the antioxidant effectiveness of α-tocopherol also decreased significantly in the presence of PC in a concentration-dependent way, but this time α-tocopherol was more effective in the presence of PC than Trolox

(Figure 4). The induction periods prior to the polymerization were coherent with the OSI values by the Rancimat test. Both tests provided the same relative order of stabilities, i.e., Tro-0 > Toc-0 > Toc-50 > Tro-50 > Toc-1000 > Tro-1000, where Tro and Toc represent Trolox and α-tocopherol, respectively, and the number indicates the PC concentration in mg $kg^{-1}$.

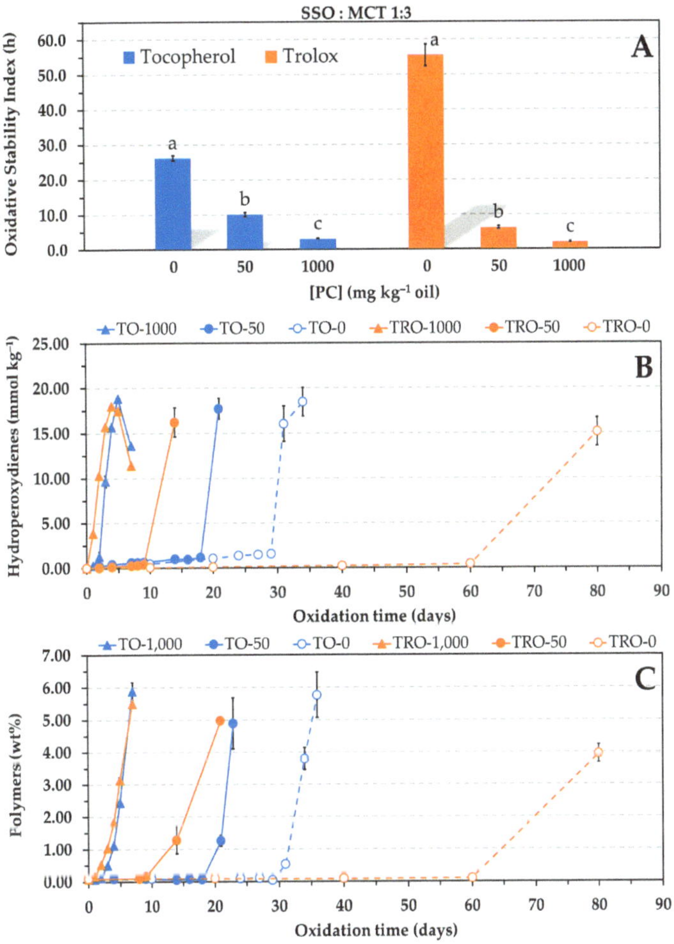


**Figure 4.** Influence of the phosphatidylcholine (PC) concentration on the antioxidant efficacy of α-tocopherol and Trolox at reduced concentration (0.06 mmol $kg^{-1}$) in a blend of SSO and MCT (1:3, *w*/*w*). Oxidative stability index determined in the Rancimat test at 100 °C (**A**). Formation of hydroperoxides (**B**) and polymers (**C**) at 60 °C in an oven. Data represent means and standard deviations ($n = 3$). For a given antioxidant, different letters in (**A**) indicate significant differences according to Duncan's test ($p < 0.05$). Some of the error bars in (**B**,**C**) are within data points.

The CMC of PC in the blend SSO-MCT was determined in the presence of 0.06 mmol $kg^{-1}$ α-tocopherol. The result obtained was 80 mg $kg^{-1}$ at 60 °C and 100 °C. Thus, it was similar to that found in SSO containing 1.16 mmol $kg^{-1}$ Trolox. A sharp increase in absorbance was observed this time in the presence of tocopherol. Therefore, similar to Trolox, the protecting effect of α-tocopherol was also drastically reduced even at a PC concentration below the CMC.

To assess how the relative concentrations of the antioxidants and PC affected the AE, the molar concentration of PC was estimated using the molecular weight of the dioleylphosphatidylcholine (786.1). In the range of PC solubility (up to 1000 mg kg$^{-1}$), the molar concentration ratios between the antioxidants and PC in Figures 1 and 2 were 18, 3, and 1, respectively. Thus, the concentration of the antioxidants was nearly 20 times higher than that of the PC when the phospholipid was tested at 50 mg kg$^{-1}$. When the concentration of α-tocopherol was ten times reduced in Figure 3, the concentration of the PC was approximately 2 and 10 times higher than that of the antioxidant when it was assayed at 200 and 1000 mg kg$^{-1}$, respectively. At these conditions, the PC only had a marginal effect. However, a drastic decrease in the AE of α-tocopherol was found in the blend SSO-MCT when it was assayed at a lower concentration and even when this was approximately the same as that of the PC (50 mg kg$^{-1}$) (Figure 4). This effect was not observed in SSO when the molar concentration of α-tocopherol was similar to that of PC (1000 mg kg$^{-1}$) in Figure 1. Similarly, when the molar concentration of PC was ca. 10 times higher compared to α-tocopherol in SSO in Figure 3 (50T-1000 or 250T-5000), the decrease in the OSI was minimal. Consequently, these results suggested that the lipid medium also played a role.

### 3.3. *Effect of PC on the Antioxidant Effectiveness of α-Tocopherol in Different Lipid Substrates*

The effect of PC on the antioxidant effectiveness of 500 mg kg$^{-1}$ α-tocopherol was studied in samples containing different proportions of SSO and MCT. The phospholipid was tested at 1000 mg kg$^{-1}$ to ensure the presence of reverse micelles. The relative molar concentrations of PC and α-tocopherol was ca. 1:1. Results obtained by the Rancimat test showed that when the SSO was diluted with MCT, the PC caused a significant decrease in the AE of α-tocopherol and this was more pronounced in the sample containing more MCT (Figure 5).

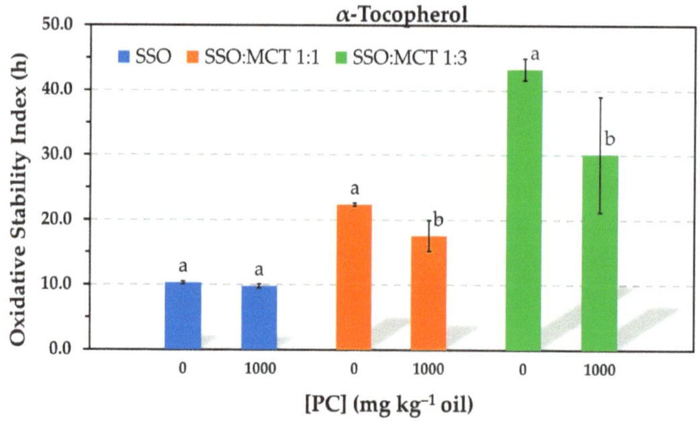


**Figure 5.** Influence of the addition of 1000 mg kg$^{-1}$phosphatidylcholine (PC) on the antioxidant efficacy of 1.16 mmol kg$^{-1}$ α-tocopherol in stripped sunflower oil (SSO), and 1:1 (*w*/*w*) and 1:3 (*w*/*w*) blends of SSO with MCT, as determined by the Rancimat test at 100 °C. Data represent means ($n$ = 3) and standard deviations. For a given lipid medium, different letters indicate significant differences according to Student's $t$-test ($p < 0.05$).

Similar results were also observed at 50 °C in the oven test, but the effect of PC, as determined by hydroperoxides measurements, was not as pronounced as the Rancimat test showed. Results for hydroperoxides only exhibited clear differences between the control sample and that containing the PC in the lipid mixture containing more MCT (Figure 6). Oxidation occurred more rapidly in the presence of PC, indicating a reduction in the AE of α-tocopherol when the SSO was diluted with MCT.

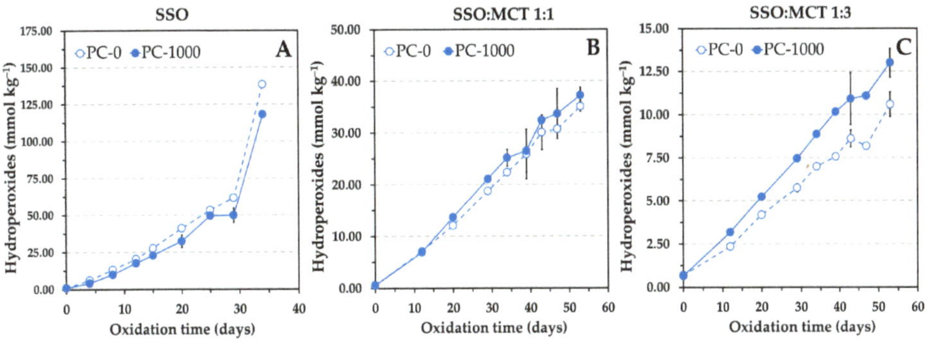


**Figure 6.** Influence of the addition of 1000 mg kg$^{-1}$ phosphatidylcholine (PC) on the formation of hydroperoxides at 50 °C in different lipid substrates containing 1.16 mmol kg$^{-1}$ α-tocopherol, namely, stripped sunflower oil (SSO) (**A**), and 1:1 (*w*/*w*) (**B**) and 1:3 (*w*/*w*) (**C**) blends of SSO with MCT. Data represent means and standard deviations (*n* = 3). Some of the error bars are within data points.

### *3.4. Effect of PC on the Antioxidant Efficacy of Gallic Acid, Propyl Gallate, and Dodecyl Gallate in SSO*

Gallic acid (GA) is a natural phenolic antioxidant found in a number of fruits and in teas [29]. Its molecular structure includes three hydroxy groups at the 3, 4, and 5 positions of the aromatic ring and, similar to Trolox, a carboxylic group. These structural elements contribute to its high polarity and water solubility [30]. GA was also tested in the present study to investigate the significant reductions in AE observed for Trolox in the presence of PC. Specifically, the examination aimed to determine if the carboxylic group played a role in the pronounced effect of PC on Trolox, and GA was employed as a model for this evaluation. For comparative purposes, propyl gallate (PG) and dodecyl gallate (DG) were also examined. In both of these antioxidants, the carboxylic group is esterified and, consequently, blocked with alkyl alcohols, which reduces the polarity of GA. In the case of DG, the alcohol also imparts amphiphilic properties to it. All three antioxidants were evaluated in the Rancimat test at an equivalent molar concentration of 0.59 mmol kg$^{-1}$. Considering the same CMC found for the PC in SSO containing Trolox (60–80 mg kg$^{-1}$), the impact of PC on the AE of GA and its esters was assessed using PC concentrations below (50 mg kg$^{-1}$) and above (1000 mg kg$^{-1}$) the CMC.

Significant reductions in AE were observed for all three antioxidants when the PC concentration exceeded the CMC. However, when the concentration of PC was below the CMC, significant differences were only observed for GA. The OSI also decreased by the addition of 50 mg kg$^{-1}$ PC, showing a similar pattern to that found for Trolox in Figure 1. In contrast, the OSI of the samples containing the ester derivatives of GA was unaffected by the addition of 50 mg kg$^{-1}$ of PC (Figure 7).

### *3.5. Influence of PC Concentration on the Oxidative Stability of Neat Sunflower Oil (Non-Purified)*

To contrast the fact that the addition of PC over a wide range of concentrations had no impact on the AE of α-tocopherol when this was tested at a concentration normally found in sunflower oils (Figures 1 and 2), the effect of PC on the oxidative stability of neat or non-purified sunflower oil was investigated by the Rancimat test. Results showed a small effect on OSI in the concentration range in which the PC was soluble in the oil (up to 1000 mg kg$^{-1}$). A very slight but significant increase was observed (Figure 8). In contrast, the OSI was reduced when the PC concentration was as high as 5000 mg kg$^{-1}$. Similarly, when the effect of PC was studied on SSO in the absence of antioxidants, the OSI also decreased moderately, showing a prooxidant role of PC reverse micelles, as reported in various studies [3,5,6].

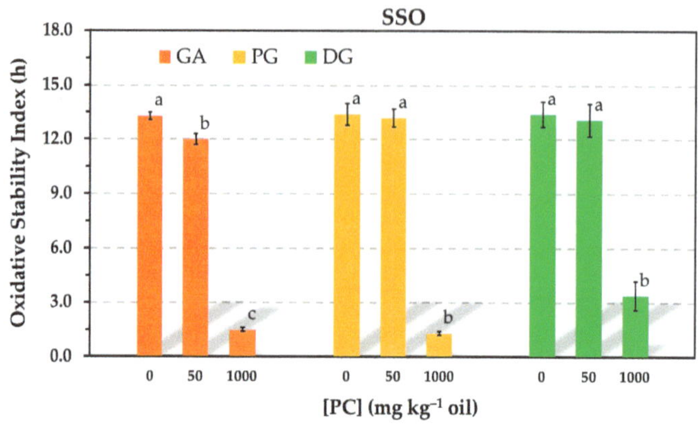


**Figure 7.** Influence of the phosphatidylcholine (PC) concentration on the oxidative stability index determined in the Rancimat test at 100 °C in stripped sunflower oil containing 0.59 mmol $kg^{-1}$ gallic acid (GA), propyl gallate (PG) or dodecyl gallate (DG). Data represent means and standard deviations ($n$ = 3). Different letters for a given antioxidant indicate significant differences according to Duncan's test ($p < 0.05$).

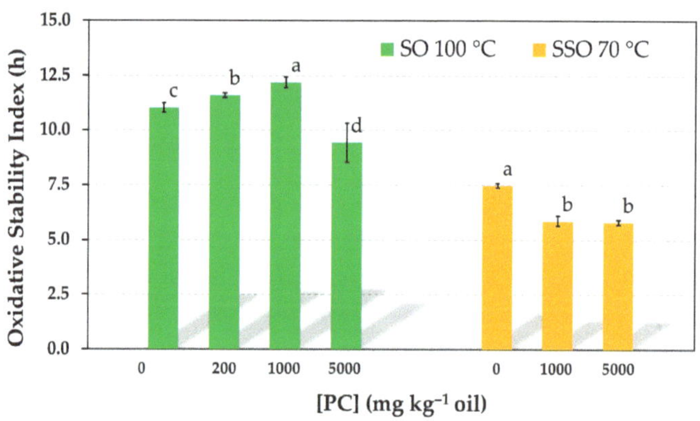


**Figure 8.** Influence of the phosphatidylcholine (PC) concentration on the oxidative stability index of refined sunflower oil (SO) and stripped sunflower oil in the absence of antioxidants determined in the Rancimat test at 100 °C and 70 °C, respectively. Data represent means and standard deviations ($n$ = 3). Different letters indicate significant differences according to Duncan's test ($p < 0.05$).

## 4. Discussion

The prooxidative role of reverse micelles in bulk oils has been demonstrated in numerous studies (e.g., [3,5,6]). It has been suggested that association colloids increase the prooxidative activity of transition metals [7]. Phospholipids can concentrate at the oil–water interface and attract metals owing to their capability of producing negatively charged interfacial regions [31]. These interfacial metals can interact with lipid hydroperoxides that can migrate and locate at the interface owing to their interfacial activity, catalyzing the hydroperoxide decomposition into free radicals [11]. Therefore, reverse micelles appear to facilitate the interactions between metals and hydroperoxides.

The results obtained in this study for α-tocopherol, as shown in Figures 1 and 2, where the AE remained unchanged in the presence of PC, suggest that the antioxidant somehow prevented the interactions between hydroperoxides and metals. In addition, these results also indicate that the polyunsaturated fatty acyl chains of the PC did not have a significant effect on the oxidation rate of the SSO, probably due to the relatively low PC concentrations.

When Trolox was tested, losses of AE were observed in the presence of PC, but their efficacy was still higher compared to α-tocopherol. To explain the different effects observed for the PC on the effectiveness of these antioxidants, we postulate the hypothesis that main oxidation reactions did not necessarily occur at pre-existing PC reverse micelles, but at reverse micelles mainly formed by hydroperoxides and other oxidation compounds. Lipid hydroperoxides could mainly form in the continuous lipid phase, in a similar way to when the PC is not present in a purified oil. Hydroperoxides are also amphiphilic molecules with interfacial activity that can associate with other interfacial components like the PC to form mixed reverse micelles and/or associate themselves to form pure micelles, as it has been suggested by other authors [3,14,32,33]. Budilarto and Kamal-Eldin [34] demonstrated the formation of colloids during the course of oxidation of purified sunflower or canola oil, i.e., in the absence of pre-existing reverse micelles. They found that the water content and the size of reverse micelles formed by hydroperoxides and other oxidation compounds increased during the induction period, defined as the period prior to an exponential increase of hydroperoxides. The increase in the water content was attributed to the bimolecular decomposition of hydroperoxides.

The capability of hydroperoxides and other lipid oxidation products to self-aggregate in the presence of pre-existing PC reversed micelles could be justified by their less surface activity [3,11]. It is thus evident that such interactions are going to depend on the hydroperoxide concentrations and that they would be more favored as oxidation proceeds. The greater affinity of polar antioxidants for these new interfacial colloids, where lipid oxidation compounds are concentrated and hypothetically lipid radicals are mainly formed, would explain their greater antioxidant effectiveness compared to lipophilic antioxidants. This was definitely the case of Trolox compared to α-tocopherol when these antioxidants were assayed in the absence of PC in the present study and in other works [6,13].

When the PC was added, the results of this study point out that the more polar antioxidant was less available to inhibit lipid oxidation, probably due to a greater affinity for PC compared to the less polar antioxidant. In this regard, Chen et al. [6] demonstrated by fluorescence measurements the incorporation of Trolox into DOPC reverse micelles. It was stated that Trolox partitioned into the water core of the micelles due to its water solubility. The partitioning of Trolox between micelles of PC and micelles of triacylglycerol-hydroperoxides appears to account for the loss of its AE.

Regarding α-tocopherol, its lower affinity for the PC micelles seemed not to affect its location to a great extent and so its availability to inhibit lipid oxidation. The low affinity of α-tocopherol for DOPC reverse micelles has been demonstrated in different studies. Compared to Trolox, Chen et al. [6] found a lower interaction of α-tocopherol with a fluorescence probe, namely, *N*-(7-nitrobenz-2-oxa-1,3-diazol-4-yl)-1,2-dihexadecanoyl-*sn*-glycero-3-phosphoethanolamine (NBD-PE), embedded into DOPC micelles. They stated that α-tocopherol has virtually no water solubility and, unlike Trolox, it cannot partition into the water phase. At best, it would align at the oil–water interface. Similarly, Rokosik et al. [14] found no interactions of α-tocopherol with DOPC reverse micelles by fluorescence measurements. The fluorescence intensity of α-tocopherol did not change in the presence of DOPC. In addition, the fluorescence intensity of the probe NBD-PE was not affected by the presence of α-tocopherol in a concentration range of 1–500 μmol $kg^{-1}$.

The losses in AE found for α-tocopherol in the presence of PC when the antioxidant concentration was reduced in Figure 4 could have been due to its low and ineffective concentration in the continuous lipid phase as a result of its association with PC. At the antioxidant concentration assayed (0.06 mmol $kg^{-1}$), the PC concentration had a greater effect on the effectiveness of Trolox (Figure 4), and α-tocopherol was more effective. This fact can be attributed to the greater affinity of Trolox for the PC. It has been suggested that partitioning of α-tocopherol and Trolox into DOPC micelles could allow the antioxidants to reduce the transition metals to their more prooxidative states and increase the oxidation rate, which would be seen as a decrease in the antioxidant effectiveness [6]. However, this effect was not observed when α-tocopherol was tested at high concentration in Figures 1 and 2.

Most of the studies on the effect of reverse micelles on lipid oxidation employ low antioxidant concentrations to shorten the experiments. As a result, the interactions of antioxidants with reverse micelles are usually extremely favored due to the great differences in concentrations between reverse micelle components and antioxidants. Furthermore, MCT oil is sometimes added to a stripped vegetable oil to have a greater volume of sample. In the present study, it was found that this practice also favored the prooxidative effect of reverse micelles (Figures 5 and 6). The reduction in AE of α-tocopherol in Figures 5 and 6 could be related to an increase in polarity of the lipid medium, since the dielectric constant of MCT oil is a bit larger compared to sunflower oil [35]. The differences in polarity with the lipid medium could determine the migration and location of antioxidants and lipid oxidation compounds and affect their affinities for reverse micelles.

The PC also had an impact on the AE when its concentration was below the CMC. In this regard, losses in AE were observed for Trolox in Figure 1 or for α-tocopherol and Trolox in Figure 4. These results are consistent with the reductions in AE reported for both α-tocopherol and Trolox in SSbO:MCT (1:3, *w*/*w*) due to the addition of 1,2-dibutyryl-*sn*-glycero-3-phosphocholine ($DC_4PC$), a synthetic phospholipid that cannot form reverse micelles [6]. Furthermore, it was hypothesized in this study that Trolox could form a complex with PC through the carboxylic group, similar to what has been described for free fatty acids [36]. Results obtained for GA and its esters when the PC was added at a concentration below the CMC, considered to be 60–80 mg $kg^{-1}$, pointed out a role of the carboxylic group (Figure 7). While the AE of GA was reduced in the presence of 50 mg $kg^{-1}$ of PC, no effect was observed for its ester derivatives. As for α-tocopherol, the reduction in its AE caused by the addition of 50 mg $kg^{-1}$ PC in Figure 4 remains unclear. It should be noted that the PC concentration assayed (50 mg $kg^{-1}$) was not much lower than the CMC (60–80 mg $kg^{-1}$), as well as that the CMC was determined by the TCNQ method, which is based on an imprecise determination subjected to the number and selection of data points. Therefore, further research is needed to clarify this matter.

The reductions in AE observed for GA and its alkyl esters (Figure 7) suggested that the polarity of the antioxidant to partition into PC reverse micelles appears to be the predominant factor involved. When the carboxylic group was blocked, the polarity provided by the three hydroxyl groups was still elevated so that strong associations with the PC micelles occurred.

The addition of PC to a commercial sunflower oil had a minor effect on the oil oxidative stability (Figure 8). The prooxidative effect of PC was only observed when its concentration was 5000 mg $kg^{-1}$, significantly exceeding its solubility limit in the oil (1000 mg $kg^{-1}$). Therefore, this fact confirms the low impact of PC reverse micelles on the AE of the α-tocopherol naturally present in sunflower oil.

## 5. Conclusions

At concentrations occurring in refined sunflower seed oils, the AE of α-tocopherol remained unaffected by the presence of PC reverse micelles in SSO. In contrast, the AE of Trolox substantially decreased with the addition of PC in a concentration-dependent way under the same experimental conditions. The location of α-tocopherol appeared not to be significantly affected by the presence of PC reverse micelles, and its availability to inhibit lipid oxidation did not change. However, Trolox seemed to be less available to inhibit lipid oxidation, probably due to its partitioning into PC reverse micelles, as observed in previous studies [6]. The reduction in the AE of Trolox was still greater than that of α-tocopherol when the PC was assayed in a range of concentrations up to 1000 mg $kg^{-1}$, which was its solubility limit in the oil. These results point out that new water–oil interfaces were created by hydroperoxides and other oxidation compounds. Even though Trolox was partitioned into PC reverse micelles losing effectiveness, its greater affinity for micelles consisting of hydroperoxides in comparison with the lipophilic antioxidant could account for its higher effectiveness.

A significant reduction in the AE of α-tocopherol was observed in a blend of SSO and MCT (1:3, *w/w*) when the antioxidant concentration was very low, comparable to the concentrations used in most studies in this field [6,13,14], and the molar concentration of PC was similar or 20 times higher than that of the antioxidant. Similarly, Trolox also showed the same pattern, but this time α-tocopherol was more effective in the presence of PC, which can be attributed to its lesser affinity for the PC and so higher effective concentration in the continuous phase to inhibit lipid oxidation. In addition, the prooxidative effect of PC in the mixture of oils, SSO and MCT, was larger compared to SSO. The slightly higher polarity of MCT could have affected the migration and location of antioxidants, and/or lipid oxidation products, as well as their associations with reverse micelles.

It can be concluded that the prooxidative effects of PC observed in different studies that employed model systems consisting of blends of stripped vegetable oils with MCT, along with antioxidants assayed at significantly lower concentrations than PC, are overstated. The effect of PC on the stability of a commercial vegetable oil, such as sunflower oil, which naturally contains substantial amounts of tocopherol, was found to be moderate.

**Supplementary Materials:** The following supporting information can be downloaded at: https://www.mdpi.com/article/10.3390/antiox12111993/s1, Figure S1: Determination of the critical micelle concentration (CMC) of phosphatidylcholine (PC) in stripped sunflower oil (SSO) in the presence of 1.16 mmol kg$^{-1}$ α-tocopherol (Toc) or Trolox at 100 °C. Toc 1 and Toc 2 refer to duplicate experiments made in different days.

**Author Contributions:** Conceptualization, J.V.; methodology, J.V.; validation, J.V. and M.-V.R.-M.; formal analysis, M.-J.G., Y.-Q.W., A.G.-G. and J.V.; investigation, J.V.; resources, J.V.; data curation, J.V.; writing—original draft preparation, J.V.; writing—review and editing, J.V. and M.-V.R.-M.; visualization, J.V.; supervision, J.V.; project administration, J.V.; funding acquisition, J.V. All authors have read and agreed to the published version of the manuscript.

**Funding:** This research was funded by MCIN/AEI/10.13039/501100011033 through project with reference PID2020-119304RB-I00.

**Institutional Review Board Statement:** Not applicable.

**Informed Consent Statement:** Not applicable.

**Data Availability Statement:** Data are contained within the article or Supplementary Materials.

**Conflicts of Interest:** The authors declare no conflict of interest. The funders had no role in the design of the study; in the collection, analyses, or interpretation of data; in the writing of the manuscript; or in the decision to publish the results.

## References

1. Chaiyasit, W.; Elias, R.J.; McClements, D.J.; Decker, E.A. Role of physical structures in bulk oils on lipid oxidation. *Crit. Rev. Food Sci. Nutr.* **2007**, *47*, 299–317. [CrossRef]
2. Chen, B.; McClements, D.J.; Decker, E.A. Minor components in food lipids: A critical review of their roles on lipid oxidation chemistry in bulk oils and emulsions. *Crit. Rev. Food Sci. Nutr.* **2011**, *51*, 901–916. [CrossRef] [PubMed]
3. Villeneuve, P.; Bourlieu-Lacanal, C.; Durand, E.; Lecomte, J.; McClements, D.J.; Decker, E.A. Lipid oxidation in emulsions and bulk oils: A review of the importance of micelles. *Crit. Rev. Food Sci. Nutr.* **2023**, *63*, 4687–4727. [CrossRef]
4. Frankel, E.N. Hydroperoxide Decomposition. In *Lipid Oxidation*, 2nd ed.; Frankel, E.N., Ed.; Woodhead Publishing Limited: Cambridge, UK, 2005; pp. 67–98.
5. Chen, B.C.; Han, A.; McClements, D.J.; Decker, E.A. Physical structures in soybean oil and their impact on lipid oxidation. *J. Agric. Food Chem.* **2010**, *58*, 11993–11999. [CrossRef]
6. Chen, B.C.; Han, A.; Laguerre, M.; McClements, D.J.; Decker, E.A. Role of reverse micelles on lipid oxidation in bulk oils: Impact of phospholipids on antioxidant activity of alpha-tocopherol and Trolox. *Food Funct.* **2011**, *2*, 302–309. [CrossRef] [PubMed]
7. Chen, B.C.; Panya, A.; McClements, D.J.; Decker, E.A. New insights into the role of iron in the promotion of lipid oxidation in bulk oils containing reverse micelles. *J. Agric. Food Chem.* **2012**, *60*, 3524–3532. [CrossRef] [PubMed]
8. Kittipongpittaya, K.; Panya, A.; McClements, D.J.; Decker, E.A. Impact of free fatty acids and phospholipids on reverse micelles Formation and lipid oxidation in bulk oil. *J. Am. Oil Chem. Soc.* **2014**, *91*, 453–462. [CrossRef]

9. Cui, L.Q.; Kittipongpittaya, K.; McClements, D.J.; Decker, E.A. Impact of phosphoethanolamine reverse micelles on lipid oxidation in bulk oils. *J. Am. Oil Chem. Soc.* **2014**, *91*, 1931–1937. [CrossRef]
10. Kittipongpittaya, K.; Panya, A.; Cui, L.Q.; McClements, D.J.; Decker, E.A. Association colloids formed by multiple surface active minor components and their effect on lipid oxidation in bulk oil. *J. Am. Oil Chem. Soc.* **2014**, *91*, 1955–1965. [CrossRef]
11. Homma, R.; Suzuki, K.; Cui, L.; McClements, D.J.; Decker, E.A. Impact of association colloids on lipid oxidation in triacylglycerols and fatty acid ethyl esters. *J. Agric. Food Chem.* **2015**, *63*, 10161–10169. [CrossRef]
12. Koga, T.; Terao, J. Phospholipids increase radical-scavenging activity of vitamin E in a bulk oil model system. *J. Agric. Food Chem.* **1995**, *43*, 1450–1454. [CrossRef]
13. Cui, L.; McClements, D.J.; Decker, E.A. Impact of phosphatidylethanolamine on the antioxidant activity of α-tocopherol and Trolox in bulk oil. *J. Agric. Food Chem.* **2015**, *63*, 3288–3294. [CrossRef]
14. Rokosik, E.; Siger, A.; Rudzinska, M.; Dwiecki, K. Antioxidant activity and synergism of canolol and α-tocopherol in rapeseed oil is affected by the presence of phospholipid association colloids. *LWT Food Sci. Technol.* **2020**, *133*, 110095. [CrossRef]
15. FAO; WHO. Codex Alimentarius Commission. In *Standard for Named Vegetable Oils CXS 210-1999*; World Health Organization, Food and Agriculture Organization of the United Nations: Rome, Italy, 1999. Available online: https://www.fao.org/fao-who-codexalimentarius/sh-proxy/en/?lnk=1&url=https%253A%252F%252Fworkspace.fao.org%252Fsites%252Fcodex%252FStandards%252FCXS%2B210-1999%252FCXS_210e.pdf (accessed on 4 September 2023).
16. Fuster, M.D.; Lampi, A.-M.; Hopia, A.; Kamal-Eldin, A. Effects of α- and γ-tocopherols on the autooxidation of purified sunflower triacylglycerols. *Lipids* **1998**, *33*, 715–722. [CrossRef]
17. *ISO 9936:2016*; Animal and Vegetable Fats and Oils—Determination of Tocopherol and Tocotrienol Contents by High-Performance Liquid Chromatography. Standard ISO/TC 34/SC 11. International Organization for Standardization: Geneva, Switzerland, 2016. Available online: https://www.iso.org/standard/69595.html (accessed on 2 November 2023).
18. Velasco, J.; Morales-Barroso, A.; Ruiz-Méndez, M.V.; Márquez-Ruiz, G. Quantitative determination of major oxidation products in edible oils by direct NP-HPLC-DAD analysis. *J. Chromatogr. A* **2018**, *1547*, 62–70. [CrossRef] [PubMed]
19. *ISO 16931:2009*; Animal and Vegetable Fats and Oils—Determination of Polymerized Triacylglycerols by High-Performance Size-Exclusion Chromatography (HPSEC). Standard ISO/TC 34/SC 11. International Organization for Standardization: Geneva, Switzerland, 2009. Available online: https://www.iso.org/standard/44881.html (accessed on 4 September 2023).
20. Singleton, W.S.; Gray, M.S.; Brown, M.L.; White, J.L. Chromatographically homogeneous lecithin from egg phospholipids. *J. Am. Oil Chem. Soc.* **1965**, *42*, 53–56. [CrossRef] [PubMed]
21. IUPAC. *Standard Methods for the Analysis of Oils, Fats and Derivatives, 7th ed*; International Union of Pure and Applied Chemistry, Blackwell Scientific: Oxford, UK, 1992.
22. Kanamoto, R.; Wada, Y.; Miyajima, G.; Kito, M. Phospholipid-phospholipid interaction in soybean oil. *J. Am. Oil Chem. Soc.* **1981**, *58*, 1050–1053. [CrossRef]
23. AOCS. *Official Methods and Recommended Practices of the American Oil Chemists' Society*, 7th ed.; AOCS Press: Champaign, IL, USA, 2022.
24. Carelli, A.A.; Eci, L.N.; Crapiste, G.H. Phosphorous-to-phospholipid conversion factors for crude and degummed sunflower oils. *J. Am. Oil Chem. Soc.* **2002**, *79*, 1177–1180. [CrossRef]
25. Velasco, J.; Andersen, M.L.; Skibsted, L.H. Evaluation of oxidative stability of vegetable oils by monitoring the tendency to radical formation. A comparison of electron spin resonance spectroscopy with the Rancimat method and differential scanning calorimetry. *Food Chem.* **2004**, *85*, 623–632. [CrossRef]
26. Laguerre, M.; Chen, B.; Lecomte, J.; Villeneuve, P.; McClements, D.J.; Decker, E.A. Antioxidant properties of chlorogenic acid and its alkyl esters in stripped corn oil in combination with phospholipids and/or water. *J. Agric. Food Chem.* **2011**, *59*, 10361–10366. [CrossRef] [PubMed]
27. Martín-Polvillo, M.; Márquez-Ruiz, G.; Dobarganes, M.C. Oxidative stability of sunflower oils differing in unsaturation degree during long-term storage at room temperature. *J. Am. Oil Chem. Soc.* **2004**, *81*, 577–583. [CrossRef]
28. Velasco, J.; Marmesat, S.; Dobarganes, C.; Márquez-Ruiz, G. Heterogeneous aspects of lipid oxidation in dried microencapsulated oils. *J. Agric. Food Chem.* **2006**, *54*, 1722–1729. [CrossRef] [PubMed]
29. Roche, A.; Ross, E.; Walsh, N.; O'Donnell, K.; Williams, A.; Klapp, M.; Fullard, N.; Edelstein, S. Representative literature on the phytonutrients category: Phenolic acids. *Crit. Rev. Food Sci. Nutr.* **2017**, *57*, 1089–1096. [CrossRef] [PubMed]
30. Jacobsen, C.; Schwarz, K.; Stockmann, H.; Meyer, A.S.; Adler-Nissen, J. Partitioning of selected antioxidants in mayonnaise. *J. Agri. Food Chem.* **1999**, *47*, 3601–3610. [CrossRef] [PubMed]
31. Shaw, L.A.; McClements, D.J.; Decker, E.A. Spray-Dried multilayered emulsions as a delivery method for ω-3 fatty acids into food systems. *J. Agric. Food Chem.* **2007**, *55*, 3112–3119. [CrossRef] [PubMed]
32. Budilarto, E.S.; Kamal-Eldin, A. The supramolecular chemistry of lipid oxidation and antioxidation in bulk oils. *Eur. J. Lipid Sci. Technol.* **2015**, *117*, 1095–1137. [CrossRef] [PubMed]
33. Ghnimi, S.; Budilarto, E.; Kamal-Eldin, A. The new paradigm for lipid oxidation and insights to microencapsulation of omega-3 fatty acids. *Compr. Rev. Food Sci. Food Saf.* **2017**, *16*, 1206–1218. [CrossRef] [PubMed]
34. Budilarto, E.S.; Kamal-Eldin, A. Water content and micelle size change during oxidation of sunflower and canola oils. *Eur. J. Lipid Sci. Technol.* **2015**, *117*, 1971–1977. [CrossRef]

35. Chen, Y.; Cheng, H.; Liang, L. Effect of oil type on spatial partition of resveratrol in the aqueous phase, the protein interface and the oil phase of o/w emulsions stabilized by whey protein and caseinate. *Antioxidants* **2023**, *12*, 589. [CrossRef] [PubMed]
36. Lehtinen, O.-P.; Nugroho, R.W.N.; Lehtimaa, T.; Vierros, S.; Hiekkataipale, P.; Ruokolainen, J.; Sammalkorpi, M.; Österberg, M. Effect of temperature, water content and free fatty acid on reverse micelle formation of phospholipids in vegetable oil. *Colloids Surf. B Biointerfaces* **2017**, *160*, 355–363. [CrossRef]

*antioxidants*

MDPI

*Article*

# Kinetic Analysis of High-Temperature Sunflower Oil Peroxidation Inhibited by the Major Families of Phenolic Antioxidants Unveils the Extraordinary Activity of 1,4-Hydroquinones

**Fabio Mollica [1], Lucia Bonoldi [2] and Riccardo Amorati [1,*]**

[1] Department of Chemistry "G. Ciamician", University of Bologna, Via S. Giacomo 11, I-40126 Bologna, Italy
[2] Research and Technological Innovation Department, Eni SpA, Via F. Maritano 26, I-20097 San Donato Milanese, Italy
* Correspondence: riccardo.amorati@unibo.it

**Abstract:** Peroxidation of vegetable oils represents a major problem for the food and biodiesel industries, and it is greatly accelerated by oil degree of unsaturation and by temperature increase. Phenols represent the most common additives used to counteract oil peroxidation, however clear structure-activity relationships at high temperatures are not available. We report, herein, a kinetic study of $O_2$ consumption during spontaneous peroxidation of sunflower oil at 130 °C in the presence of 18 antioxidants belonging to the main families of natural and synthetic phenols, including α-tocopherol, alkylphenols (BHT, BHA), hydroquinones (TBHD), catechols (quercetin, catechin) and gallates. Results show that TBHQ provide the best protection in terms of induction period (*IP*) duration and $O_2$ consumption rate. EPR spectroscopy demonstrated that the inhibition activity is negatively correlated to the stability of the phenoxyl radical of the antioxidant ($A^\bullet$), suggesting that chain propagation with linoleate (RH) moieties $A^\bullet + RH \rightarrow AH + R^\bullet$ decreases the efficacy of those antioxidants forming persistent $A^\bullet$ radicals. These results provide important information to optimize the antioxidant activity of phenols and of novel phenol-based materials.

**Keywords:** peroxidation; kinetic; bond dissociation; tocopherol; polyphenols; EPR; sunflower oil; peroxyl radical; oxygen consumption; phenoxyl radical

## 1. Introduction

Lipid peroxidation is a main problem for edible oil utilization, and for the long-term preservation of foods rich in polyunsaturated lipids. The formation of hydroperoxides and other oxygenated products like epoxides, aldehydes, ketones, diols, etc. are correlated with off-flavor, insoluble deposits, and with the formation of potentially toxic compounds [1]. Besides, recycling used oil as feedstock for biorefineries requires the control of oxidation and polymerization processes that can cause corrosion and clogging problems [2]. It is therefore understandable why the development of novel and more effective ways to counteract lipid peroxidation is still an active research field. Phenols represent the most important class of natural and artificial antioxidants, and therefore it is not surprising that many recent studies are based on the manipulation and optimization of their characteristics. For example, in the emerging field of nano-antioxidants [3,4], natural phenolic polymers like lignin [5] and melanin-like materials [6,7] are among the most promising radical-trapping platforms. Besides, many natural-based waste products currently investigated as a source of antioxidants are particularly rich in phenols, like for instance, cashew-nut shell liquid [8], or olive mill wastewater [9].

Knowledge of the factors that determine the efficacy of phenolic antioxidants is of fundamental importance to guide the research toward significant advancements. While several experimental and theoretical studies performed in the past decades have clarified the chemico-physical basis of the antioxidant activity of phenols [10,11], the prediction of the efficacy in real systems (such as in emulsions, etc.) remains challenging [12]. The behavior of phenols as antioxidants for oils at high temperatures is particularly important because these conditions are found not only in many technological applications such as cooking and oil processing but also in accelerated stability tests such as Rancimat [13]. Previous studies reported the effect of phenols as stabilizers toward oil (or fatty acids methyl esters, FAME) peroxidation at high temperatures, but rarely they are performed by comparing a significant number of different phenols and, to the best of our knowledge, never report a chemical explanation of the results. For example, Dodos et al. compared the activity of ten phenolic antioxidants in the stability at 110 °C of FAME from soybean oil (rich in linoleic acid) by using Rancimat equipment and found the longest induction period (*IP*) in the case of *tert*-butyl hydroquinone (TBHQ), while monophenols like BHT and BHA did not show significant effects [14]. R. Becker and A. Knorr, in a study of the oxidation of rapeseed oil at 130 °C, showed that, among several investigated phenols, 2,5-di-*tert*-butyl-hydroquinone had the best activity in terms of *IP* [15]. However, the reason for this activity ranking remained unclear.

We, therefore, decided to study the antioxidant activity of 18 phenols, chosen among the main families of natural and synthetic phenols, with the aim of deriving a structure/activity relationship and understanding the mechanism of inhibition of lipid peroxidation. We used a recently developed method consisting of real-time measuring of the $O_2$ consumption in the headspace of a purified oil sample, vigorously stirred at 130 °C [16,17]. This approach minimizes the loss of the antioxidant by evaporation and allows determining not only the *IP* but also the slope of the $O_2$ consumption during the *IP* ($-d[O_2]_{inh}/dt = R_{inh}$). While IP is a measure of the lifetime of the antioxidants in the selected environment, $R_{inh}$ correlates with the rate of reaction of the antioxidant with free radicals, i.e., its efficacy in slowing down oxidation [16]. The collected results are analyzed by considering the accepted mechanisms underlying lipid peroxidation. With the help of electron paramagnetic resonance (EPR) spectroscopy, we were able to detect, in specific cases, the presence of stable radicals, affording a plausible explanation of the efficacy order. Our results, besides confirming the previously unexplained high activity of hydroquinones, provide a predictive rationalization based on the behavior of the radicals derived from phenols under peroxidation conditions.

## 2. Materials and Methods

### 2.1. Materials

All chemicals and solvents were commercially available (Merk, Sigma-Aldrich, Milan, Italy). *Para*-methoxyphenol (**1**), 2-*tert*-butyl-4-methoxyphenol (**2**), 2,6-di-*tert*-butyl-4-methylphenol (BHT) (**3**), 2,4,6-trimethylphenol (**4**), 2-methoxy-4-ethylphenol (**5**), 2,6-dimethoxy-4-allylphenol (**6**), (±)-α-Tocopherol (**7**), 2,2,5,7,8-pentamethyl-6-chromanol (**8**), *para*-hydroquinone (**9**), *tert*-butylhydroquinone (**10**), 2,5-di-*tert*-butylhydroquinone (**11**), *para*-benzoquinone (**12**), *tert*-butylbenzoquinone (**13**), 2,5-di-*tert*-butylbenzoquinone (**14**), catechol (**15**), 4-methylcatechol (**16**), 4-*tert*-butylcatechol (**17**), caffeic acid (**18**), catechin (**19**), quercetin (**20**), propyl gallate (**21**), triacetin were of the highest available purity and were used as received. Antioxidant concentrations are % *w*/*w*. Solvents were HPLC-grade and were used without further purification. High-linoleic sunflower oil was purchased from a local market. Stripped sunflower oil (SSO) was prepared as previously reported by percolation on basic alumina and by treatment with activated carbon [17]. The SSO composition was determined by $^1$H-NMR [18]: polyunsaturated chains 59.4 mol %, monounsaturated chains 27.3 mol %, saturated chains 13.3 mol %.

### 2.2. Autoxidation Experiments

Oxygen consumption during the sunflower oil autoxidation was measured by an $O_2$-sensitive optical probe (Pyroscience GmbH, Aachen, Germany) in a sealed apparatus previously described [16]. Briefly, 2.00 g of oil containing variable amounts of the antioxidants is introduced into a 10 mL round-bottomed flask with a magnetic stir bar connected to a miniaturized water-cooled condenser to keep the temperature of the $O_2$-sensitive probe below 40 °C (total internal volume 29 mL). The sample is kept under vigorous stirring in a silicone bath at 130 °C. The $O_2$ probe is calibrated in air before each experiment. The amount of consumed $O_2$ is calculated from the internal volume of the apparatus. The experiments with peroxidized SSO were performed by oxidizing SSO at 130 °C measuring $O_2$ uptake. Then the sample was aerated, and a small volume of a concentrated octanol solution of **10** was added.

### 2.3. Analysis of Antioxidant Stability

The experiments were performed under the same conditions as the autoxidation experiments, except for the use of triacetin, which is resistant to oxidation at 130 °C, in place of SSO. The concentrations of the antioxidants were determined by HPLC separation coupled to a photodiode array (DAD) UV-vis detection and the flow was set to 0.9 mL/min. The column was a Sepachrom $C_{18}$, 5 μm, 250 mm × 4.1 mm. The mobile phase was an isocratic mixture of $MeCN/H_2O$ 60:40 for **9**, **10** and **11**; $MeCN/H_2O$ 90:10 for **2**; $MeCN/H_2O$ + 3% formic acid 50:50 for **21**; MeCN/MeOH 90:10 for **8**.

### 2.4. EPR Analysis

The determination of the concentration of the radicals of **2**, **3**, **7**, and **10** was performed by an X-band electron paramagnetic resonance (EPR) spectrometer (Elexsys 500, Bruker, Billerica, MA, USA) equipped with a variable temperature dewar placed in the resonant cavity. Suprasil quartz EPR tubes were filled with the pure oil samples (no dilution) up to 4 cm and thermostatted at 40 °C by nitrogen gas flowing in the dewar. The small quantity of oil ensured a homogeneous temperature inside the analyzed sample. Typical instrumental settings were: microwave power = 10 mW, amplitude = 2 G, 64 accumulations. The radical concentration was obtained from the double integration of the EPR signal multiplied by the inverse temperature in Kelvin degrees to correct for the Curie temperature dependence and calibrated with the signal of the strong-pitch standard from Bruker. The spectra were simulated with the software Winsim [19], freely available from the Public Electron paramagnetic resonance Software Tools (PEST) from NIEHS (National Institue of Environmental Health Sciences). The radicals were identified as the phenoxyl radicals of the investigated antioxidants from the agreement of their spectral parameters (hyperfine coupling constants, *a*, and *g*-factors) with those reported in the literature [20,21] or with those expected from their structure [22].

### 2.5. Statistical Analysis

$O_2$-consumption experiments and HPLC analysis were performed in duplicate. The values are expressed as the average ± SD (standard deviation). Student's *t*-test was applied for comparisons between two mean values. Differences were considered to be of significance when $p$ was $\leq 0.05$.

## 3. Results

### 3.1. Autoxidation Experiments

Stripped high-linoleic sunflower oil (SSO), in the absence of antioxidants, showed an initial linear $O_2$ consumption, followed by a decrease in the oxidation rate with the decline of $O_2$ concentration in the headspace (Figure 1, trace a).

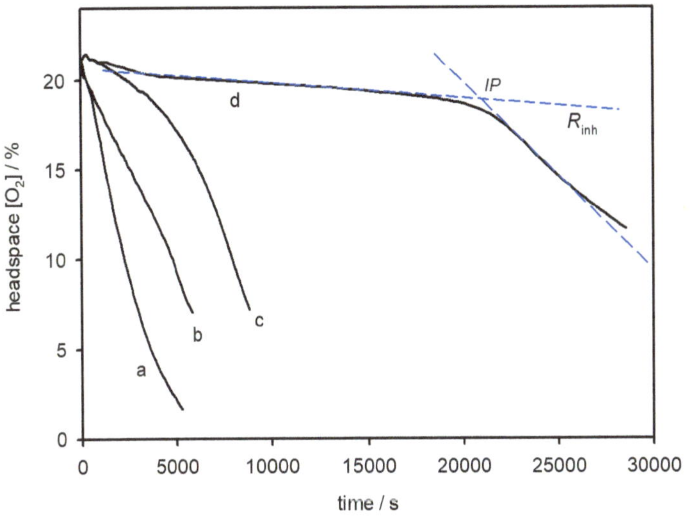


**Figure 1.** $O_2$ consumption during the autoxidation of SSO at 130 °C without antioxidants (**a**) or in the presence of 0.1% (*w*/*w*) of α-tocopherol (**b**), BHT (**c**) and TBHQ (**d**).

In the presence of the antioxidants, the autoxidation onset of SSO was retarded for a variable period of time (induction period, *IP*), after which the $O_2$ consumption was faster. The minimum slope of $O_2$ consumption during the inhibited or retarded period (inhibited rate, $R_{inh}$) and the duration of the inhibition *IP* were taken as indicators of the efficacy of the antioxidants. In Figure 1, some typical $O_2$ consumption curves and the derivation of $R_{inh}$ and *IP* are reported. By following this procedure, we analyzed several natural and synthetic phenolic antioxidants all at the identical concentration of 0.1% *w*/*w*, and the results are reported in Scheme 1. The antioxidants were divided into three families depending on the number of OH groups and their relative position because, as will be evident in the discussion of the results, this feature influences the fate of the radical of the antioxidant and the overall antioxidant efficacy.

Monophenols, characterized by the presence of only one phenolic OH group, include α-tocopherol (**7**) and its synthetic analog **8**, two *ortho*-methoxy phenols (**5** and **6**) having a structural relationship with natural compounds like sinapic acid and eugenol, the well-known antioxidant additives BHT (**3**) and BHA (**2**), and other alkylated or methoxylated phenols used for comparison purposes (**1** and **4**).

Hydroquinones are characterized by the presence of two OH groups in *para* position, and a variable number of alkyl substituents. They include unsubstituted hydroquinone (**9**), *tert*-butyl hydroquinone (TBHQ, **10**), and 2,5-di-*tert*-butyl hydroquinone (**11**). The corresponding oxidized *para*-quinones (**12–14**) were also investigated to ascertain the influence of these compounds on the radical chain, as it was reported that they may act as alkyl ($R^•$) [23] or $HOO^•$ radical traps [24].

The last group is represented by catechols, having at least two OH groups in *ortho* position. They include the natural antioxidants caffeic acid (**18**), catechin (**19**), and quercetin (**20**) as well as unsubstituted catechol (**15**) and the alkylated catechols **16** and **17**, while gallates are represented by the antioxidant additive propyl gallate (**21**).

With the exception of 4-methoxyphenol (**1**) and of benzoquinones **12–14**, all the investigated compounds had some retarding effect on SSO peroxidation. However, the only antioxidants that were able to afford a strong inhibition for a long time were the hydroquinones **9** and **10**. In fact, they were able to reduce 30 and 100 times the slope of $O_2$ consumption, respectively, while the *IP* was about $2.5 \times 10^4$ s, that is, considerably longer than that of the best non-hydroquinone antioxidants like propyl gallate. If considered on a molar basis, **10** displayed the highest radical trapping stoichiometry, that is the number of radicals trapped by each antioxidant molecule, among all investigated phenols. Because of its exceptional characteristics, the *IP* dependence on the concentration of **10** was investigated in detail, as shown in Figure 2. The inset in this figure shows that in the concentration range considered, *IP* is linearly dependent on **10** concentrations.

Monophenols

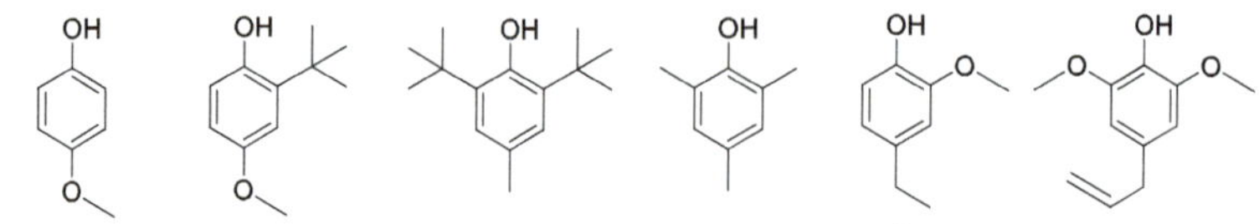

| | **1** | **2** | **3** (BHT) | **4** | **5** | **6** |
|---|---|---|---|---|---|---|
| $R_{inh}$ (nmol/s) | 69±2 | 22±2 | 16±2 | 34±3 | 32±2 | 38±3 |
| *IP* (s) | undefined | 6700±200 | 7800±300 | undefined | 3200±300 | undefined |

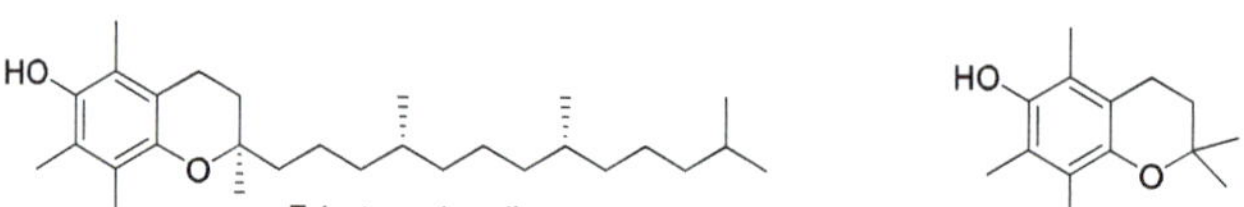

| | **7** (α-tocopherol) | **8** |
|---|---|---|
| $R_{inh}$ (nmol/s) | 21± 2 | 25± 2 |
| *IP* (s) | 4200 ± 300 | 6100 ± 300 |

Hydroquinones quinones

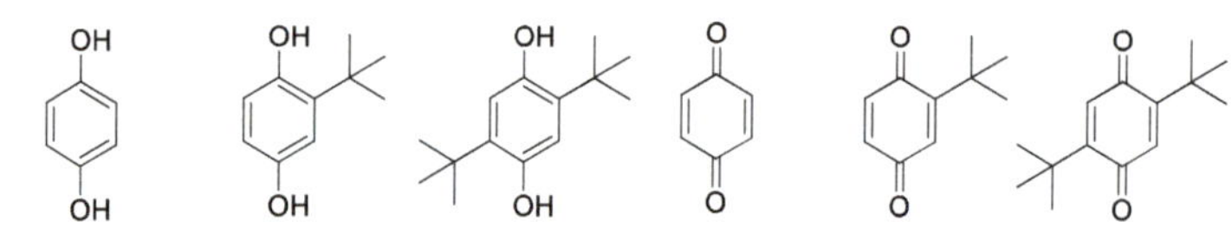

| | **9** | **10** (TBHQ) | **11** | **12** | **13** | **14** |
|---|---|---|---|---|---|---|
| $R_{inh}$ (nmol/s) | 2.6 ± 0.3 | 0.8 ± 0.1 | 35 ± 1 | 62 ± 0.8 | 70 ± 10 | 65 ± 4 |
| *IP* (s) | 25,500 ± 400 | 21,300 ± 700 | 7800 ± 400 | undefined | undefined | undefined |

Catechols gallates

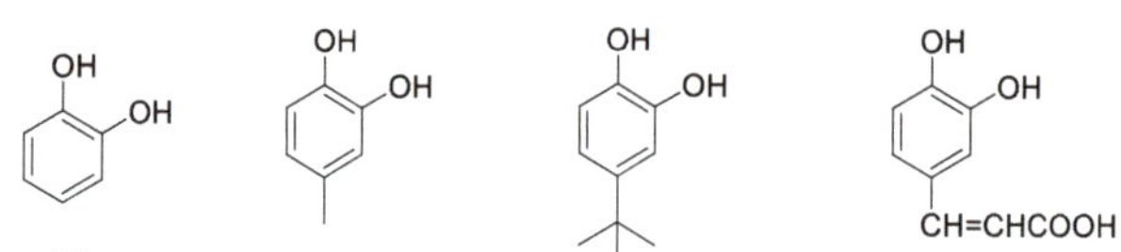

| | **15** | **16** | **17** | **18** (caffeic acid) |
|---|---|---|---|---|
| $R_{inh}$ (nmol/s) | 19 ± 2 | 9.4 ± 0.6 | 10 ± 1 | 12 ± 1 |
| *IP* (s) | 7100 ± 400 | 8900 ± 300 | 11,200 ± 600 | 11,400 ± 600 |

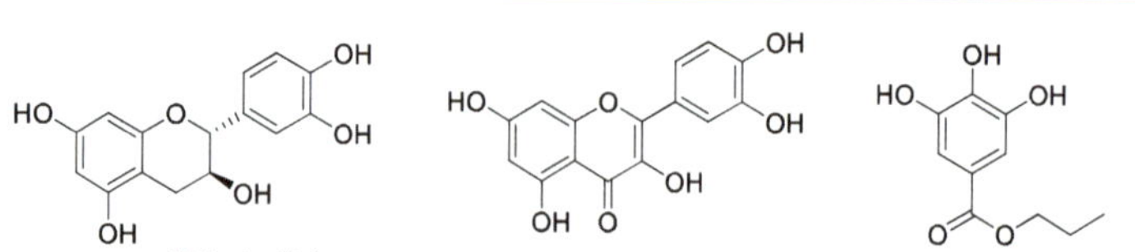

| | **19** (catechin) | **20** (quercetin) | **21** (propyl gallate) |
|---|---|---|---|
| $R_{inh}$ (nmol/s) | 23 ± 2 | 39 ± 4 | 8.1 ± 0.7 |
| *IP* (s) | 6200 ± 500 | 3700 ± 300 | 10,200 ± 800 |

**Scheme 1.** Antioxidant effect of the mono-phenols: oxygen consumption rate in the inhibited period ($R_{inh}$) and duration of the inhibited period *IP*. The slope without antioxidants was 70 ± 8 nmol/s. Undefined *IP* indicates that no induction period is present.

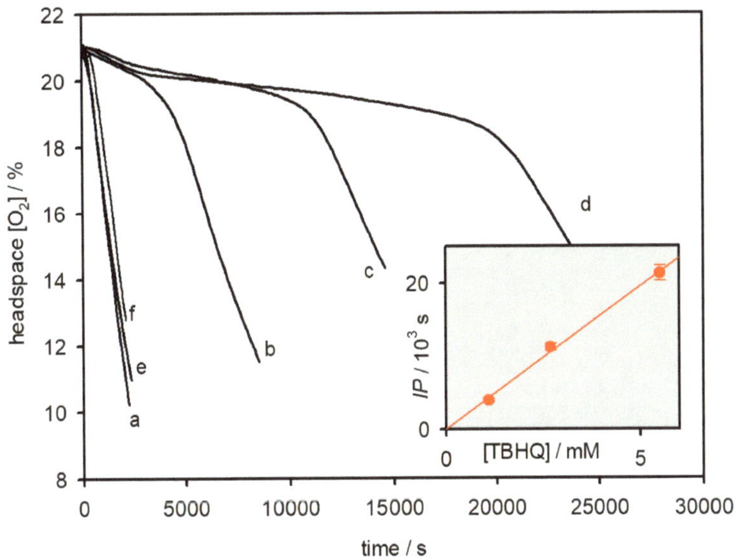


**Figure 2.** $O_2$ consumption during the autoxidation of SSO at 130 °C without antioxidants (**a**) or in the presence of **10**: 0.02% (**b**), 0.05% (**c**) and 0.1% (**d**), or **13**, 0.1% (**e**). $O_2$ consumption of peroxidized SSO in the presence of **10**, 0.02% (**f**). The inset reports the relationship between the **10** concentrations and *IP*. Some error bars are smaller than the symbol size.

In order to test the effect of hydroperoxides on the *IP* of **10**, we studied the effect of **10** (0.02%) in an SSO sample that was previously oxidized without any antioxidant, until reaching a [ROOH] ≈ 200 mM, assuming that all consumed $O_2$ is transformed into ROOH. The results show that the antioxidant effect of **10** almost disappears (compare plot f with b, Figure 2), indicating a major effect of peroxide accumulation on *IP*.

### 3.2. EPR Experiments

To investigate the relationship between the stability of the radicals and the antioxidant activities of the investigated phenols, we measured the electron paramagnetic resonance (EPR) spectra of **2**, **3**, **4**, and **10** at 0.1% in SSO under air. Aerated oil samples containing the antioxidants were introduced in the EPR cavity and heated at 40 °C. This temperature was chosen as the best compromise to optimize the signal-to-noise ratio. In the case of α-tocopherol (**7**) and BHT (**3**), the signal arising from their phenoxyl radicals could be clearly detected (see Figure 3a,b), while in the case of **2** the signal was weak, and in the case of **10** only traces of the radical could be evidenced. The identity of the radicals was ascertained by analyzing the experimental spectra by numerical simulation as shown by the red lines in Figure 3. This procedure afforded coupling constants (*a*) and *g* factors that were in good agreement with those reported in the literature for the respective phenoxyl radicals [19–22]. The radical concentrations (see Figure 3) were in the nanomolar range and, considering the experimental error, the radical stability decreased in the order: **3** ≈ **7** > **2** >> **10**.

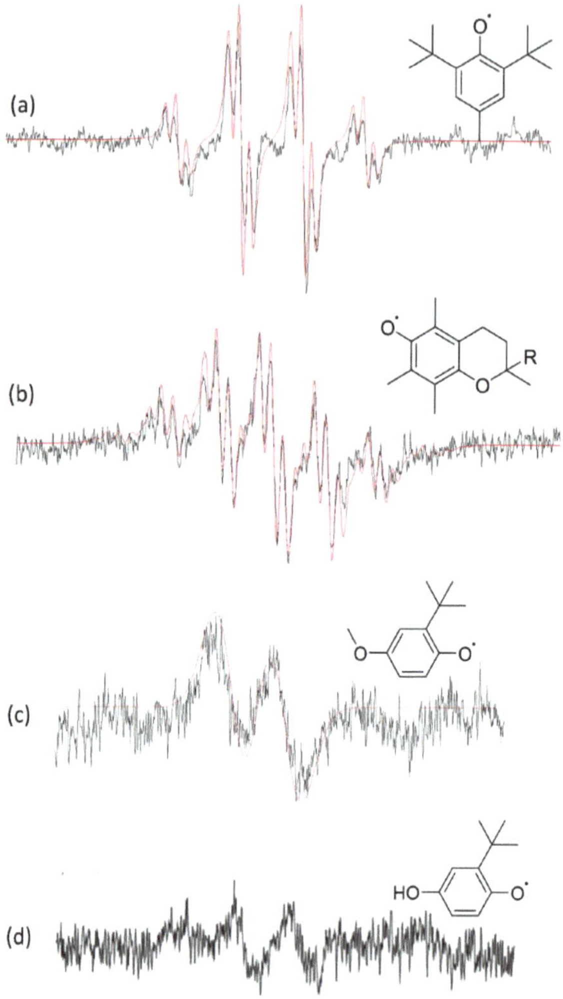


(a)

| | Lit. (gauss) | Exp. (gauss) |
|---|---|---|
| a(2H) | 1.54 | 1.80 |
| a(H,CH3) | 11.61 | 11.15 |
| g | | 2.0045 |
| [radical] = 34 nM | | |

(b)

| | Lit. (gauss) | Exp. (gauss) |
|---|---|---|
| a(H,CH3) | 5.94 | 5.89 |
| a(H,CH3) | 4.41 | 4.77 |
| a(H,CH2) | 1.39 | 1.15 |
| a(H,CH2) | 1.39 | 1.14 |
| a(H,CH3) | 0.75 | 1.51 |
| g | 2.00469 | 2.0047 |
| [radical] = 24 nM | | |

(c)

| | Exp. (gauss) |
|---|---|
| a(H) | 5.66 |
| g | 2.005 |
| [radical] = 4.5 nM | |

(d)

| [radical] = traces |
|---|

**Figure 3.** Experimental and simulated (red lines) EPR spectra of SSO after the addition of 0.1% at 40 °C of **3** (**a**), **7** (**b**), **2** (**c**), **10** (**d**). The coupling constants (*a*), the *g*-factor, and the radical concentrations (estimated error ± 50%) are also reported.

*3.3. Stability of Antioxidants*

The stability at 130 °C was investigated to understand the role of phenol degradation on the duration of the inhibited period. Selected antioxidants (see Figure 4) were dissolved in triacetin, an oxidation-resistant triacylglycerol having no unsaturations, and were placed in the same reactor used in the peroxidation experiments. Then, their concentration was measured by HPLC-DAD analysis at time intervals for 6 h. The hydroquinones **9–11** were considered in order to clarify the excellent antioxidant behavior of **9** and **10** in SSO autoxidation experiments, as opposed to the relatively small efficacy of the dialkylated analog **11**. The monophenol **2** and propyl gallate (**21**) were included for comparison purposes, while **8** was selected to investigate the behavior of α-tocopherol.

We found that in all cases the concentration remained approximately constant, except for the dialkylated hydroquinone **11** and the α-tocopherol analog **8**, whose concentrations, however, decreased only by 20% and 15%, respectively, after 6 h ($21.6 \times 10^3$ s). If considering that in the SSO autoxidation experiments their *IP* was 7.7 × $10^3$ s and 6.6 × $10^3$ s, respectively, even in the case of the least stable phenols **8** and **11**, thermal degradation plays a negligible role in the duration of the inhibition. In addition, these experiments also demonstrated that the antioxidant loss by evaporation was negligible even in the case of the low molecular weight phenols **2** and **8**.

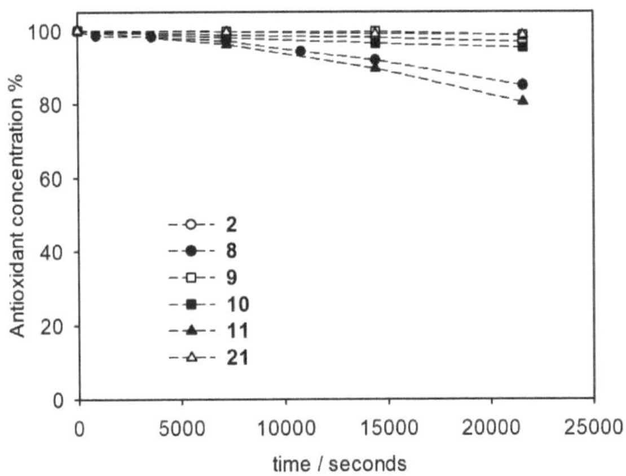


**Figure 4.** Stability at 130 °C of selected phenols in triacetin under air. Error bars are smaller than the symbol size.

## 4. Discussion

The results collected herein show, in a comprehensive manner, the effect of the major families of phenolic antioxidants on the oxidation of high-linoleic sunflower oil at 130 °C. Despite their structural similarity, phenolic antioxidants display highly variable inhibition capability, as expressed by the reduction of the $R_{inh}$ rate, and by the duration of the inhibited period *IP*. Hydroquinones **9** and **10** display the highest inhibition, while other phenols having a very similar structure, such as their methylated analogs **1** and **2**, are much less active. To understand these results, reactions 1–9 occurring during SSO peroxidation have to be considered (Scheme 2) [10,25].

The initiation step (reactions 1 and 2) consists of the formation of radicals in the system with an overall rate $R_i$. Reactions 3–5 describe the processes occurring in not-inhibited autoxidation, where hydrogen-atom abstraction (reaction 4) occurs (mainly) at the bis-allylic position of the linoleate residues and the radical chain propagates with rate constant $k_p$ [10].

The initiation process may occur either through the direct reaction of activated C-H (usually bis-allylic ones) with $O_2$ (reaction 1) or by decomposition of hydroperoxides ROOH (reaction 2) [25,26]. Considering that only a small fraction of bis-allylic groups is consumed, the first mechanism is expected to decline only with $O_2$ concentration, while the second mechanism is marginal at the beginning of the reaction, or in the presence of strong antioxidants, but becomes more and more important as the autoxidation proceeds and hydroperoxides are accumulated.

The $R_i$ parameter has a great influence on *IP* because the faster radicals are produced in the system, the faster the antioxidants are consumed, and the *IP* is shortened. By applying the steady state approximation, which states that in radical chains the concentration of radical species is constant and that the overall rate of the formation of radicals must be equal to that of their disappearance, and considering $R_i$ is constant during the inhibited period, the following relation applies: $IP = n[AH]/R_i$, where [AH] is the initial concentration of the antioxidant and $n$ is the number of radicals trapped by each AH. At low temperatures, phenols quench ROO• by reactions 6–8 and are usually characterized by $n \approx 2$ [10,25].

Initiation

$$RH + O_2 \longrightarrow R^{\bullet} + HOO^{\bullet} \xrightarrow{RH} R^{\bullet} \quad (1)$$

$$ROOH \longrightarrow RO^{\bullet} + HO^{\bullet} \xrightarrow{RH} R^{\bullet} \quad (2)$$

$$\text{rate} = R_i$$

Propagation

$$R^{\bullet} + O_2 \longrightarrow ROO^{\bullet} \quad (3)$$

$$ROO^{\bullet} + RH \xrightarrow{k_p} ROOH + R^{\bullet} \quad (4)$$

Termination

$$2\ ROO^{\bullet} \xrightarrow{k_t} \text{non-radical products} \quad (5)$$

Termination by the antioxidant

$$ROO^{\bullet} + AH \xrightarrow{k_{inh}} ROOH + A^{\bullet} \quad (6)$$

$$ROO^{\bullet} + A^{\bullet} \longrightarrow \text{non-radical products} \quad (7)$$

$$A^{\bullet} + A^{\bullet} \longrightarrow \text{non-radical products} \quad (8)$$

propagation by the antioxidant

$$A^{\bullet} + RH \longrightarrow AH + R^{\bullet} \quad (9)$$

**Scheme 2.** Reactions occurring during SSO autoxidation in the presence of antioxidants. RH = polyunsaturated triacylglycerols, AH = antioxidant.

In the case of **10**, which is one of the best antioxidants analyzed here, the *IP* increases linearly with **10** concentrations (Figure 2) suggesting that $R_i$ is constant throughout the duration of the inhibited period of all three experiments shown in Figure 2, otherwise reduction in the *IP* should have been observed caused by the increase of hydroperoxides. On the other hand, when **10** is used as an inhibitor of an SSO sample already containing a high level of hydroperoxides, no antioxidant effect is observed, suggesting a major role of ROOH in increasing $R_i$.

In the case where $R_i$ would be only due to the bulk oil's original composition (i.e., mainly the bis-allilic hydrogen concentration) in all samples, *IP* would be approximately the same throughout the investigated phenols. However, it turned out not to be so: *IP* varied significantly depending on the antioxidant (see values in Scheme 1): antioxidants causing a small reduction of $O_2$ consumption rate (i.e., those having high $R_{inh}$) were also characterized by low *IP*. To explain this effect, we hypothesized that in the case of weak antioxidants also, hydroperoxides may contribute to $R_i$, increasing its value with respect to the pristine bulk phase. The $O_2$ consumed during the inhibited period of weak antioxidants is converted into hydroperoxides, that in turn increase $R_i$, causing a shortening of *IP*.

Reactions 6–8 account for the action of phenolic radical-trapping antioxidants, consisting in H-atom donation to the chain-carrying peroxyl radicals $ROO^{\bullet}$, with a kinetic rate constant $k_{inh}$, followed by the decay of $A^{\bullet}$ with another radical (reactions 7 and 8) [10,25]. The slope of the inhibited period, $R_{inh}$, is inversely proportional to $k_{inh}$. In the steady state approximation, $R_{in} = -d[O_2]/dt = -(k_p[RH]R_i)/(n\ k_{inh}[AH])$ [10].

In addition, it must also be considered that, if the decay of the $A^{\bullet}$ radical is not efficient, it can react with the substrate, propagating the oxidative chain via reaction 9. This reaction, although regenerating an active antioxidant molecule AH, has, however, the undesired effect of opposing reaction 6, thus contributing to the formation of hydroperoxides and increasing $R_{inh}$.

Overall, this reaction scheme shows that good antioxidants must have a high $k_{inh}$ value (reaction 6) and must decay efficiently to avoid the onset of reaction 9.

It is well known that, in order to have high $k_{inh}$ values, phenols must possess electron-donating (ED) substituents in *ortho* and *para* positions with respect to the reactive OH group (see Scheme 3) [27]. ED substituents decrease the bond dissociation enthalpy (*BDE*) of the OH group and thus more easily render the H-atom abstraction by ROO• radicals. The substituent effect is proportional to the ED power of the substituents (OMe = OH > Me = $^t$Bu > H) and is modulated by the bulkiness of the substituents and by the presence of intramolecular H-bonds [27]. For example, *ortho*-methoxy phenols have a low reactivity toward ROO• because the OH group is engaged in a strong H-bond with the OMe substituent [28]. Accordingly, phenols having only one ED substituent (like **1**) or *ortho*-methoxy substituents (like **5** and **6**), were poor inhibitors of SSO autoxidation.

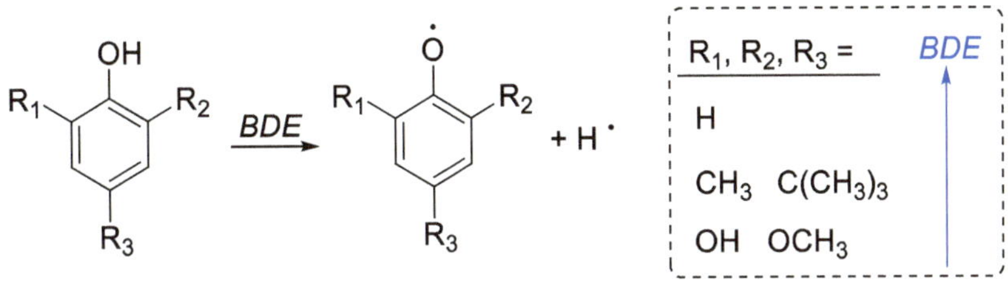


**Scheme 3.** Substituent effect on the O-H bond dissociation enthalpy (*BDE*) of phenols.

However, these rules seem not to apply in every case. For instance, **2** and **10** which have the same number of substituents with the same ED ability, show instead very different antioxidant effects, both in terms of $R_{inh}$ and of *IP*. In addition, α-tocopherol (**7**) and its analog **8** which have the lowest *BDE*(OH) among the investigated phenol [27] are not the most active ones.

As stability experiments showed that the spontaneous decay of the antioxidants at 130 °C is negligible (see Figure 4), these differences must be explained by considering the interplay between reactions 6–9. EPR experiments were performed to investigate this stability. They showed that the stability of the radicals of the antioxidants increases with the number and bulkiness of the substituents in *ortho* and *para* positions to the phenolic OH in the order **3** ≈ **7** > **2**, while the stability of the radical from **10** is extremely low. This result can be explained by the tendency of phenoxyl radicals to decay mainly by radical-radical addition by using the *ortho* and *para* positions, as shown in Scheme 4A [22]. If these positions contain alkyl or methoxy substituents, the decay is slow and possibly reversible.

On the other hand, the 4-hydroxylphenoxyl radical formed by **10** (see Scheme 4B) is very unstable because it decays by H-atom transfer with another radical forming the *para*-benzoquinone **13** and is therefore unable to propagate the oxidation via reaction 9 Interestingly, **13** is the main product that is formed in oxidized soybean oil stabilized by **10** [29].

The instability ranking experimentally observed by EPR follows the antioxidant ranking, proving that persistent radicals can behave detrimentally, because of their contribution to the chain propagation mechanism.

In this context, the disappointing behavior of catechol and gallates was unexpected, because also in their case, the phenoxyl radical may potentially decay by forming an *ortho*-quinone. However, the present results rather suggest that the OH groups in the phenoxyl radicals from catechols and gallates are not easily transferred to another radical, likely because they are involved in a strong intramolecular H-bond with the -O• group (see Scheme 4C) [30].

In agreement with this view of reasoning, the plot of *IP* versus $R_{inh}$, both divided by the molar concentration of the antioxidants to take into account their different molecular weights, shows the presence of a general trend consisting of an inverse relation between *IP* and $R_{inh}$ (see Figure 5), although the rather scattered data points suggest that other factors contribute to the *IP* vs. $R_{inh}$ relationship.

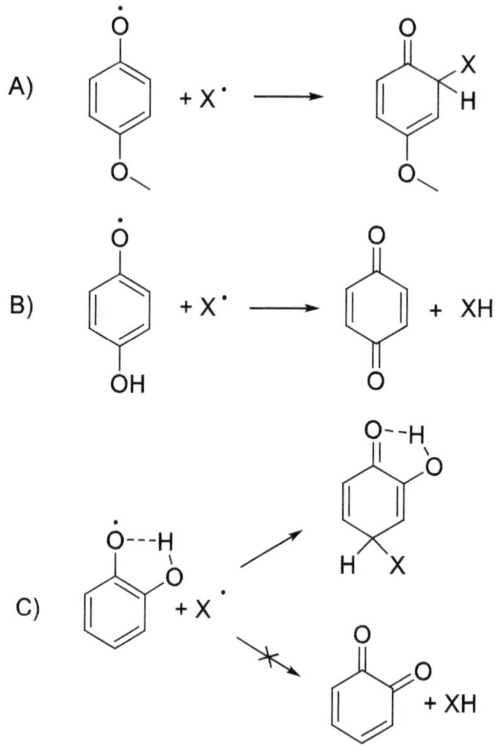


**Scheme 4.** Decay pathways of the phenoxyl radicals formed by monophenols (**A**), *para*-hydroquinones (**B**), and catechols (**C**) upon reaction with a generic radical X•.

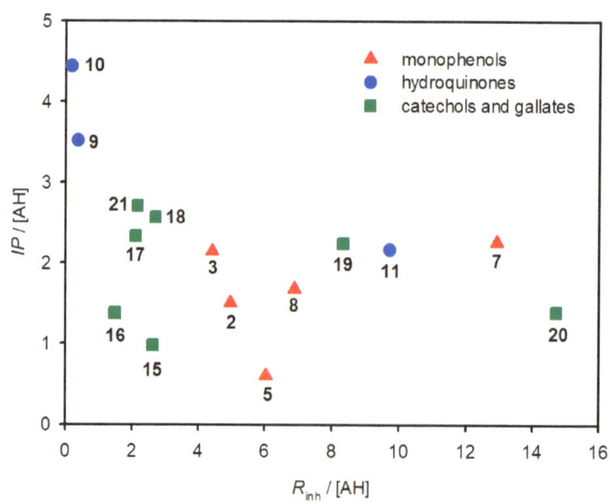


**Figure 5.** Relation between $R_{inh}$ and *IP* divided by the molar antioxidant concentration (in arbitrary units). Antioxidants with undefined *IP* were omitted.

It should also be pointed out that the phenomenon of chain propagation by the phenoxyl radicals of the antioxidants A• is not observed at low temperatures when the autoxidation of SSO is accelerated by the decomposition of azoinitiators, probably because the high flux of ROO• radicals ensures an efficient decay of A• for all antioxidants, while the low temperature slows down reaction 9 [31]. On the other hand, the low *IP* of α-tocopherol compared to γ-tocopherol in soybean and sunflower oils at 100 °C has been in part attributed to reaction 9 [32]. Moreover, the chain propagation by the α-tocopheroxyl radicals is responsible for the tocopherol-mediated peroxidation in low-density lipoproteins [33].

The low efficacy of the dialkylated hydroquinone **11** is instead to be ascribed to the tendency of the electron-rich semiquinones to react with $O_2$ forming *para*-benzoquinones and HOO•, which propagates the oxidative chain (Scheme 5) [34].

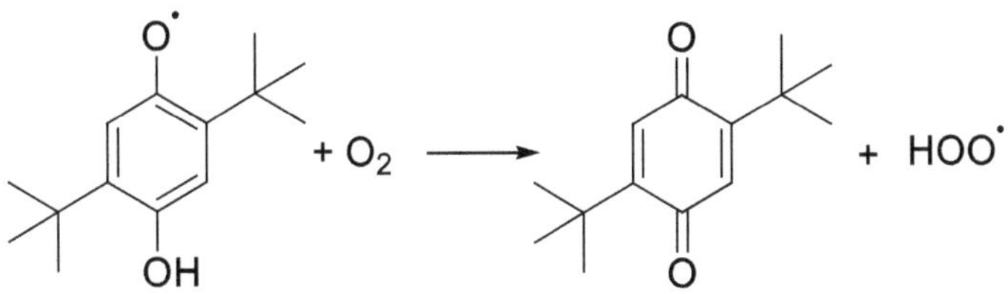

**Scheme 5.** Chain propagation of dialkylated semiquinone radicals.

## 5. Conclusions

Autoxidation experiments at 130 °C of stripped sunflower oil in the presence of various phenolic antioxidants, having monophenolic, hydroquinone, or catechol active moieties, showed that hydroquinones with one alkyl substituent display the best activity in terms of inhibition degree of $O_2$ consumption and duration of the inhibited period. By using EPR spectroscopy, we could demonstrate that the stability of the radical of the antioxidants plays a counterintuitive role in determining the overall activity. Antioxidants like α-tocopherol, which form relatively persistent radicals, are not very effective in counteracting oil peroxidation, because their radicals can abstract an H-atom from polyunsaturated lipids. On the other hand, hydroquinones form very unstable semiquinone radicals that disappear quickly by H-atom donation to another radical, and thus are incapable of propagating the oxidative chain.

Although many efforts are still needed to fully clarify the mechanisms of the inhibition of lipid peroxidation, these results are important to develop novel antioxidant strategies for the protection of oils with a high polyunsaturated fatty acid content at high temperatures, and possibly of lipid-containing foods in general. Considering the interest in finding new sources of natural phenolic antioxidants, knowledge of the structural factors underlying their activity can drive the research toward new (poly)phenolic materials having hydroquinone functionality [35] with practical applications for food and biodiesel stabilization.

**Author Contributions:** Conceptualization, R.A.; methodology, R.A. and F.M.; investigation, F.M. and L.B.; writing—original draft preparation, R.A.; writing—review and editing, R.A., F.M. and L.B. All authors have read and agreed to the published version of the manuscript.

**Funding:** This research was funded by Eni S.p.A. by providing a scholarship to F.M.

**Institutional Review Board Statement:** Not applicable.

**Informed Consent Statement:** Not applicable.

**Data Availability Statement:** Data is contained within the article.

**Conflicts of Interest:** The funders had no role in the design of the study; in the collection, analyses, or interpretation of data; in the writing of the manuscript; or in the decision to publish the results.

## References


1. Feng, J.; Berton-Carabin, C.C.; Fogliano, V.; Schroën, K. Maillard reaction products as functional components in oil-in-water emulsions: A review highlighting interfacial and antioxidant properties. *Trends Food Sci. Technol.* **2022**, *121*, 129–141. [CrossRef]
2. Romola, C.V.J.; Meganaharshini, M.; Rigby, S.P.; Moorthy, I.G.; Kumar, R.S.; Karthikumar, S. A comprehensive review of the selection of natural and synthetic antioxidants to enhance the oxidative stability of biodiesel. *Renew. Sustain. Energy Rev.* **2015**, *145*, 111109. [CrossRef]
3. Baschieri, A.; Amorati, R. Methods to determine chain-breaking antioxidant activity of nanomaterials beyond DPPH•. A review. *Antioxidants* **2021**, *10*, 1551. [CrossRef]
4. Borges, A.; de Freitas, V.; Mateus, N.; Fernandes, I.; Oliveira, J. Solid lipid nanoparticles as carriers of natural phenolic compounds. *Antioxidants* **2020**, *9*, 998. [CrossRef]
5. Piccinino, D.; Capecchi, E.; Tomaino, E.; Gabellone, S.; Gigli, V.; Avitabile, D.; Saladino, R. Nano-structured lignin as green antioxidant and UV shielding ingredient for sunscreen applications. *Antioxidants* **2021**, *10*, 274. [CrossRef] [PubMed]
6. Roy, S.; Rhim, J.W. New insight into melanin for food packaging and biotechnology applications. *Crit. Rev. Food Sci. Nutr.* **2022**, *62*, 4629–4655. [CrossRef]
7. Mollica, F.; Lucernati, R.; Amorati, R. Expanding the spectrum of polydopamine antioxidant activity by nitroxide conjugation. *J. Mater. Chem. B* **2021**, *9*, 9980–9988. [CrossRef] [PubMed]
8. Amorati, R.; Attanasi, O.A.; Favi, G.; Menichetti, S.; Pedulli, G.F.; Viglianisi, C. Amphiphilic antioxidants from "cashew nutshell liquid" (CNSL) waste. *Org. Biomol. Chem.* **2011**, *9*, 1352–1355. [CrossRef]
9. Russo, E.; Spallarossa, A.; Comite, A.; Pagliero, M.; Guida, P.; Belotti, V.; Caviglia, D.; Schito, A.M. Valorization and potential antimicrobial use of olive mill wastewater (OMW) from Italian olive oil production. *Antioxidants* **2022**, *11*, 903. [CrossRef]
10. Helberg, J.; Pratt, D.A. Autoxidation vs. antioxidants–the fight for forever. *Chem. Soc. Rev.* **2021**, *50*, 7343–7358. [CrossRef]
11. Guo, Y.; Pizzol, R.; Gabbanini, S.; Baschieri, A.; Amorati, R.; Valgimigli, L. Absolute antioxidant activity of five phenol-rich essential oils. *Molecules* **2021**, *26*, 5237. [CrossRef] [PubMed]
12. Ten Klooster, S.; Boerkamp, V.; Lazaridi, E.; Yang, S.; Takeuchi, M.; Berton-Carabin, C.; Schroën, K.; Janssen, H.; Friedrich, H.; Hohlbein, J.; et al. Lipid oxidation in food emulsions: Analytical challenges and recent developments. In *Lipid Oxidtion in Food and Biological Systems*; Bravo-Diaz, C., Ed.; Springer: Cham, Germany, 2022; pp. 3–29.
13. Farhoosh, R.; Niazmand, R.; Rezaei, M.; Sarabi, M. Kinetic parameter determination of vegetable oil oxidation under Rancimat test conditions. *Eur. J. Lipid Sci. Technol.* **2008**, *110*, 587–592. [CrossRef]
14. Dodos, G.S.; Tsesmeli, C.E.; Zannikos, F. Evaluation of the antimicrobial activity of synthetic and natural phenolic type antioxidants in biodiesel fuel. *Fuel* **2017**, *209*, 150–161. [CrossRef]
15. Becker, R.; Knorr, A. An Evaluation of antioxidants for vegetable oils at elevated termperatures. *Lubr. Sci.* **1996**, *8*, 95–117. [CrossRef]
16. Mollica, F.; Lucarini, M.; Passerini, C.; Carati, C.; Pavoni, S.; Bonoldi, L.; Amorati, R. Effect of antioxidants on high-temperature stability of renewable bio-oils revealed by an innovative method for the determination of kinetic parameters of oxidative reactions. *Antioxidants* **2020**, *9*, 399. [CrossRef] [PubMed]
17. Mollica, F.; Gelabert, I.; Amorati, R. Synergic antioxidant effects of the essential oil component γ-terpinene on high-temperature oil oxidation. *ACS Food Sci. Technol.* **2022**, *2*, 180–186. [CrossRef]
18. Guillén, M.D.; Ruiz, A. Rapid Simultaneous Determination by proton NMR of unsaturation and composition of acyl groups in vegetable oils. *Eur. J. Lipid Sci. Technol.* **2003**, *105*, 688–696. [CrossRef]
19. Duling, D.R. Simulation of multiple isotropic spin-trap EPR spectra. *J. Magn. Reson. Ser. B* **1994**, *104*, 105–110. [CrossRef]
20. Eloranta, J.; Hamalainen, E.; Salo, E.; Makela, R.; Kekalainen, U. Radicals formed from α-tocopherol under oxidizing and reducing conditions. An EPR and ENDOR study. *Acta Chem. Scand. A* **1983**, *37*, 383–391. [CrossRef]
21. Tuner, H.; Korkmaz, M. Radiostability of butylated hydroxytoluene (BHT): An ESR study. *Nucl. Instrum. Methods Phys. Res.* **2007**, *258*, 388–394. [CrossRef]
22. Lucarini, M.; Pedrielli, P.; Pedulli, G.F.; Cabiddu, S.; Fattuoni, C. Bond dissociation energies of O−H bonds in substituted phenols from equilibration studies. *J. Org. Chem.* **1996**, *61*, 9259–9263. [CrossRef]
23. Denisov, E.T. Reactivity of quinones as alkyl radical acceptors. *Kinet. Catal.* **2006**, *47*, 662–671. [CrossRef]
24. Guo, Y.; Baschieri, A.; Mollica, F.; Valgimigli, L.; Cedrowski, J.; Litwinienko, G.; Amorati, R. Hydrogen atom transfer from HOO• to *ortho*-quinones explains the antioxidant activity of polydopamine. *Angew. Chem. Int. Ed.* **2021**, *60*, 15220–15224. [CrossRef] [PubMed]
25. Denisov, E.T.; Khudyakov, I.V. Mechanisms of action and reactivities of the free radicals of inhibitors. *Chem. Rev.* **1987**, *87*, 1313–1357. [CrossRef]
26. Jensen, R.K.; Korcek, S.; Zinbo, M.; Johnson, M.D. Initiation in hydrocarbon autoxidation at elevated temperatures. *Int. J. Chem. Kinet.* **1990**, *22*, 1095–1107. [CrossRef]
27. Lucarini, M.; Pedulli, G.F. Free radical intermediates in the inhibition of the autoxidation reaction. *Chem. Soc. Rev.* **2010**, *39*, 2106–2119. [CrossRef]
28. Amorati, R.; Menichetti, S.; Mileo, E.; Pedulli, G.F.; Viglianisi, C. Hydrogen-atom transfer reactions from *ortho*-alkoxy-substituted phenols: An experimental approach. *Chem. Eur. J.* **2009**, *15*, 4402–4410. [CrossRef]

29. Li, J.; Bi, Y.; Yang, H.; Wang, D. Antioxidative properties and interconversion of *tert*-butylhydroquinone and *tert*-butylquinone in soybean oils. *J. Agric. Food Chem.* **2017**, *65*, 10598–10603. [CrossRef]
30. Lucarini, M.; Mugnaini, V.; Pedulli, G.F. Bond dissociation enthalpies of polyphenols: The importance of cooperative effects. *J. Org. Chem.* **2002**, *67*, 928–931. [CrossRef]
31. Baschieri, A.; Pizzol, R.; Guo, Y.; Amorati, R.; Valgimigli, L. Calibration of squalene, p-cymene, and sunflower oil as standard oxidizable substrates for quantitative antioxidant testing. *J. Agric. Food Chem.* **2019**, *67*, 6902–6910. [CrossRef]
32. Yanishlieva, N.V.; Kamal-Eldin, A.; Marinova, E.M.; Toneva, A.G. Kinetics of antioxidant action of α- and γ-tocopherols in sunflower and soybean triacylglycerols. *Eur. J. Lipid Sci. Technol.* **2002**, *104*, 262–270. [CrossRef]
33. Bowry, V.W.; Stocker, R. Tocopherol-mediated peroxidation. The prooxidant effect of vitamin E on the radical-initiated oxidation of human low-density lipoprotein. *J. Am. Chem. Soc.* **1993**, *115*, 6029–6044. [CrossRef]
34. Barsukova, T.V.; Loshadkin, D.; Pliss, E. Substituted p-hydroquinones as inhibitors of lipid peroxidation. *Chem Phys Lipids* **2003**, *125*, 49–58.
35. Giner, R.M.; Ríos, J.L.; Máñez, S. Antioxidant activity of natural hydroquinones. *Antioxidants* **2022**, *11*, 343. [CrossRef]

*antioxidants*

MDPI

*Article*

# Synthesis, Characterization and Evaluation of a Novel Tetraphenolic Compound as a Potential Antioxidant

**Mengqi Xu [1,2], Pengcheng Meng [1], Hongyan Wang [3], Jun Liu [1], Tao Guo [3], Zhenjie Zhu [1,2] and Yanlan Bi [1,2,*]**

[1] College of Food Science and Technology, Henan University of Technology, Zhengzhou 450001, China; 2021920059@stu.haut.edu.cn (M.X.); pengchengmeng@haut.edu.cn (P.M.); liujun@haut.edu.cn (J.L.); 2021920086@stu.haut.edu.cn (Z.Z.)

[2] Food Laboratory of Zhongyuan, Luohe 462300, China

[3] College of Chemistry and Chemical Engineering, Henan University of Technology, Zhengzhou 450001, China; why@haut.edu.cn (H.W.); taoguo@haut.edu.cn (T.G.)

* Correspondence: byl@haut.edu.cn

**Abstract:** A novel antioxidant containing four hydroxyl groups, namely 2,2′-(2-methylpropane-1,3-diyl)bis(hydroquinone) (MPBHQ), was synthesized using hydroquinone and methylallyl alcohol as the raw materials, phosphoric acid as the catalyst, and toluene as the solvent system. The structure of MPBHQ was characterized by mass spectrometry, nuclear magnetic resonance, ultraviolet spectroscopy, and infrared spectroscopy. The results showed that MPBHQ has a good radical scavenging effect, as measured by the ORAC assay, DPPH radical scavenging assay, ABST radical scavenging assay, and Rancimat test. In fatty acid methyl ester and lard without exogenous antioxidants, MPBHQ showed better antioxidant performance than butylated hydroxytoluene (BHT), hydroquinone (HQ), tert-butyl hydroquinone (TBHQ), and propyl gallate (PG), meeting the need for a new antioxidant with better properties to ensure the oxidative stability of lipids and biodiesel.

**Keywords:** tetrahydroxyl-phenol antioxidants; synthesis and purification; free radical scavenging; lipid antioxidant

## 1. Introduction

Phenolic antioxidants are the main types of antioxidants, most of which are hydroxyl groups directly linked to the aromatic ring [1]. These compounds can provide hydrogen protons to block free-radical-dominated chain reactions [2], which gives them good antioxidant activity in oil and fat products and other chemical products. Although consumers currently prefer natural phenolic antioxidants, such as tocopherols, tea polyphenols, syringic acid, and rosmarinic acid [3], they are usually expensive due to their restricted sources and difficulty in purification [4]. Moreover, they have undesirable flavors for specific products or people [5]. The mainstream of the market is still synthetic phenolic antioxidants, which are widely used in food [1], medicine [6], cosmetics [7], fuel [8], feed [9], and other fields [10], by virtue of their low price, high yield, and good antioxidant effect.

Synthetic phenolic antioxidants have a similar structure, which is usually a phenolic hydroxyl group and an alkyl group attached to the benzene ring [11]. Commercial synthetic phenolic antioxidants, including butyl hydroxyanisole (BHA), butylated hydroxytoluene (BHT), and tert-butyl hydroquinone (TBHQ), are commonly used to inhibit the oxidative degradation of plastics and rubber and to delay the oxidative rancidity of food when used as food additives [12,13]. TBHQ is a common synthetic antioxidant containing two hydroxyl groups, which has a significant antioxidant effect on vegetable oils with high unsaturated fatty acids [14,15]. Propyl gallate (PG) is also a common synthetic polyphenol antioxidant, which is commonly used in oil-rich foods, such as fried foods and cookies [16]. However, most of the existing synthetic phenolic antioxidants are usually volatile at high temperatures (frying) and are prone to decomposition during the storage of oil and fat products [17], which is mainly because of their small molecular weight [18], and these defects reduce their antioxidant properties. Moreover, the application of synthetic phenolic antioxidants mainly focuses on mono-hydroxy and di-hydroxy species [19], and there is a lack of research on and application of synthetic polyphenolic antioxidants. Therefore, new antioxidants of a larger molecular weight with excellent thermostability are favored, and they may have better antioxidant properties and more extended application fields.

Hydroquinone (HQ) is an important raw material additive and intermediate in the synthesis of pharmaceuticals [20] and other fine chemicals. But HQ is less used in the oil and fat processing industry due to its poor oil solubility, small molecular weight, and easy volatility [21]. For the synthesis of phenolic antioxidants, HQ is an important reactant for the synthesis of TBHQ. TBHQ is synthesized from hydroquinone and tert-butyl alcohol by the Foucault alkylation reaction, which enhances the oil solubility and the antioxidant capacity of a synthetic antioxidant. It is well-known that the antioxidant capacity of natural phenolic antioxidants containing four phenolic hydroxyl groups, such as epicatechin and rosmarinic acid, is significantly stronger than that of tocopherol containing one hydroxyl group and TBHQ containing two phenolic hydroxyl groups [11,22,23]. However, TBHQ has a molecular weight of 164 and only contains two phenolic hydroxyl groups, which leads to its limited antioxidant capacity and the universality of the use environment [17]. Based on this design theory, 2-(*tert*-butyl)-5-methylbenzene-1,4-diol (TBMHQ) containing four phenolic hydroxyl groups was synthesized [24], and the DPPH radical scavenging test and oxidative decay test results showed that it has enhanced antioxidant properties. Therefore, thermal stability and antioxidant activity can be improved by increasing the molecular weight of the substance and introducing more hydroxyl groups. In addition, the introduction of alkyl terminations improves the oil solubility of the antioxidant. The alkylating reagent chosen is methyl allyl alcohol, which has one more double bond than the usual alkylating reagent used in the synthesis of antioxidants [25]. It is designed to increase the molecular weight of hydroquinone by the Friedel–Crafts alkylation reaction.

In this work, a novel synthetic polyphenol antioxidant named 2,2′-(2-methylpropane-1,3-diyl)bis(hydroquinone) (MPBHQ) containing four hydroxyl groups was successfully synthesized based on the Friedel–Crafts reaction. The molecular weight and main chemical properties of MPBHQ were identified by high-resolution mass spectrum (HRMS), retention factor (Rf) value, and UV spectroscopy, and the chemical structure was also confirmed by HRMS, nuclear magnetic resonance (NMR), and Fourier-transform infrared (FTIR) spectroscopy. The antioxidant properties of MPBHQ were studied by the ORAC assay, DPPH radical scavenging assay, ABTS radical scavenging assay, and Rancimat test. MPBHQ exhibits strong antioxidant activity in its edible oil system and mixed fatty acid methyl ester. Therefore, MPBHQ, a novel synthetic antioxidant with a definite chemical structure and excellent antioxidant capacity, has broad application prospects in the food and chemical industries.

## 2. Materials and Methods

### 2.1. Materials

Toluene, hydroquinone, 2-methylallyl alcohol, phosphoric acid, potassium persulfate, C18 column chromatography silica gel (40–60 μm), iodine (AR), and other non-specifically indicated chemicals of analytical purity grade were from Shanghai Aladdin Biochemical Technology Co., Ltd. (Shanghai, China). Silica-gel-coated plates (100 × 100 mm, 0.25 mm) were purchased from Qingdao Ocean Chemical Co., Ltd. (Qingdao, China). ABTS (>98%), DPPH (>98%), mixed fatty acid methyl ester (fatty acid methyl ester, ≥99%), 2,2′-azobis(2-amidinopropane) dihydrochloride (AAPH), Trolox (6-hydroxy-2,5,7,8-tetramethylchroman-2-carboxylic acid), which is a hydrophilic analogue of vitamin E, fluorescein sodium salt, tert-butylhydroquinone (TBHQ, >98%), butylated hydroxytoluene (BHT, >98%), and propyl gallate (PG, >98%) were also purchased from Shanghai Aladdin Biochemical Technology Co., Ltd. Refined lard without antioxidants was purchased from Tianjin Jiuyuan Oils and Fats Co., Ltd. (Tianjin, China). Chromatographically pure methanol was obtained from

VBS Biologic Inc. (New York, NY, USA). Deuterated methanol (≥99.8%) was obtained from Cambridge Isotope Laboratories, Inc. (Andover, MA, USA).

### 2.2. Synthesis of MPBHQ

MPBHQ was synthesized through a liquid-phase reaction strategy using phosphoric acid as the catalyst. The main factors, including reaction temperature, reaction time, and reactant ratio, were chosen to investigate the effect on the synthesis of MPBHQ. Typically, 4.4 g of hydroquinone and 4 mL of phosphoric acid were added to 20 mL of toluene with magnetic stirring in a thermostatic water bath. When the reaction temperature reached 90 °C, 4 mL of 2-methylallyl alcohol was added dropwise to the above mixture. After the reaction was performed for 20 min at 90 °C with magnetic stirring, the reaction products were transferred into a pear-shaped separatory funnel. Then, the aqueous phase containing phosphoric acid was separated, and the unreacted p-phenol was also separated by washing with hot water. The retained organic phase was transferred into a 100 mL round bottom flask, and the toluene was evaporated at 70 °C using a rotary evaporator. Eventually, a white powder was obtained after drying in a vacuum oven at 70 °C for 6 h. The synthesis of MPBHQ and the harvest process of the crude products are shown in Scheme 1.

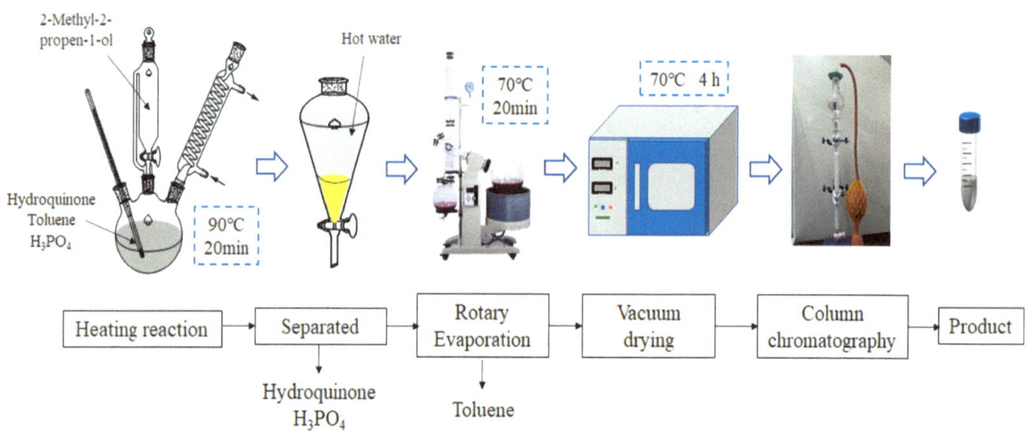


**Scheme 1.** Flowchart on the synthesis and harvest process of MPBHQ crude products.

### 2.3. Content of MPBHQ

The content of MPBHQ was analyzed by high-performance liquid chromatography (HPLC). The sample was dissolved in methanol, and this solution was purified by a 0.45 μm filter membrane. The sample concentration was finally fixed at 1 mg·mL$^{-1}$. The signal of the compound was collected on a high-performance liquid chromatograph (Waters 2695, Waters Corp., Milford, MA, USA) equipped with a C18 column (Waters SunFire, 4.6 mm × 250 mm, 5 μm) and an ultraviolet (UV)/visible detector (Waters 2489, Waters Corp., Milford, MA, USA). The detection wavelength was 280 nm, the column temperature was 35 °C, and the injection volume was 20 μL. The mobile phase was a mixture of methanol and water, and the phase A was aqueous solution containing 0.5% acetic acid, phase B was methanol. The flow rate of the mobile phase was 0.8 mL·min$^{-1}$. The samples were analyzed according to the following gradient procedure. The elution profile was as follows: 0 min, 60% A (aqueous solution containing 0.5% acetic acid); 1–8 min, 60–20% A; 8–15 min, 20–0% A; 15–20 min, 0% A; 20–28 min, 0–60% A. The content of MPBHQ was calculated according to the area normalization method.

### 2.4. Separation and Purification

Reversed-phase column chromatography coupled with a laboratory self-loaded C18 column (17 × 300 mm, 40–60 μm) was selected for the separation and purification of the crude products containing MPBHQ. After the C18 silica-gel was activated by using methanol for 8 h, it was loaded using a wet-process column-loading strategy. Before the separation process, the column was kept flat with quartz sand and equilibrated with the eluent until the column bed was uniform and bubble-free. After dissolving the crude products containing MPBHQ in methanol completely, the solution was loaded in the pre-column and separated in the following elution program. The mobile phase was a mixture of methanol and water (methanol:water, 4:6, $v/v$), the flow rate was 5 $mL \cdot min^{-1}$, and the sample loading mass was 0.2 g. Finally, the eluate was collected into a 10 mL test tube for subsequent analysis.

### 2.5. Identification of the Structure of MPBHQ

The structure of MPBHQ was identified based on the information from the HRMS, NMR, and FTIR. The chemical bond information of MPBHQ was recorded from a Brucker ALPHA FTIR spectrometer (Bruker, Karlsruhe, Germany) using ATR with the wavenumber range of 400–4000 $cm^{-1}$. For LC-MS analysis, the conditions for this part of the liquid chromatography (LC) were the same as those in Section 2.3., but the injection volume was 10 μL. The ionization of substances depended on the Electron Spray Ionization (ESI). The executive mass spectrometer was performed in the negative ion mode. The mass range was 110–500 $m/z$. The high-purity nitrogen was used as a nebulizer and drying gas, and the desolvation gas flow rate was 380 L/h. The ToF-MS functions were set as follows: the desolvation temperature was 180 °C, the source temperature was 80 °C, the capillary voltage was 2.9 kV, and the sampling cone voltage was 30 eV. $^{1}$H NMR, $^{13}$C NMR, 2D COSY, 2D HSQC, and 2D HMBC spectra were recorded at 25 °C on a Bruker AVANCE III 500 MHz NMR spectrometer (500.13 MHz proton frequency) equipped with a BBI liquid probe and Z-gradient system. Deuterated methanol was used as a lock solvent, and the concentration of the working solution for NMR was 30 $mg \cdot mL^{-1}$. All the pulse programs used for NMR experiments were the standard ones equipped in Bruker software Topspin 3.2. For 1D−$^{1}$H NMR and 1D−$^{13}$C NMR spectra, and both of the pulse programs "zg" and "zgpg30" were used with a delay time of 2.0 s. The 2D COSY spectra were acquired with the pulse program "cosygpmfqf" with a relaxation delay of 2.0 s. A data matrix of 2048 × 128 points was recorded with 8 scans for each increment. Two-dimensional HSQC spectra were acquired with the pulse program "hsqcetgpsisp2.2" with a relaxation delay of 2.0 s and a J(CH) constant of 145 Hz. Two-dimensional HMBC spectra were acquired with the pulse program "hmbcgpndqf" with a relaxation delay of 1.5 s and a J(XH) constant of 8 Hz. The data matrix of 1024 × 256 points were recorded in HSQC and HMBC with 8 scans for each increment. $^{1}$H NMR (500 MHz, Methyl alcohol-d) δ 6.55 (d, J = 1.9 Hz, 1H, H-3), 6.35 (d, J = 2.0 Hz, 1H, H-1), 0.90 (d, 3H, H-6), 2.52 (m, H, H-5), 3.98 ppm (d, 2H, H-4). $^{13}$C NMR (500 MHz, Methyl alcohol-d) δ (118.2, C-5), δ (116.4, C-3), δ (115.2, C-2), δ (23.6, C-9), δ (33.6, C-8), δ (46.2, C-7). HRMS (ESI): Calcd for $C_{16}H_{18}O_4^-$ $[M\text{-}H]^-$ $m/z$ 273.0934, found 273.0934. The purity of MPBHQ sample was determined by HPLC as 98.0%.

### 2.6. Retention Factor Value

The retention factor (Rf) value of selected compounds was determined by thin layer chromatography (TLC). Typically, 5 μL of methanol solution containing 0.15 mg of the samples were spotted on a silica-gel plate. Then, the plate was developed in the chloroform methanol mixture (19:1, $v/v$), and the total expansion distance of TLC plate was 9 cm. After the expansion was completed and the solvent was evaporated, iodine was used to develop the color. The Rf value was calculated based on the shift distance of the sample and using the equation:

$$\mathrm{Rf} = \mathrm{d_c}/\mathrm{d_o}$$

where $d_c$ and $d_o$ represent the distance moved from the origin by component and the distance moved from the origin by solvent, respectively.

### 2.7. Free-Radical Scavenging Assay

DPPH radical scavenging rate and ABTS radical scavenging rate were selected as two indexes to evaluate the antioxidant capacity of MPBHQ. The DPPH radical scavenging assay of BHT, HQ, TBHQ, PG, and MPBHQ was performed as described in Jiang et al. [19]. Briefly, 0.5 mL of antioxidant methanol solution containing 1.5, 3.0, 6.0, 12.0, 24.0, and 48.0 $\mu g \cdot mL^{-1}$ was mixed with 3.0 mL of 0.1 mM DPPH methanol solution, respectively. This mixture was sufficiently shaken and reacted under darkness at room temperature for 30 min. The absorption was measured on a TU-1810 UV-visible spectrophotometer (PuXi Tongyong, Beijing, China) at 517 nm. The methanol was used as the blank test sample. A total of 0.5 mL of methanol was added in 3.0 mL DPPH solution, and this mixture served as the control.

ABTS radical scavenging activity of MPBHQ BHT, HQ, TBHQ and PG was determined according to the previous method reported by Feng [26]. A total of 9 mL of 7 mM ABTS aqueous solution was mixed with 1 mL of 24.5 mM potassium persulfate solution, and this mixture was diluted 50 times after it was left to react in the dark at room temperature for 16 h. Then, 0.2 mL of each antioxidant methanol solution was dispersed in 4.0 mL of ABTS solution and reacted in the dark for 6 min. The absorbance value was collected on a TU-1810 UV-visible spectrophotometer at 734 nm. Methanol was used as the blank test sample, and the mixture containing 0.2 mL of methanol and 4.0 mL of DPPH solution served as the control.

The radical scavenging activity (RSA) was calculated using the equation:

$$RSA\ (\%) = [(A_2 - A_1)/A_2] \times 100\%$$

where $A_1$ is the absorbance value of the tested sample, and $A_2$ is the absorbance value of control. The free radical scavenging activity of antioxidants is expressed as $EC_{50}$, which represents the effective concentration required to achieve 50% antioxidant capacity [27].

### 2.8. ORAC Assay

The oxygen radical absorption capacity (ORAC) assay was measured by a modified method, and this method used an enzyme marker (SPARK, Tecan, Männedorf, Switzerland) with an excitation wavelength of 485 nm and an emission wavelength of 530 nm. In each well, 150 μL of FL (78 nM) and 25 μL of sample, blank (PBS), or standard (Trolox, 6.25–100 μmol/L) were placed, and then 25 μL of AAPH (73 mM) were added. The fluorescence was measured immediately after the addition, and the measurements were then taken every 5 min until the relative fluorescence intensity (FI%) was less than 5% of the value of the initial reading. The measurements were taken in triplicate. The area under the fluorescence decay curve (AUC) was calculated by applying the following equations:

$$ORAC\ (\mu m\ TE) = (C_{Trolox}/C_{sample}) \cdot [(AUC_{sample} - AUC_{blank})/(AUC_{Trolox} - AUC_{blank})]$$

$$AUC = 0.5 \times [2 \times (f_0 + f_1 + \ldots + f_{n-1} + f_n) - f_0 - f_n]\triangle t$$

where $f_0$ is the initial fluorescence and $f_n$ is the fluorescence at time n.

### 2.9. Rancimat Test

The antioxidant activities of MPBHQ, BHT, HQ, TBHQ, and PG in lard system and mixed fatty acid methyl ester system were determined according to the AOCS Official Method Cd 12b-92 Oil Stability Index, and the data were collected on a Rancimat 892 Apparatus (Metrohm, Herisau, Switzerland). A total of 3.00 ± 0.01 g of lard containing antioxidants were subjected to accelerated oxidation under the following conditions. The heating temperature was 120, 130, and 140 °C with an airflow rate fixed at 20 $L \cdot h^{-1}$, and the measuring cell distilled water dosage was 60 mL. The oxidation induction time of BHT,

HQ, TBHQ, PG, and MPBHQ on lard at different concentrations (0.01%, 0.02%, 0.04%) was determined, and lard without antioxidant was the control. The protection factor (Pf) values were calculated following the equation:

$$Pf = IP_{sample}/IP_{control}$$

where $IP_{sample}$ and $IP_{control}$ represent the induction period (IP) of lard samples with and without antioxidants, respectively.

### 2.10. Statistical Analysis

All experiments were performed in duplicate, and data were presented as mean ± standard deviation (SD). The data were analyzed by Duncan's multiple range tests using ver. 26.0 SPSS (IBM SPSS, Almonk, NY, USA).

## 3. Results and Discussion

### 3.1. Synthesis, Purification and Characterization of MPBHQ

The novel tetra phenolic antioxidant MPBHQ was obtained as illustrated in Scheme 1. The synthesis of MPBHQ was based on the phosphoric acid ($H_3PO_4$) catalyzed Friedel–Crafts alkylation of hydroquinone (A) and 2-methylallyl alcohol (B). Toluene as the solvent provided the reaction environment. The composition of the crude products containing MPBHQ were analyzed, and the possible products were shown in Figure 1a. The effect of reaction temperature, molar ratio of hydroquinone to methyl allyl alcohol, and reaction time on the synthesis of MPBHQ was investigated, and the results were shown in Table 1. When the molar ratio of hydroquinone, methyl allyl alcohol and $H_3PO_4$ was 1:1.2:0.5, the reaction temperature was 90 °C, and the reaction time was 20 min, the content of MPBHQ reached 41%. After the separation and purification according to the steps shown in Section 2.4 and Scheme 1, the MPBHQ content in the purified product can reach 98.0%, which was determined by a high-performance liquid chromatograph.

The properties and structure of MPBHQ were identified and confirmed by TLC, ultraviolet spectroscopy, NMR, HRMS (ESI), FTIR, etc. For the Rf values of HQ and MPBHQ obtained by TLC, and Rf value can reflect the polarity of the substance [25]. The Rf values of HQ and MPBHQ were 0.20 and 0.45, respectively, which indicated that MPBHQ exhibited stronger lipophilicity than HQ. From (Figure 1b), MPBHQ showed a strong UV absorption peak centered at 297 nm. After mixing MPBHQ with KOH solution, the absorption peak is enhanced and red-shifted, indicating that MPBHQ is a phenolic compound. The chemical structure of MPBHQ was further confirmed by FTIR. From Figure 1c, the absorption peak centered at 3316.88 $cm^{-1}$ is assigned to the stretching vibration of the phenolic hydroxyl group on the benzene ring [28]. The peaks located at 2949.95 and 1020.42 $cm^{-1}$ are assigned to the stretching vibration and in-plane bending vibration of C-H on the benzene ring, respectively. The peaks centered at 2947.88 and 2834.50 $cm^{-1}$ are attributed to the stretching vibration of C-H on alkanes, and at 1422.41 $cm^{-1}$ is attributed to the bending vibrations of C-H on alkanes. In order to further determine the structure of the MPBHQ, its molecular weight was determined by HRMS. From Figure 2a, the primary-ion mass spectra of MPBHQ showed the fundamental peak $m/z$ 273.0934 $[M\text{-}H]^-$, which inferred its molecular weight to be 274. As shown in Figure 2b, the secondary-ion mass spectra of MPBHQ shows more detailed structural information, such as the benzene ring structure. The chemical structure of MPBHQ was identified by NMR spectra. The NMR spectra of MPBHQ was shown in Figure 3. As shown in Figure 3a, the chemical shifts of hydrogen on the benzene ring were located at around 6–7 ppm. The hydrogen on C9 has the biggest electron shielding effect in the molecule structure, so it shows the smallest chemical shift in the high field region at 0.90 ppm. The chemical shifts of hydrogen at H4 and H5 are 3.98 ppm and 2.52 ppm, respectively.

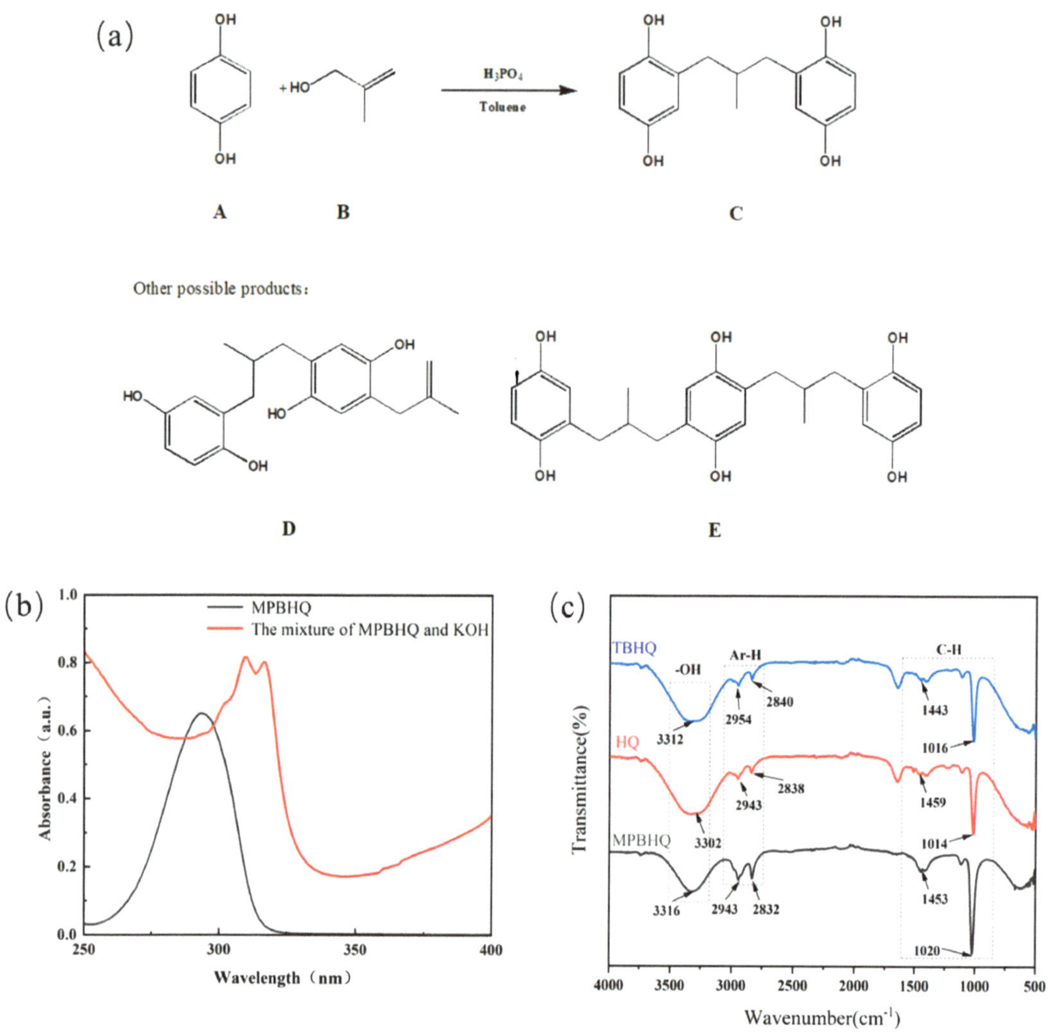


**Figure 1.** (**a**) Possible products in the synthesis reaction. (**b**) Ultraviolet spectrogram of MPBHQ solution and the mixture of MPBHQ and KOH. (**c**) The FTIR spectra of MPBHQ, HQ and TBHQ.

**Table 1.** The reaction results of several typical reactions.

| | Temperature (°C) | Molar Ratio of Phenol to Alcohol | Time (min) | Composition of the Reaction Products (%) | | |
|---|---|---|---|---|---|---|
| | | | | HQ | MPBHQ | Others |
| 1 | 90 | 1:2 | 20 | 25 | 20 | 55 |
| 2 | 120 | 1:1 | 40 | 37 | 10 | 53 |
| 3 | 60 | 1:2 | 20 | 52 | 20 | 28 |
| 4 | 90 | 1:1.2 | 20 | 22 | 41 | 37 |

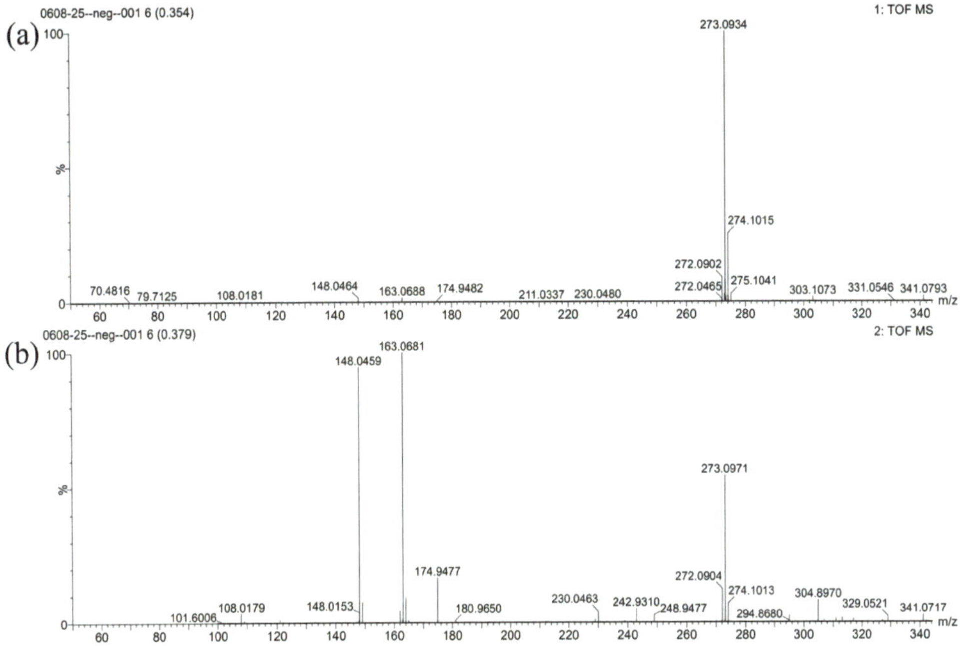


**Figure 2.** The primary (**a**) and secondary (**b**) ion mass spectra of MPBHQ.

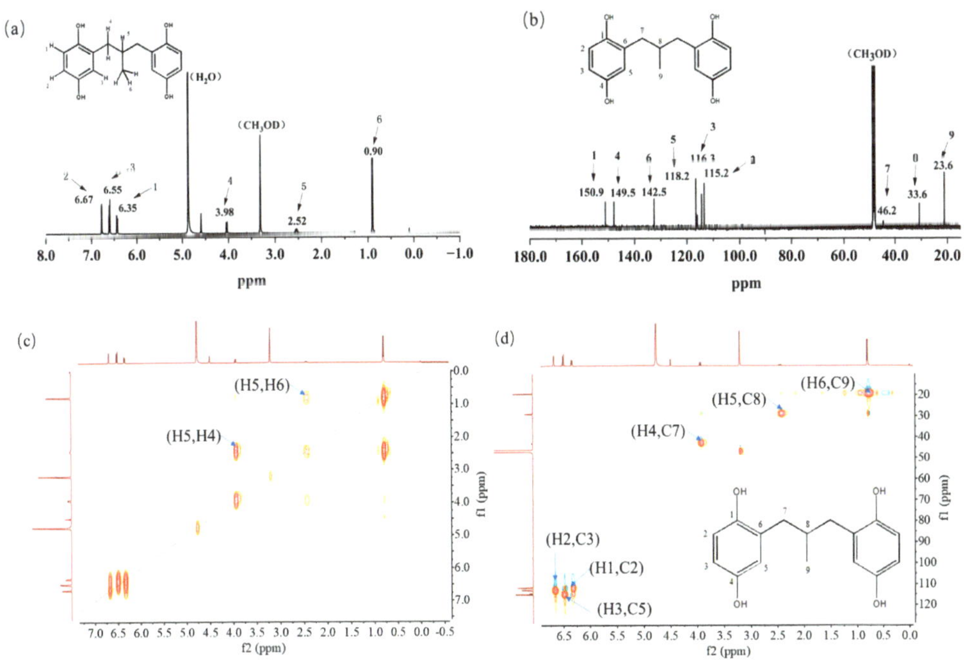


**Figure 3.** *Cont.*

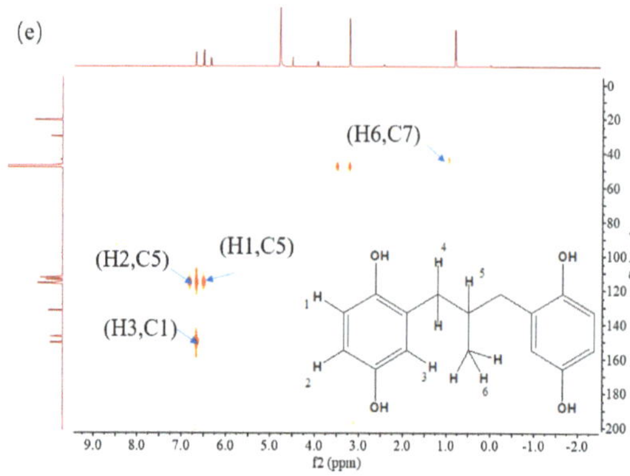


**Figure 3.** NMR spectra of MPBHQ, including $^{1}$H NMR (**a**), $^{13}$C NMR (**b**), $^{1}$H-$^{1}$H COSY-NMR (**c**), $^{13}$C-$^{1}$H HSQC-NMR (**d**), and $^{13}$C-$^{1}$H HMBC-NMR (**e**).

All the assignments of hydrogen chemical shifts in Figure 3a were further determined by the COSY spectra (Figure 3c). The information on the coupling of each group was obtained from the relevant peaks outside the diagonal of the COSY spectra. We can easily obtain the coupling information of H5 and H6 for H4 and H5. Based on the correlation signals of C-H in HSQC spectra (Figure 3d), all the assignments of carbons in MPBHQ molecules were shown in Figure 3b. The chemical shifts of carbons connecting the two phenolic hydroxyl positions (C1 and C4) are determined at 150.9 ppm and 149.5 ppm, respectively. The chemical shifts of carbons attached to hydrogens on the benzene ring were determined at the range of 110–120 ppm. For the alkyl fraction, the chemical shift of the methyl carbon was determined at 23.6 ppm. The C8 of methylene was assigned to the chemical shift of 33.6 ppm. Then, from Figure 3d, we found a correlation between the three hydrogen atoms on the benzene ring except the hydrogen atoms on the hydroxyl group and the carbon atoms of the benzene ring, which are H1, H2, and H3. The correlations of hydrogen atoms H4, H5, and H6 on alkyl corresponding to C7, C8, and C9 are also marked in the figures. In Figure 3e, the correlations of H and C in adjacent or interphase positions are marked, such as the correlations of H6 and C7, and H1 and C5. In summary, the chemical information of MPBHQ was determined, and it is an alkylbenzene compound containing four hydroxyl groups, as shown in Figure 3.

### 3.2. DPPH Radical Scavenging Assay

The DPPH radical scavenging assay has the advantages of being efficient, convenient, sensitive, and reproducible. Moreover, the $EC_{50}$ value of the sample concentration corresponding to a free radical scavenging rate of 50% was calculated to reflect the scavenging efficiency of the antioxidants on free radicals [29]. MPBHQ contains four phenolic hydroxyl groups, which is presumed to have a strong antioxidant effect. MPBHQ and the compounds containing different amounts of phenolic hydroxyl groups, BHT, HQ, TBHQ, and PG, were selected to examine their scavenging ability against DPPH radicals, and the antioxidant capacity of MPBHQ was reflected by comparing the DPPH free radical scavenging rate. From Figure 4a, the DPPH radical scavenging rate of MPBHQ and other antioxidants increased with the increase in antioxidant concentrations. $EC_{50}$ values of TBHQ, HQ, PG, and MPBHQ were 22.20, 11.34, 8.74, and 7.93 μg/mL, respectively. Due to the weak ability of BHT, the $EC_{50}$ values of BHT cannot be calculated. Among all of the selected compounds, the DPPH radical scavenging ability of HQ, PG, and MPBHQ increased rapidly with increasing concentration from 1.5 to 24 μg/mL. Moreover, when the compound concentration was 24 μg/mL, their free radical scavenging rate reached 86.10%, 86.21%, and 93.85%, respectively. Overall, the DPPH radical scavenging ability was MPBHQ > PG > HQ > TBHQ > BHT. That is to say, the more phenolic hydroxyl groups of the compound, the better the scavenging activity on DPPH radicals. The scavenging ability of MPBHQ containing tetrahydroxyl groups was almost twice as high as that of TBHQ. This is mainly because of the enhanced hydrogen supply capacity from the presence of two hydroquinones in the structure and more DPPH capture by the released [H] from the stable resonators [18]. MPBHQ combined the hydrogen donating capacity from the two hydroquinones. Therefore, the DPPH radical scavenging capacity is almost twice that of TBHQ. Based on the selected method in this work, the DPPH radical scavenging assay in methanol solution is mainly dominated by single electron transfer (SET) mechanism. However, ABTS radicals can be eliminated by both SET and hydrogen atom transfer (HAT) mechanism. In order to evaluate the antioxidant activity of MPBHQ in detail, ABTS radical scavenging assay was carried out.

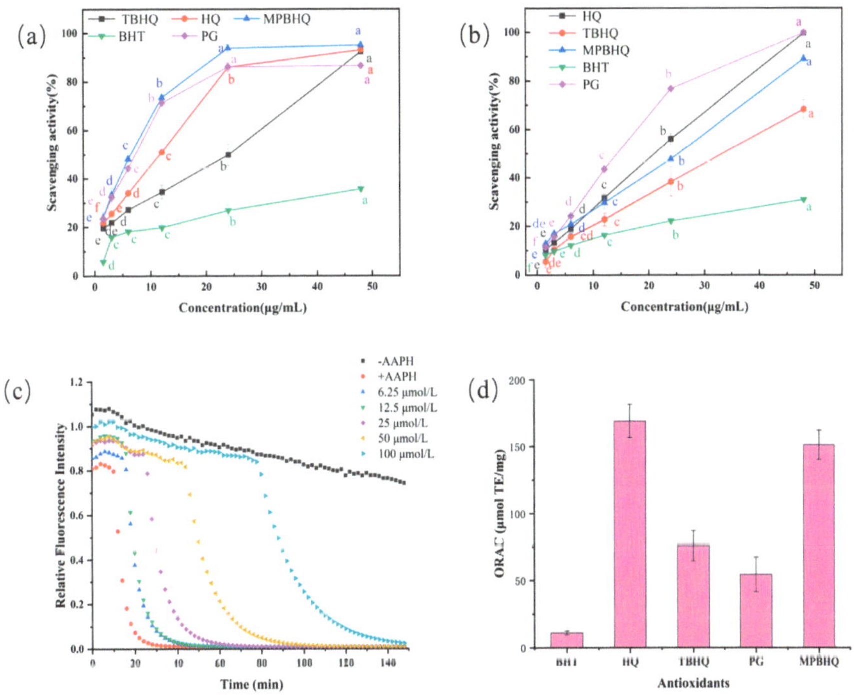


**Figure 4.** (**a**) DPPH radical scavenging activity of different antioxidants. (**b**) ABTS radical scavenging activity of different antioxidants. (**c**) The attenuation curve of fluorescence intensity of Trolox solution. (**d**) ORAC values of different antioxidants. The different lowercase letters in the graph, including a, b, c, d, e, and f, indicate the significant difference of $EC_{50}$ values of the same substance at different concentrations ($p < 0.05$).

### 3.3. ABTS Radical Scavenging Assay

The ABTS radical scavenging activity of the selected antioxidant compounds was shown in Figure 4b. The scavenging rate of ABTS radicals increased with the increase in the antioxidant concentration (1.5–48 μg/mL). $EC_{50}$ values of TBHQ, MPBHQ, HQ, and PG were 33.34, 24.35, 21.81, and 18.17 μg/mL, respectively. BHT still showed weak scavenging activity, and the scavenging ability of ABTS radical was PG > HQ > MPBHQ > TBHQ > BHT. Different from the results of the DPPH radical scavenging assay, the ABTS radical scavenging activity of MPBHQ with four hydroxyl groups was higher than that of TBHQ with two hydroxyl groups and BHT with one hydroxyl group, but lower than that of PG with three hydroxyl groups and HQ with two hydroxyl groups. This may be due to the amphiphilic nature of benzothiazole structure of the ABTS radical rather than the hydrophobic aromatic ring structure of the DPPH radical, which makes it easier to undergo electron transfer or hydrogen transfer with phenolics [30,31]. Therefore, it is speculated that the alkyl groups attached to MPBHQ and TBHQ are the main reason for their lower scavenging effect on ABTS radical than PG and HQ. Overall, it can be demonstrated that MPBHQ has strong radical scavenging ability in non-oil systems, indicating its strong antioxidant ability. Whether MPBHQ can exhibit strong antioxidant activity in oil and fat with a more complex composition, it is necessary to carry out research on the antioxidant capacity of MPBHQ in oil and fat systems.

From the results of DPPH and ABTS assay (Figure 4a,b), MPBHQ exhibited strong single electron transfer ability for scavenging free radicals. However, it is generally believed that the method based on the HAT mechanism is similar to the classical lipid peroxidation method, which is closest to the real oxidation process of the physiological environment. Therefore, in order to detect the antioxidant activity of MPBHQ in lipids, ORAC assay was carried out.

### 3.4. ORAC Assay

The antioxidant capacity of phenolic compounds mainly relies on the hydroxyl group on their basic backbone to provide [H], thus achieving the purpose of scavenging free radicals and terminating free radical chain reactions. The substitution position and number of hydroxyl groups are important for their antioxidant activity. The antioxidant capacity of phenolics is enhanced with the degree of hydroxylation. However, the ORAC value of HQ is larger than both TBHQ with two hydroxyl groups and PG with three hydroxyl groups, which is attributed to the fact that HQ is more hydrophilic in comparison, which causes it to release [H] and then scavenge free radicals more easily in aqueous solutions. MPBHQ has side chain alkyl groups, but it also contains a higher number of hydroxyl groups. Thus, MPBHQ exhibited a higher ability to scavenge free radicals than that of TBHQ, PG and BHT. The ORAC assay responded to the ability of MPBHQ to provide [H] to convert free radicals into more stable products to stop free radical chain reactions. The results of SET (DPPH and ABTS assay) and HAT (ORAC assay) assay shown that MPBHQ has a stronger ability to transfer hydrogen and single electrons. This will make it have strong application potential in the food and chemical industry. Furthermore, we selected a lard system and mixed fatty acid methyl ester system to evaluate the value and effect of MPBHQ as an antioxidant in edible oil and biodiesel using the Rancimat test.

### 3.5. Antioxidant Activity in Lard and Fatty Acid Methyl Esters—Rancimat Test

The Rancimat test was used to comparatively study the antioxidant activity of BHT, HQ, TBHQ, PG, and MPBHQ in the pure oil system, and the antioxidant activity of MPBHQ was comprehensively evaluated. Lard without antioxidants was selected as raw material, and the antioxidant concentrations were set to 0.01, 0.02, and 0.04% (*w*/*w*). The execution temperature of Rancimat tests was set in the range of 120–140 °C, and the tests were carried out at air-saturated conditions. The maximum concentration of antioxidant compounds was set as 400 mg/kg to study whether the antioxidant activity of MPBHQ changed with increasing concentration. Additionally, the oxidation induction time (IP) and antioxidant protection factor (Pf) were used as evaluation indices. Generally, higher Pf values indicated stronger antioxidant activity. Specifically, Pf $<$ 1 indicated pro-oxidant effect, Pf = 1 indicated no antioxidant activity, 1 $<$ Pf $<$ 2 showed weak antioxidant activity, 2 $<$ Pf $<$ 3 indicated effective antioxidant activity, and Pf $>$ 3 showed that the compound had strong antioxidant activity [32,33]. According to the trend of Pf values shown in Figure 5a, when the addition amount of antioxidant compounds is 0.02%, the antioxidant capacity is MPBHQ > PG > TBHQ > HQ > BHT at 120, 130, and 140 °C. When the execution temperature is 120 °C, the Pf value of lard with MPBHQ is 1.4, 1.5, 2.5, and 4 times than that of lard with PG, TBHQ, HQ, and BHT, respectively. When the execution temperature rises to 140 °C, the Pf value of lard with MPBHQ is about 2.0, 2.0, 3.0, and 4.2 times than that of lard with PG, TBHQ, HQ, and BHT, respectively. This indicates that MPBHQ has better antioxidant capacity than that of PG, TBHQ, HQ, and BHT, and this may be mainly because MPBHQ has a higher relative molecular mass and evaporates less at relatively high temperatures [18]. Furthermore, the higher number of phenolic hydroxyl groups is the main reason for the strong antioxidant capability of MPBHQ, and the presence of the methyl propyl electron-giving substituent in the structure further enhances its antioxidant activity [19,34].

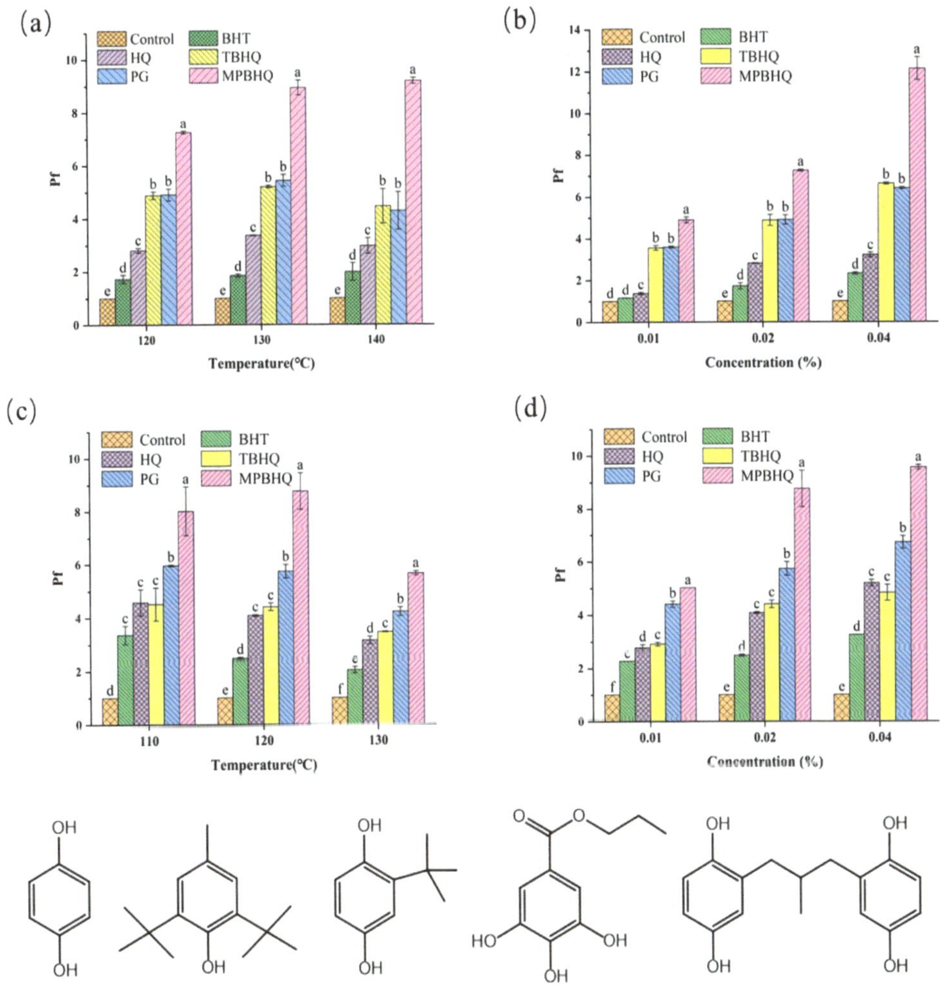


**HQ** **BHT** **TBHQ** **PG** **MPBHQ**

**Figure 5.** (**a**) Changes in Pf values of lard containing 0.02 wt.% of antioxidants at different temperatures. (**b**) Changes in Pf values of lard containing different antioxidants concentrations at 120 °C. (**c**) Changes in Pf values of fatty acid methyl ester containing 0.02 wt.% of antioxidants at different temperatures. (**d**) Changes in Pf values of fatty acid methyl ester containing different antioxidants concentrations at 120 °C. The chemical structure of the five antioxidants is shown below. The different lowercase letters in the graph, including a, b, c, d, e, and f, indicate the significant difference of Pf values of the different substances at same condition ($p < 0.05$).

According to the trend of Pf values shown in Figure 5b, when the heating condition was 120 °C, all five antioxidant compounds showed stronger antioxidant performance with an increase in additive amount (0–0.04 wt.%). When the additive amount of MPBHQ was 0.01 wt.%, the Pf value was 4.90. However, the addition of PG or TBHQ should be doubled to achieve similar antioxidant effects, and their corresponding Pf values were 4.90 and 4.88, respectively. It is even less optimistic that when the addition amount was 0.04%, the Pf values of HQ and BHT were only 3.21 and 2.32, respectively. In general, for the oil system, MPBHQ has the strongest antioxidant capacity. Therefore, MPBHQ has a broad application prospect due to its strong antioxidant capacity and thermal stability, including the antioxidant protection of industrial fats, amongst others.

In order to further explore the performance of MPBHQ and broaden its application scenarios, its antioxidant capacity in mixed fatty acid methyl esters was evaluated. According to the trend of Pf values shown in Figure 5c, when the addition amount of antioxidant compounds was 0.02%, the antioxidant capacity was MPBHQ > PG > TBHQ > HQ > BHT at 120, 130 and 140 °C of fatty acid methyl esters. According to the trend of Pf values shown in Figure 5d, when the heating condition was 120 °C, all five antioxidant compounds showed stronger antioxidant performance with the increase in additive amount (0–0.04 wt.%). In summary, MPBHQ also exhibits excellent antioxidant properties in fatty acid methyl esters, which contain polyhydroxyl and alkane structures that give it both good oil solubility and good antioxidant properties. Fatty acid methyl esters are the main components of biodiesel, so MPBHQ can be used as a potential antioxidant for biodiesel.

Table 2 shows the antioxidant properties of some synthetic antioxidants compared with MPBHQ. MPBHQ has the strongest advantage in DPPH free radical scavenging tests. In the Rancimat test, MPBHQ showed a strong antioxidant effect, but was second to 6,6′-(butane-1,1-diyl)bis(4-methylbenzene-1,2-diol) (BMB) studied by Olajide et al. [18]. In summary, MPBHQ has strong free radical scavenging ability and prevents oxidative rancidity of oil and fat.

**Table 2.** Comparison of the antioxidant properties of MPBHQ and several synthetic polyhydroxy antioxidants.

| Antioxidant | Structure | Condition of DPPH | $EC_{50}/\mu g \cdot mL^{-1}$ | Condition of Rancimat | Pf | Reference |
|---|---|---|---|---|---|---|
| Compound **1** | 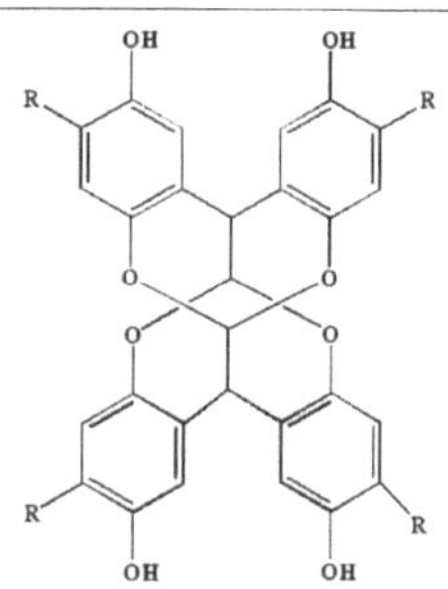 | 0.5 mL of sample, 2.5 mL of DPPH (0.1 mM) | 31.64 | 3 g of lard, 120 °C, 0.02% of antioxidant, 20 L/h air flow rate | 6.12 | [19] |
| Compound **2** | Compound 1 (R=Me) Compound 2 (R=t-Bu) | 0.5 mL of sample, 2.5 mL of DPPH (0.1 mM) | 106.95 | 3 g of lard, 120 °C, 0.02% of antioxidant, 20 L/h air flow rate | 7.28 | [19] |
| BMB | 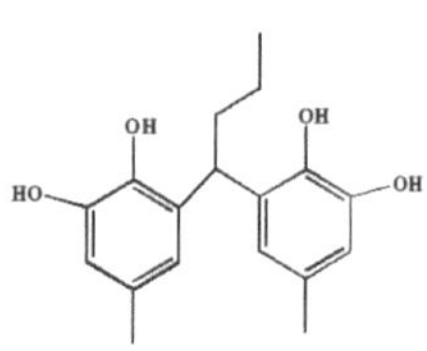 | 0.5 mL of sample, 3 mL of DPPH (0.1 mM) | 24.39 | 3 g of lard, 120 °C, 0.02% of antioxidant, 20 L/h air flow rate | 14.85 | [18] |

**Table 2.** *Cont.*

| Antioxidant | Structure | Condition of DPPH | $EC_{50}/\mu g \cdot mL^{-1}$ | Condition of Rancimat | Pf | Reference |
|---|---|---|---|---|---|---|
| MPBHQ | 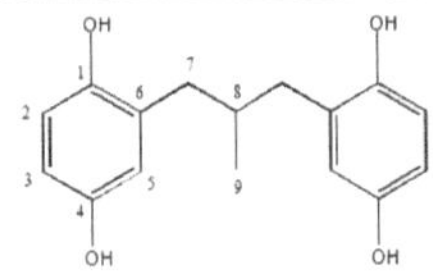 | 0.5 mL of sample, 3 mL of DPPH (0.1 mM) | 7.93 | 3 g of lard, 120 °C, 0.02% of antioxidant, 20 L/h air flow rate | 7.25 | This work |
| TBHQ | 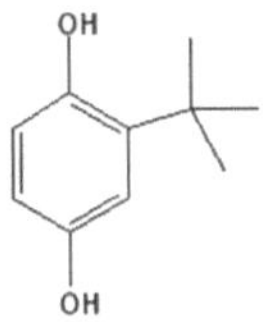 | 0.5 mL of sample, 3 mL of DPPH (0.1 mM) | 22.20 | 3 g of lard, 120 °C, 0.02% of antioxidant, 20 L/h air flow rate | 4.88 | This work |
| PG | 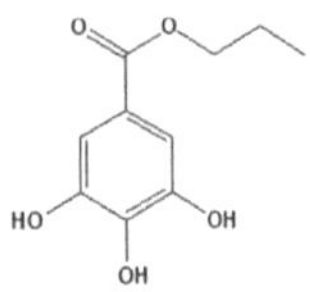 | 0.5 mL of sample, 3 mL of DPPH (0.1 mM) | 11.34 | 3 g of lard, 120 °C, 0.02% of antioxidant, 20 L/h air flow rate | 4.90 | This work |

## 4. Conclusions

In this work, a novel antioxidant named MPBHQ containing four phenolic hydroxyl groups was successfully synthesized. After optimized separation and purification, the chemical structure of MPBHQ was identified, and its antioxidant properties were evaluated by free radical scavenging assays based on the SET mechanism and HAT mechanism, as well as the oxidative rancidity tests of a lard system and a mixed fatty acid methyl ester system. The chemical structure of MPBHQ was confirmed by the HRMS, NMR, FTIR and UV spectroscopy, and its chemical name was designated as 2,2′-(2-methylpropane-1,3-diyl)bis(hydroquinone). The results of DPPH and ABST radical scavenging assay indicate that MPBHQ has an excellent ability to scavenge free radicals dominated by SET mechanism. MPBHQ also showed strong antioxidant properties in the ORAC assay, and this indicated that MPBHQ also had excellent hydrogen transfer ability. Furthermore, the results of Rancimat test show that MPBHQ has excellent ability to inhibit the oxidation of edible oil and mixed fatty acid methyl ester. By comparing the oxidation resistance test results of MPBHQ, PG, TBHQ, HQ, and BHT, the reason why MPBHQ has higher antioxidant capacity is mainly because it has more phenolic hydroxyl groups and the methyl propyl electron-giving substituent, as well as the relatively higher thermostability. In summary, MPBHQ meets the requirements of a new antioxidant with strong antioxidant properties at high temperatures, and MPBHQ will have great potential application value in industrial products such as biodiesel and edible oils.

**Author Contributions:** Data curation, formal analysis, writing—original draft, M.X.; conceptualization, methodology, writing—review and editing, supervision, P.M.; conceptualization, methodology, software, validation, H.W.; methodology, conceptualization, J.L.; methodology, supervision, T.G.; software, validation, Z.Z.; visualization, supervision, investigation, project administration, funding acquisition, Y.B. All authors have read and agreed to the published version of the manuscript.

**Funding:** This study was financially supported by the National Natural Science Foundation of China (No. 32172131).

**Institutional Review Board Statement:** Not applicable.

**Informed Consent Statement:** Not applicable.

**Data Availability Statement:** Data are contained within the article.

**Conflicts of Interest:** We declare that we do not have any commercial or associative interests that represents any conflicts of interest in connection with the submitted work.

## References


1. Farooq, S.; Abdullah; Zhang, H.; Weiss, J. A comprehensive review on polarity, partitioning, and interactions of phenolic antioxidants at oil–water interface of food emulsions. *Compr. Rev. Food Sci. Food Saf.* **2021**, *20*, 4250–4277. [CrossRef]
2. Shahidi, F.; Ambigaipalan, P. Phenolics and polyphenolics in foods, beverages and spices: Antioxidant activity and health effects. *J. Funct. Foods* **2015**, *18*, 820–897. [CrossRef]
3. Zawawi, N.; Chong, P.J.; Mohd Tom, N.N.; Saiful Anuar, N.S.; Mohammad, S.M.; Ismail, N.; Jusoh, A.Z. Establishing Relationship between Vitamins, Total Phenolic and Total Flavonoid Content and Antioxidant Activities in Various Honey Types. *Molecules* **2021**, *26*, 4399. [CrossRef]
4. Caleja, C.; Finimundy, T.C.; Pereira, C.; Barros, L.; Calhelha, R.C.; Sokovic, M.; Ivanov, M.; Carvalho, A.M.; Rosa, E.; Ferreira, I. Challenges of traditional herbal teas: Plant infusions and their mixtures with bioactive properties. *Food Funct.* **2019**, *10*, 5939–5951. [CrossRef]
5. Guo, Y.; Amorati, R.; Valgimigli, L. Synergic antioxidant activity of γ-Terpinene with phenols and polyphenols enabled by hydroperoxyl radicals. *Food Chem.* **2020**, *345*, 128468. [CrossRef]
6. Keramat, S.; Sharebiani, H.; Patel, M.; Fazeli, B.; Stanek, A. The potential role of antioxidants in the treatment of peripheral arterial disease. A Systematic Review. *Antioxidants* **2022**, *11*, 2126. [CrossRef]
7. Costa, R.; Santos, L. Delivery systems for cosmetics—From manufacturing to the skin of natural antioxidants. *Powder Technol.* **2017**, *86*, 306–320. [CrossRef]
8. Zhang, F.; Li, J.; Yang, S.; Bi, Y. Inhibitory effect of antioxidants on biodiesel crystallization: Revealing the role of antioxidants. *Fuel* **2021**, *297*, 120782. [CrossRef]
9. Sun, H.; Chen, D.; Cai, H.; Chang, W.; Wang, Z.; Liu, G.; Deng, X.; Chen, Z. Effects of fermenting the plant fraction of a complete feed on the growth performance, nutrient utilization, antioxidant functions, meat quality, and intestinal Microbiota of Broilers. *Animals* **2022**, *12*, 2870. [CrossRef]
10. Rodil, R.; Quintana, J.B.; Basaglia, G.; Pietrogrande, M.C.; Cela, R. Determination of synthetic phenolic antioxidants and their metabolites in water samples by downscaled solid-phase extraction, silylation and gas chromatography–mass spectrometry. *J. Chromatogr. A* **2010**, *1217*, 6428–6435. [CrossRef]
11. Li, Z.J.; Yang, F.J.; Yang, L.; Zu, Y.G. Comparison of the antioxidant effects of carnosic acid and synthetic antioxidants on tara seed oil. *Chem. Cent. J.* **2018**, *12*, 12–37. [CrossRef]
12. Liu, R.; Song, S.; Lin, Y.; Ruan, T.; Jiang, G. Occurrence of synthetic phenolic antioxidants and major metabolites in municipal Sewage Sludge in China. *Environ. Sci. Technol.* **2015**, *9*, 2073–2080. [CrossRef]
13. Liu, R.; Mabury, S.A. Synthetic phenolic antioxidants in personal care products in Toronto, Canada: Occurrence, human exposure, and discharge via greywater. *Environ. Sci. Technol.* **2019**, *6*, 7–12. [CrossRef] [PubMed]
14. Liu, C.; Li, J.; Bi, Y.; Wang, X.; Sun, S.; Yang, G. Thermal losses of tertiary Butylhydroquinone (TBHQ) and its effect on the qualities of palm oil. *J. Oleo Sci.* **2016**, *65*, 739–748. [CrossRef] [PubMed]
15. Yu, N.; Ye, Q.; Nie, X.; Xia, C.; Meng, X. Distribution of 2-tert-butyl-1,4-benzoquinone and its precursor, tert-butylhydroquinone, in typical edible oils and oleaginous foods marketed in Hangzhou City, China. *Food Chem.* **2021**, *361*, 130039. [CrossRef] [PubMed]
16. Cui, H.; Tao, F.; Hou, Y.; Lu, Y.; Zheng, T.; Sang, S.; Lv, L. Dual effects of propyl gallate and its methylglyoxal adduct on carbonyl stress and oxidative stress. *Food Chem.* **2018**, *265*, 227–232. [CrossRef]
17. Ye, Q.; Meng, X.; Jiang, L. Identification and assessment of residual levels of the main oxidation product of tert-butylhydroquinone in frying oils after heating and its cytotoxicity to RAW 264.7 cells. *Food Chem.* **2018**, *264*, 293–300. [CrossRef]
18. Olajide, T.M.; Pasdar, H.; Weng, X.C. A novel antioxidant: 6,6′-(butane-1,1-diyl)bis(4-methylbenzene-1,2-diol). *Grasas Aceites* **2018**, *69*, e269. [CrossRef]
19. Jiang, Z.W.; Weng, X.C.; Huang, Y.; Hou, J.P.; Liao, X.Y. Synthesis and antioxidant activity of two novel tetraphenolic compounds derived from toluhydroquinone and tertiary butylhydroquinone. *Grasas Aceites* **2014**, *65*, e16. [CrossRef]
20. Owolabi, J.O.; Fabiyi, O.S.; Adelakin, L.A.; Ekwerike, M.C. Effects of skin lightening Cream Agents—Hydroquinone and Kojic Acid, on the skin of adult female experimental rats. *Clin. Cosmet. Investig. Dermatol.* **2020**, *13*, 283–289. [CrossRef]
21. Searle, T.; Al-Niaimi, F.; Ali, F.R. Hydroquinone: Myths and reality. *Clin. Exp. Dermatol.* **2020**, *4*, 636–640. [CrossRef] [PubMed]
22. Jordán, M.J.; Lax, V.; Rota, M.C.; Lorán, S.; Sotomayor, J.A. Relevance of carnosic acid, carnosol, and rosmarinic acid concentrations in the in vitro antioxidant and antimicrobial activities of *Rosmarinus officinalis* (L.) methanolic extracts. *J. Agric. Food Chem.* **2012**, *60*, 9603–9608. [CrossRef] [PubMed]
23. Li, P.; Yang, X.; Lee, W.J.; Huang, F.; Wang, Y.; Li, Y. Comparison between synthetic and rosemary-based antioxidants for the deep frying of French fries in refined soybean oils evaluated by chemical and non-destructive rapid methods. *Food Chem.* **2021**, *335*, 127638. [CrossRef] [PubMed]
24. Huang, Y.; Jiang, Z.; Liao, X.; Hou, J.; Weng, X. Antioxidant activities of two novel synthetic methylbenzenediol derivatives. *Czech J. Food Sci.* **2014**, *32*, 348–353. [CrossRef]
25. Kodomari, M.; Nawa, S.; Miyoshi, T. Direct allylation of aromatic compounds with allylic chloride using the supported reagents system $ZnCl_2/SiO_2/K_2CO_3/Al_2O_3$. *J. Chem. Soc. Chem. Commun.* **1995**, *18*, 18–25. [CrossRef]

26. Feng, Y.; Zhang, X. Polysaccharide extracted from Huperzia serrata using response surface methodology and its biological activity. *Int. J. Biol. Macromol.* **2020**, *157*, 267–275. [CrossRef]
27. Chen, Z.; Bertin, R.; Froldi, G. EC50 estimation of antioxidant activity in DPPH assay using several statistical programs. *Food Chem.* **2013**, *138*, 414–420. [CrossRef]
28. Shi, B.; Yu, J.; Tang, T.; Yuan, L.; Tang, Y. Study on UV spectrum and antioxidant properties of 3-tert-Butyl-4-hydroxyanisole molecule. *Russ. J. Phys. Chem. A* **2021**, *95*, 343–348. [CrossRef]
29. Shi, G.; Liao, X.; Olajide, T.M.; Liu, J.; Jiang, X.; Weng, X. Butylated caffeic acid: An efficient novel antioxidant. *Grasas Aceites* **2017**, *68*, 201. [CrossRef]
30. Schaich, K.M.; Tian, X.; Xie, J. Hurdles and pitfalls in measuring antioxidant efficacy: A critical evaluation of ABTS, DPPH, and ORAC assays. *J. Funct. Foods* **2015**, *14*, 111–125. [CrossRef]
31. Tian, X.; Schaich, K.M. Effects of Molecular Structure on Kinetics and Dynamics of the Trolox Equivalent Antioxidant Capacity Assay with $ABTS^{+\bullet}$. *J. Agric. Food Chem.* **2013**, *61*, 5511–5519. [CrossRef]
32. Wen, W.; Weng, X.; Cheng, D. Antioxidant activities of natural phenolic components from *Dalbergia odorifera* T. Chen. *Food Chem.* **2000**, *71*, 45–49. [CrossRef]
33. Symoniuk, E.; Ratusz, K.; Krygier, K. Comparison of the oxidative stability of linseed (*Linum usitatissimum* L.) oil by pressure differential scanning calorimetry and Rancimat measurements. *J. Food Sci. Technol.* **2016**, *53*, 3986–3995. [CrossRef] [PubMed]
34. Kajiyama, T.; Ohkatsu, Y. Effect of para-substituents of phenolic antioxidants. *Polym. Degrad. Stab.* **2001**, *71*, 445–452. [CrossRef]

*antioxidants*

MDPI

*Article*

# Effects of the Chemical Composition on the Antioxidant and Sensory Characteristics and Oxidative Stability of Cold-Pressed Black Cumin Oils

**Aleksandra Szydłowska-Czerniak [1,*], Monika Momot [1,2], Barbara Stawicka [2] and Dobrochna Rabiej-Kozioł [1]**

1 Department of Analytical Chemistry and Applied Spectroscopy, Faculty of Chemistry, Nicolaus Copernicus University in Toruń, Gagarina 7, 87-100 Toruń, Poland
2 Bunge Polska Sp. z o.o., Niepodległości 42, 88-150 Kruszwica, Poland
* Correspondence: olasz@umk.pl; Tel.: +48-56-611-4786

**Abstract:** The antioxidant capacity (AC); amounts of tocopherols, sterols, and polycyclic aromatic hydrocarbons; oxidative parameters; fatty acid composition (FAC); and sensory quality of cold-pressed black cumin oils (CPBCOs) available on the Polish market were analyzed and compared. The AC levels of the CPBCO samples were determined using four assays, namely 2,2-diphenyl-1-picrylhydrazyl (DPPH = 226.8–790.1 µmol TE/100 g), 2,2′-azino-bis(3-ethylbenzothiazoline-6-sulfonic acid) (ABTS = 385.9–1465.0 µmol TE/100 g), cupric-reducing antioxidant capacity (CUPRAC = 975.3–19,823.3 µmol TE/100 g), and Folin–Ciocalteu assays (FC = 168.1–643.7 µmol TE/100 g). The FAC scores were typical for black cumin oil, except for the sample CPBCO4, which had a higher content of α-linolenic acid (C18:3 = 23.33%), pointing to possible oil adulteration. Additionally, the concentrations of total sterols (TSC = 372 mg/100 g) and tocopherols (TTC = 42.3 mg/100 g) in this sample were higher than those for other investigated oils (TSC = 159–222 mg/100 g, TTC = 1.9–10.4 mg/100 g respectively). The oxidative stability levels (IP = 8.21–37.34 h), peroxide values (PV = 21.36–123.77 meq $O_2$/kg), acid values (AV = 6.40–22.02 mg KOH/kg), and the sums of four specific polycyclic aromatic hydrocarbons (∑4PAHs = 4.48–46.68 µg/kg) in the studied samples differed significantly ($p < 0.05$). A sensory lexicon including 12 attributes was developed and applied for the sensory evaluation of oils using a quantitative descriptive analysis (QDA).

**Keywords:** cold-pressed black cumin oils; antioxidant capacity; tocopherols; sterols; oxidative stability; fatty acid profile; quantitative descriptive sensory analysis; principal component analysis

## 1. Introduction

Cold-pressed vegetable oils are gaining increasing interest among conscious consumers. Nowadays, with health and nutrition becoming so important, virgin oils are being chosen instead of solvent-extracted and refined oils [1]. Recently, extra virgin olive oil, linseed oil, rapeseed oil, and other oils have been widely used in food, cosmetic, and pharmaceutical applications. It is well known that cold pressing is a mechanical process in which the oil is extracted and separated from the seeds without chemical intervention or the addition of any other substances at a temperature that does not exceed 35 °C. Therefore, cold-pressed oils are recognized as natural, healthy, and abundant in nutritious compounds such as tocopherols, polyphenols, and essential fatty acids [2]. *Nigella sativa* L. seeds and cold-pressed black cumin oils (CPBCOs) are part of the diet in Asia and Africa (Egypt, Tunisia). Lately, CPBCOs have gained more extensive popularity in Europe due to their beneficial fatty acid composition (FAC) and bioactive compounds, including tocopherols, sterols, phenolic compounds, and essential oils. These phytochemicals have anti-inflammatory activity and play key roles in supporting the immune system and in the treatment of allergies and atopic dermatitis. Black cumin oils are also valued for their specific taste, and they are suitable for food seasoning [2–4].

It is known that *Nigella sativa* L. seeds can consist of 30–50% fat, depending on the cultivar and seed maturity [4,5]. Unsaturated fatty acids constitute 80–90% of the total lipids (predominantly linoleic and oleic acids), making black cumin oil a healthy product. There can be differences between the profiles of fatty acids due to the country of origin; for example, the linoleic acid amounts varied from 37 to 71% in black cumin seeds cultivated in Bangladesh, Egypt, Iran, Morocco, Syria, Tunisia, and Turkey [4–7].

Besides the favorable fatty acid composition, CPBCOs are rich in bioactive compounds such as tocopherols, dominated by $\gamma$- and $\alpha$-tocopherols, as well as phytosterols, such as $\beta$-sitosterol, stigmasterol, and campesterol [3,6,8]. The phenolic compounds such as thymoquinone and carvacrol are responsible for the antiradical activity of this oil [8–10]. These chemical compounds are the main terpenes in essential oils from *Nigella sativa* L. seeds, which on the one hand give the oil its antioxidant potential, and on the hand provide a unique sensory profile that makes the black cumin oil very aromatic. Moreover, *trans*-anethole, p-cymene, thymoquinone, and other volatile compounds with a distinctive aroma and taste were identified in essential oils from black cumin seeds [2,3]. On the other hand, various factors such as the seed varieties, oxidation processes, levels of contaminants, storage conditions, and technological processes can affect the sensory profiles of oils from black cumin seeds. The presence of essential oils and tocopherols results in powerful protection against oxidation—much higher than for other cold-pressed oils. The oxidative stability levels of CPBCOs measured as induction period values (IP = 13.45–38.22 h) using the Rancimat test were about 2–10 times higher than the IP values (3.67–13.98 h) for linseed and rapeseed oils but similar to the IP values of roasted pumpkin seed oils (IP = 20.29–34.39 h) [11].

According to the European Commission, *Nigella sativa* L. seeds and black cumin oils are considered as novel foods listed in the Novel Food Catalogue [12], because these products have just started expanding in Eastern and Western Europe. It is known that during the development process, the creation of new products is imperative to optimize parameters such as the shape, color, appearance, odor, taste, consistency, and safety and to encourage the consumer to purchase. Moreover, the interaction of all these components is critical to achieving a complete balance, leading to excellent quality and good acceptability. This approach is required due to consumers demanding food products with high-quality characteristics, while the food industry should find the right way to satisfy the consumers' requirements. A sensory analysis allows the estimation of the consumer acceptability and quality of food products, acting as an inherent part of the plan to create new products. The quantitative descriptive analysis (QDA) approach has been recognized as a tool for measuring and optimizing the sensory attributes of various food products. Sensory attribute lexicons for the evaluation of cold-pressed sunflower and rapeseed oils using the QDA methodology have been developed and implemented [13–15]. Nevertheless, to the best of our knowledge, a QDA methodology to develop the descriptive terminology and sensory profiles of CPBCO samples has not been proposed.

Although information is available on the chemical composition and oxidative stability of black cumin oils, practically nothing is known about the sensory attributes of oils extracted from *Nigella sativa* L. seeds using various techniques. Therefore, the primary aim of this study was to investigate the chemical composition—mainly via the antioxidant compound contents and oxidative stability levels—of CPBCOs available on the Polish market and to correlate them with sensory attributes. A product-specific glossary was prepared, and a QDA was applied to determine and quantify sensorial descriptors of CPBCOs. Moreover, a principal component analysis (PCA) was applied to classify and discriminate oil samples based on their antioxidant and fatty acids profiles, antioxidant potential, oxidative stability, and sensory characteristics.

## 2. Materials and Methods

### 2.1. Chemicals

All reagents of analytical or HPLC grade were purchased from Merck Sp. z o. o. (Warszawa, Poland). Redistilled water was used for the preparation of the solutions.

### 2.2. Materials

The experimental materials consisted of seven commercial cold-pressed black cumin oils originally packed into amber glass bottles (250–500 mL) purchased in shops on the Polish market. Samples were collected randomly from seven well-known Polish manufacturers who pressed black cumin oils from seeds cultivated in different localities of Poland, Egypt, and India. To maintain the discretion of the manufacturers, the oil samples were designated as CPBCO1–CPBCO7. All oil samples were within their stated shelf lives and stored in a refrigerator until analysis.

### 2.3. Determination of Tocopherols and Sterols Contents

The tocopherols were analyzed following ISO 9936:2016 standard [16] using high-performance liquid chromatography with some modifications. Briefly, each oil (0.5 g) was dissolved in hexane (5 mL), injected (5–20 μL) into a LiChrospher 100 Diol (125 × 4 mm, 5 μm particle size) column (Merck, Darmstadt, Germany), and analyzed using an Agilent 1100 HPLC system (Agilent Technologies, Palo Alto, CA, USA) with an autosampler and fluorescence detector. The mobile phase was hexane with tetrahydrofuran (96:4, $v/v$%) at a flow rate of 0.8 mL/min. Excitation and emission wavelengths at 280 and 340 nm, respectively, were used.

The sterols were measured using gas chromatography according to ISO 12228-1:2014 [17]. In brief, the oil samples were saponified using a 1 mol/L methanolic potassium hydroxide solution. The sterol fraction was extracted three times with a hexane/methyl tert-butyl ether (1:1) mixture. The separation and quantification of the silylated sterol fraction were performed using capillary gas chromatography (Agilent Technologies 6890, Wilmington, DE, USA) using a DB-35MS (J&W Scientific, Folsom, CA, USA) capillary column (25 m, 0.20 mm i.d. and 0.33 μm film thickness) and a flame ionization detector (FID). The detector and injector temperature was 300 °C, while hydrogen was used as a carrier gas with a 1.5 mL/min flow rate. Here, 5α-cholestane was used as an internal standard and individual sterols in oils were determined depending on their relative retention times according to the standards.

The total sum of the tocopherols and sterols and their percentage distribution were calculated.

### 2.4. Determination of Antioxidant Capacity

Before the antioxidant capacity (AC) determination, methanolic extracts of the studied oils were prepared according to the extraction procedure used for phenolic antioxidants from cold-pressed black cumin seed oil described by Lutterodt et al. [9] with minor modifications. Briefly, 2 g of oil was weighed into test tubes and extracted with 5 mL of methanol for 30 min using an orbital shaker (SHKA25081 CE, Labo Plus, Warszawa, Poland). Then, the test tubes were placed in a refrigerator for 30 min, and the methanolic extracts were separated from the oil samples. The extracted procedure was repeated three times. The methanolic extracts were collected and stored in glass bottles. The extracts were kept in a cold dark place until analysis.

The AC levels of the CPBCOs were analyzed using four modified spectrophotometric methods, namely 2,2-diphenyl-1-picrylhydrazyl (DPPH), 2,2′-azino-bis(3-ethylbenzothiazoline-6-sulfonic acid) (ABTS), cupric-reducing antioxidant capacity (CUPRAC), and Folin–Ciocalteu (FC) methods, as described previously [18].

DPPH method: The 0.5 mL of DPPH solution in methanol (304.0 μmol/L) was added to 0.1–0.2 mL of methanolic oil extracts diluted by methanol to 1.4–1.3 mL, respectively. The obtained mixtures were shaken vigorously and allowed to stand at room temperature in darkness for 15 min. Then, the absorbance of each solution was measured at 517 nm against a reagent blank (0.5 mL of DPPH methanolic solution mixed with 2 mL of methanol).

ABTS method: To generate the $ABTS^{\bullet+}$ solution, potassium persulfate (2.45 mmol/L) and ABTS (7 mmol/L) were mixed (1:0.5) and kept for 16 h in dark conditions. After this, the initial absorbance $ABTS^{\bullet+}$ solution was adjusted to 0.70 at 734 nm through dilution with ethanol. Next, 0.05–0.10 mL of each methanolic oil extract was allowed to react with the $ABTS^{\bullet+}$ solution (2.45–2.40 mL). After 5 min incubation at 30 °C, the absorbance of the reaction mixture was measured at 734 nm against a reagent blank (2.5 mL of $ABTS^{\bullet+}$ solution).

CUPRAC method: Firstly, 2 mL of $CuCl_2$ (0.01 mol/L), 2 mL of neocuproine solution (0.0075 mol/L), and 2 mL of ammonium acetate buffer (1 mol/L, pH = 7.0) were added into a 10 mL volumetric flask. Then, 0.1–0.2 mL of oil methanolic extracts were mixed, and the total volume was brought up to 10 mL with redistilled water. The mixture absorbance was recorded against a blank (all reagents without oil extracts) at 450 nm after 30 min incubation at room temperature.

FC method: In a 10 mL volumetric flask, 1.0 mL of methanolic oil extract and 0.5 mL of FC reagent were combined and mixed until homogeneous at room temperature for 3 min. Next, the reaction was neutralized with 1 mL of $Na_2CO_3$ (22.0%) made up to the mark with redistilled water and then allowed to react in darkness for 1 h. The absorbance measurements of the blue-colored solutions were carried out at 765 nm against a reagent blank.

The absorbance of each reaction mixture in a 1 cm glass cell was read with a Hitachi U-2900 spectrophotometer (Hitachi, Tokyo, Japan). The AC levels of methanolic extracts are expressed as micromoles of Trolox equivalents (TE) per 100 g sample.

### 2.5. Determination of Oxidative Stability

The oxidative stability index values of the analyzed oils were determined using a 743 Rancimat instrument made by Metrohm, applying the AOCS Official Method Cd 12b-92 [19]. Here, 3 g of each CPBCO was measured in the Rancimat test instrument at a temperature of 100 °C ± 0.3 °C with a gas flow rate of 20 L/h and 60 mL of water in the measuring vessel.

The peroxide value (PV) of each oil was measured according to ISO 27107 (2010) [20] via potentiometric titration with the use of a 905 Titrando high-end titrator from Metrohm (Warszawa, Poland) after dissolving the oil in the chloroform–glacial acetic acid mixture in the presence of a saturated solution of potassium iodide. The iodide was then oxidized using oil peroxides, and the formed iodine was titrated with sodium thiosulfate to an equivalence point. The result was reported in milliequivalents of oxygen per kilogram of oil (meq $O_2$/kg). However, the acid value (AV) was analyzed according to ISO 660 (1996) [21] via the directly acid–base titration of oil in an alcoholic medium against standard potassium hydroxide. The AV is defined as the number of milligrams of potassium hydroxide required to neutralize the free fatty acids (FFA) present in one gram of oil.

### 2.6. Determination of Water and Volatile Matter Content

The water and volatile matter contents in the studied cold-pressed black cumin oils were measured gravimetrically according to ISO 662 (2016) [22]. The amounts of both the moisture and volatile matter were determined in triplicate by drying 10 g of each oil in an oven at the temperature of 103 °C. The drying process was repeated until reaching a constant weight.

### 2.7. Determination of Polycyclic Aromatic Hydrocarbons

Four polycyclic aromatic hydrocarbons (PAHs) limited by EU regulations (benzo(a)pyrene (B(a)P), chrysene (Chry), benzo(a)anthracene (B(a)A), and benzo(b)fluoranthene (B(b)F)) in the analyzed CPBCO samples were determined according to the internal method using high-performance liquid chromatography with a fluorescence detector (HPLC-FLD, Shimadzu, Kyoto, Japan), Zorbax Eclipse PAH column (particle size 3.5 μm, length 150 mm, diameter 4.60 mm, Agilent Technologies, Santa Clara, CA, USA), and Eclipse XDB-C18 precolumn (3.5 μm, 4.6 × 150 mm, Agilent Technologies) at an oven temperature of 30 °C. The oil samples were dissolved in cyclohexane and extracted to dimethyl formaldehyde. Benzo(b)chrysene was used as an internal standard. The calibration standards were prepared via the dilution of the PAH standards into acetonitrile. These four compounds were quantified using calibration curves plotted for each PAH in the concentration range of 0.25–8.50 μg/kg. The limit of detection (LOD) values (0.08, 0.06, 0.04, and 0.07 μg/kg) and limit of quantification (LOQ) values (0.26, 0.20, 0.14, and 0.23 μg/kg) were calculated for B(a)P, Chry, B(a)A, and B(b)F, respectively. The LOD is the lowest concentration of each PAH in the studied oils that can be detected by the proposed method, while the LOQ is the lowest concentration at which an analyte can be quantitated with a linear response and acceptable precision and accuracy.

### *2.8. Determination of Fatty Acid Composition*

The fatty acid profile of each CPBCO was determined according to the official method ISO 5508 [23]. Fatty acid methyl esters prepared using the ISO 5509 standard [24] method were separated on a gas chromatograph (HP 5890 GC) equipped with a flame ionization detector (FID) (Hewlett-Packard, Avondale, PA, USA) and a high polar capillary column BPX 70 (60 m × 0.25 mm, 0.25 μm). The temperatures on the injector and detector were adjusted to 250 °C, while the oven temperature was as follows: heating from 150 to 210 °C at 1.3 °C/min, holding at 210 °C for 5 min. The carrier gas was helium at a flow rate of 0.6 mL/min. The identification of fatty acids was accomplished using external fatty acid methyl ester (FAME) standards, and the results are presented as weight percentages of the total fatty acids.

### *2.9. Sensory Evaluation and Quality Assessment*

Two sensory leaders with experience in the sensory field of cold-pressed oils sensory evaluations screened differences in the studied samples and reviewed the previous literature in order to identify a sensory flavor lexicon for the CPBCOs [25,26]. The initial list of attributes was presented during the sensory training session (6 sessions of 45 min) with reference samples and definitions. The selection of attributes was conducted according to ISO 11035 (1994) [27]. As can be seen in Table 1, the final lexicon for CPBCOs consisted of 12 attributes, including 2 basic tastes (sweet, bitter), 6 flavors (overall flavor intensity (OFI), herb-like, flower-like, medical-like, fuel-like, spicy-like), 3 mouth feel sensory terms (astringency, pungency, painty), and the color intensity.

The expert panel consisted of ten (three males and seven females, ages 22–52 years), well-trained individuals (100 h of training) experienced in the sensory evaluation of cold-pressed oils. The selection, training, and monitoring of the assessors were conducted according to ISO 8586:1 [28] and ISO 8586:2 [29].

The CPBCO samples were evaluated in triplicate by each panelist in a randomized, balanced black design. The panelists were instructed to assess the intensity of each of the attributes using a 10-cm intensity scale with an anchoring point from 0 (not perceived) to 10 (very high intensity). Additionally, the overall sensory quality (OSQ) was determined using a 5-point scoring scale ranging from 1 (very poor quality) to 5 (very good quality) based on the DGF Standard [26]. The quality level and characteristics used to determine the OSQ values of the oils are presented in Table 2.

**Table 1.** Sensory lexicon used for descriptive analysis of cold-pressed black cumin oils.

| Sensory Attributes | Definition | Reference Product | Score |
|---|---|---|---|
| OFI | The intensity of all flavor and taste attributes taken together | 1—Freshly refined rapeseed oil<br>8—Cold-pressed black cumin oil | 1; 8 |
| Color intensity | The intensity of brown color | Brown color intensity wheel | 0; 5; 10 |
| Sweet taste | A basic taste simulated by such as sugar | Fresh roasted cold-pressed oil | 9.5 |
| Bitter taste | A basic taste simulated by such substances as quinine and caffeine | 0.2% of caffeine in water or good quality of extra virgin oil | 7 |
| Herb-like flavor | The flavor reminiscent of herbs | Fresh cold-pressed black cumin oil | 6 |
| Flower-like flavor | The flavor reminiscent of fresh flowers | Freshly cold-pressed peanut oil | 8.5 |
| Medical-like flavor | The flavor reminiscent of medical, hospital, pharmacy | Cold-pressed black cumin oil stored for 24 months in room temperature | 7 |
| Fuel-like flavor | The flavor reminiscent of fuel | Bad quality cold-pressed black cumin oil | 9 |
| Spicy-like flavor | The flavor reminiscent of cumin | Cumin seed | 7 |
| Astringency | Shrinking or drying effect on the tongue surface caused by substances such as tannins | Slices of a green banana | 10 |
| Pungency | The biting sensation that can be perceived throughout the mouth cavity | A good quality extra virgin olive oil | 10 |
| Painty | The flavor reminiscent of oils such as linseed and rapeseed oils containing linolenic acid; it cannot be noted in non-linolenic acid oils such as peanut oil | Good quality canola oil aged 4–8 days at 60 °C or until PV = 10.0 meq $O_2$/kg is reached | 10 |

**Table 2.** Overall sensory quality. General characteristics of cold-pressed black cumin oils.

| Score | Quality Level | Characteristic |
|---|---|---|
| 5 | Very good | Extra fresh taste characteristic for cold-pressed black cumin oil, high intensity taste with positive attributes such as herb-like, flower-like, spicy-like, and sweet taste. |
| 4 | Good | Positive tastes such as herb-like, flower-like, and sweet-like flavor at medium intensity. Slightly bitter and astringency is acceptable. |
| 3 | Fair | Negative taste attributes such as fuel-like, medical like, and bitter taste and mouth feeling attributes such as pungency, astringency, and painty at low intensity were identified. |
| 2 | Bad | Negative taste attributes such as fuel-like, medical-like, and bitter taste and mouth feeling attributes such as pungency, astringency, and painty at moderate intensity were identified. |
| 1 | Very bad | Negative taste attributes such as fuel-like, medical-like, and bitter taste and mouth feeling attributes such as pungency, astringency, and painty at high/very high intensity were identified. |

The industrial laboratories used the 5-point quality scale to determine the OSQ values to check if oils meet the product specification requirements. In this study, we based our assessment on internal industry requirements, where the minimum acceptable scores for fresh cold-pressed and stored oils were 4.0 and 3.5, respectively.

The oil samples were coded with three-digit code numbers and served at room temperature in the amount of 80 mL in a blue jar. The panelists were instructed to clean their palates by rinsing their mouths with a weak warm black tea or eating a piece of apple. All sensory sessions were conducted in sensory laboratory designs according to ISO 8586:1 [28] and ISO 8586:2 [29].

### 2.10. Statistical Analysis

All measurements were conducted in triplicate, while the AC results were taken in five repetitions. All measurements were reported as means ± standard deviation (SD) and were processed using an analysis of variance (ANOVA) and post hoc Duncan test to determine the differences among means. A probability of $p > 0.05$ was deemed significant. The Pearson correlation and principal component analyses (PCA) were performed using Statistica (Windows software package, version 8.0; StatSoft Inc., Tulsa, OK, USA), while Fizz software (Biosystems, Courtenon, France) was applied for the collection of all sensory data.

## 3. Results and Discussion

### 3.1. Tocopherol and Phytosterol Contents in Cold-Pressed Black Cumin Oils

It is well known that tocopherols exhibit potent antioxidant properties due to their ability to donate phenolic hydrogen. Therefore, these lipophilic antioxidants are crucial for the protection of unsaturated fatty acids against peroxidation. In addition, the tocopherol profile is an important indicator of the authenticity of vegetable oils.

The results of the qualitative and quantitative analyses of tocopherols in CPBCOs are presented in Table 3.

**Table 3.** The compositions and contents of tocopherols and sterols in cold-pressed black cumin oils.

| Compound | Content * ± SD (mg/100 g) | | | | | | |
|---|---|---|---|---|---|---|---|
| | CPBCO1 | CPBCO2 | CPBCO3 | CPBCO4 | CPBCO5 | CPBCO6 | CPBCO7 |
| α-Tocopherol | 6.9 ± 0.1 [e] | 0.6 ± 0.01 [a] | 3.0 ± 0.5 [c] | 6.5 ± 0.1 [e] | 1.6 ± 0.03 [b] | 5.5 ± 0.1 [d] | 3.4 ± 0.05 [c] |
| β-Tocopherol | 2.9 ± 0.05 [c] | <LOD | <LOD | <LOD | <LOD | 0.4 ± 0.02 [b] | 0.2 ± 0.01 [a] |
| γ-Tocopherol | <LOD | 1.3 ± 0.03 [a] | 1.1 ± 0.02 [a] | 35.1 ± 0.5 [d] | 2.3 ± 0.04 [c] | 1.3 ± 0.02 [a] | 1.8 ± 0.03 [b] |
| δ-Tocopherol | 0.6 ± 0.01 [a] | <LOD | <LOD | 0.7 ± 0.02 [a] | <LOD | <LOD | <LOD |
| Total tocopherols | 10.4 ± 0.1 [e] | 1.9 ± 0.03 [a] | 4.1 ± 0.05 [b] | 42.3 ± 0.5 [f] | 3.9 ± 0.05 [b] | 7.2 ± 0.01 [d] | 5.4 ± 0.06 [c] |
| Brassicasterol | 2 ± 0.03 [b] | <LOD | <LOD | 18 ± 0.3 [c] | <LOD | 1 ± 0.04 [a] | <LOD |
| Campesterol | 30 ± 0.5 [d] | 19 ± 0.3 [a] | 24 ± 0.4 [b] | 99 ± 1.5 [e] | 24 ± 0.4 [b] | 27 ± 0.4 [c] | 25 ± 0.4 [b] |
| Δ5-Avenasterol | 28 ± 0.4 [a] | 29 ± 0.4 [a,b] | 33 ± 0.5 [c,d] | 33 ± 0.6 [c,d] | 35 ± 0.5 [d] | 31 ± 0.7 [b,c] | 31 ± 0.5 [b,c] |
| β-Sitosterol | 110 ± 1.7 [c] | 81 ± 1.2 [a] | 118 ± 1.8 [d] | 182 ± 2.7 [f] | 103 ± 1.5 [b] | 121 ± 1.8 [e] | 119 ± 2.0 [d,e] |
| Total phytosterols | 208 ± 3 [c] | 159 ± 5 [a] | 211 ± 9 [d] | 372 ± 14 [f] | 201 ± 4 [b] | 222 ± 2 [e] | 212 ± 6 [d] |

Note: * $n = 3$; different letters ([a–f]) within the same row indicate significant differences between amounts of tocopherols and sterols in the studied oils (one-way ANOVA and Duncan test, $p < 0.05$). Abbreviations: SD—standard deviation; LOD—detection limit; CPBCO—cold-pressed black cumin oil.

In the studied CPBCO samples, the total tocopherol content (TTC) values ranged from 1.9 to 42.3 mg/100 g (Table 3). The highest TTC (42.3 mg/100 g) for CPBCO4 indicates pressing this oil from an entirely different seed variety or the adulteration of this oil with another vegetable oil.

For comparison, the total concentrations of tocopherols in black cumin oils pressed from the seeds grown in Macedonia, Turkey, and Egypt were 4.5 mg/100 g, 6.5 mg/100 g and 34.0 mg/100 g, respectively [6,30,31].

Concerning the tocopherol composition of the studied oils, six samples of CPBCO2-CPBCO7 (except CPBCO1) contained high percentages of the two homologous α- + γ-tocopherols varying from 94% up to 100%. A somewhat lower sum of α- + γ-tocopherol levels (57–85%) in cold-pressed and extracted black cumin oils from different countries was observed by other authors [6,8,30,31]. However, a similar percentage of α- + γ-tocopherols (98.7%) for black cumin oil extracted with chloroform–methanol from a local *Nigella sativa* L. seed variety was reported by Hassanein et al. [32].

It is noteworthy that among all analyzed samples, CPBCO4 had the lowest amount of α-tocopherol (15% of the total tocopherol content) and the highest level of γ-tocopherol (83% of the total tocopherol content). Additionally, the CPBCO1 sample had the lowest level of sum α- + γ-tocopherols (66% of total tocopherol content), whereas this oil contained the highest amount of β-tocopherol (2.9 mg/100 g). For comparison, CPBCO6 and CPBCO7 had significantly ($p < 0.05$) lower concentrations of β-tocopherol (0.4 and 0.2 mg/100 g, respectively). Interestingly, similar levels of δ-tocopherol (0.6–0.7 mg/100 g) were found only in two samples, CPBCO1 and CPBCO4 (Duncan test, $p > 0.05$, Table 3).

The significant differences indicated by the Duncan test ($p < 0.05$) for the tocopherol composition and concentration results for seven commercial CPBCOs can be explained by the influences of genetic, agronomic, environmental (seed varieties, seed growing location, etc.), and technological factors [6,8,30–32].

It is evident that the TTC values in all studied CPBCOs were lower than the tocopherol levels in oils pressed from other seeds, such as sunflower (28.1–383.0 mg/100 g), rapeseed (41.9–71.4 mg/100 g), camelina (97.2 mg/100 g), flaxseed (41.0–58.9 mg/100 g), hemp (69.3 mg/100 g), sesame (52.1 mg/100 g), pumpkin (29.1 mg/100 g), walnut (42.3 mg/100 g), rosehip (103.6 mg/100 g), and milk thistle (26.2 mg/100 g) [6,30,33].

Moreover, sterols are bioactive compounds that naturally occur in cold-pressed oils. They have a tendency to lower LDL cholesterol levels in the blood, reducing the risk of cardiovascular diseases. Apart from this hypocholesterolemic function, these phytochemicals and their derivatives can also be potent antioxidants [34]. Their antioxidant activities can be attributed to the formation of an allylic free radical and its isomerization to other relatively stable free radicals. In addition, the sterol composition can be considered as an important parameter for determining the adulteration or authenticity of cold-pressed oils due to each seed variety having a specific sterol profile.

It is noteworthy that the sterol compositions of the analyzed CPBCOs from the seven different manufacturers differed significantly (Duncan test, $p < 0.05$, Table 3). The total concentration of phytosterols (TSC) in CPBCO samples varied from 159 to 372 mg/100 g. As seen in Table 3, β-sitosterol was the most predominant sterol in all oil samples, followed by Δ5-avenasterol, campesterol, and brassicasterol. CPBCO4 and CPBCO2 revealed the highest and lowest β-sitosterol amounts, which were 182 and 81 mg/100 g, respectively. However, insignificant differences ($p > 0.05$) in this sterol concentrations were observed between the CPBCO3, CPBCO6, and CPBCO7 samples. The determined high β-sitosterol (48.9–56.1%) was in close agreement with those for black cumin oils from different origins (32.3–59.1%), as reported by other authors [8,30–32].

However, the Δ5-avenasterol and campesterol concentrations were similar for all studied oils (except CPBCO4). In addition, brassicasterol was found only in three oil samples (CPBCO1, CPBCO4, and CPBCO6), while a significantly higher brassicasterol content was present in CPBCO4 (Duncan test, $p < 0.05$, Table 3).

It can be noted that the sterol composition of CPBCO4 was not typical for black cumin oils. The high amounts of β-sitosterol, campesterol, and brassicasterol found in CPBCO4 suggest that it was adulterated with cold-pressed linseed oil.

### 3.2. Antioxidant Capacity of Cold-Pressed Black Cumin Oils

The AC results for the seven commercial CPBCOs are summarized in Table 4.

The CPBCO samples had their antioxidative properties confirmed by four in vitro antioxidant assays, namely radical scavenging (ABTS, DPPH) and reducing power (CUPRAC, FC) methods. It is noteworthy that the AC results measured using different analytical methods differed significantly ($p < 0.05$). The different mechanisms of the used analytical methods—namely single electron transfer (SET), hydrogen atom transfer (HAT), or a combination of both—probably caused these discrepancies between the AC results. It can be noted that the CUPRAC and ABTS values of the CPBCOs were higher than the DPPH and FC results. This can be explained by the fact that CUPRAC and ABTS are reactive with both hydrophilic and hydrophobic antioxidants. However, the DPPH test only allows the determination of antioxidants that can quench the purple DPPH free radicals in the alcoholic solution by providing hydrogen atoms or via electron donation and conversion to the yellow-colored non-radical form (DPPH-H). In contrast, the FC method is performed for the analysis of phenolic antioxidants capable of reducing a mixture of phosphomolybdic/phosphotungstic acid complexes in an alkaline medium, yielding a blue-colored product. Unfortunately, some non-phenolic compounds (reducing sugars and amino acids) with an FC-reductive ability can overestimate the AC values determined by FC assays [35].

**Table 4.** Antioxidant capacity levels of cold-pressed black cumin oils.

| Oil Sample | Antioxidant Capacity * ± SD (μmol TE/100 g) | | | |
|---|---|---|---|---|
| | DPPH | ABTS | CUPRAC | FC |
| CPBCO1 | 677.3 ± 16.0 [d] | 971.6 ± 20.9 [c] | 17,440.6 ± 749.8 [f] | 526.7 ± 23.7 [d] |
| CPBCO2 | 790.1 ± 11.4 [f] | 1465.0 ± 34.0 [f] | 19,823.3 ± 915.9 [g] | 643.7 ± 25.6 [e] |
| CPBCO3 | 276.9 ± 13.5 [b] | 385.9 ± 11.4 [a] | 975.3 ± 43.7 [a] | 206.8 ± 8.4 [b] |
| CPBCO4 | 355.8 ± 11.1 [c] | 653.9 ± 18.3 [b] | 7402.9 ± 379.0 [d] | 354.7 ± 15.3 [c] |
| CPBCO5 | 757.3 ± 13.2 [e] | 1270.8 ± 36.8 [e] | 8932.2 ± 549.4 [e] | 337.8 ± 13.4 [c] |
| CPBCO6 | 229.8 ± 5.3 [a] | 1098.6 ± 13.7 [d] | 2939.9 ± 413.9 [b] | 168.1 ± 12.8 [a] |
| CPBCO7 | 226.8 ± 6.7 [a] | 993.2 ± 32.2 [c] | 4144.0 ± 377.7 [c] | 177.0 ± 9.6 [a] |

Note: * $n = 5$; different letters ([a–g]) within the same column indicate significant differences between antioxidant activity of cold-pressed black cumin oils (one-way ANOVA and Duncan test, $p < 0.05$). Abbreviations: SD—standard deviation; CPBCO—cold-pressed black cumin oil; DPPH—2,2-diphenyl-1-picrylhydrazyl method; ABTS—2,2′-azino-bis(3-ethylbenzothiazoline-6-sulfonic acid) method; CUPRAC—cupric-reducing antioxidant capacity method; FC—Folin–Ciocalteu method.

The AC variability was visible between the different CPBCO samples (Duncan test, $p < 0.05$, Table 4). Definitely, the CPBCO2 had the highest AC values determined by all analytical methods (DPPH = 790.1 μmol TE/100 g, ABTS = 1465.0 μmol TE/100 g, CUPRAC = 19,823.3 μmol TE/100 g, FC = 643.7 μmol TE/100 g). The AC (except by CUPRAC value) for CPBCO2 was around four times higher than the AC of the oil with the lowest antioxidant potential (Table 4). Moreover, CPBCO5 had high radical scavenging activity against ABTS and DPPH radicals, whereas the reducing ability of CPBCO1 was found to be high in both CUPRAC and FC methods.

The Duncan test indicated that insignificant differences ($p > 0.05$) in DPPH and FC results were observed between the CPBCO6 and CPBCO7, while the CPBCO1 and CPBCO7 had similar ABTS values. As can be seen, the mean CUPRAC results significantly differed ($p < 0.05$) from each other, but the CPBCO4 and CPBCO5 samples revealed the same power to reduce Mo(VI) to Mo(V), with the subsequent formation of a green phosphate/Mo(V) complex.

The variability in AC between different the oils can be explained by genetic (seed varieties), agronomic (plant growing conditions), and technological factors, as well as the storage conditions and time. The agronomic influence on the antioxidant properties of crude oils cold-pressed from *Nigella sativa* L. seed varieties from Tunisia and Iran was confirmed by Cheikh-Rouhou and co-workers [5]. The polyphenol content in the oil from Tunisian variety was lower (245 mg gallic acid (GA)/kg) than the total phenolic content (TPC) in the oil from Iran (309 mg GA/kg).

The extraction technique affected the amounts of antioxidants in oils from *Nigella sativa* L. seeds. The TPC was the highest in cold-pressed black cumin oil (36.05 mg GA/kg), followed by Soxhlet-extracted oil (21.44 mg GA/kg), and was the lowest in the oil obtained from black cumin seeds using microwave-assisted extraction (15.19 mg GA/kg) [31]. However, the cold-pressed oil from Nigella seeds was a poorer source of phenolic compounds (TPC = 94.40 mg GA/100 mL), and had lower antiradical activity as determined by the DPPH method ($IC_{50}$ = 2.30 mg/mL) and lower reducing ability as analyzed via the FRAP assay (329 mmol/100 mL) than the oil extracted using supercritical fluid extraction (TPC = 160.51 mg GA/100 mL, $IC_{50}$ = 1.58 mg/mL, 538.67 mmol/100 mL) [36].

However, among the different cold-pressed oils (flaxseed, walnut, rapeseed, pumpkin, evening primrose, black cumin), the black cumin oil exhibited the highest lipid-soluble antioxidant capacity as determined by photochemiluminescence (7.7 mM Trolox/L oil) and DPPH (1.2 mM TE/L oil) assays [37]. The antiradical properties (DPPH = 1.85–3.18 mM TE/kg and 76.43–83.52 μmol/100 μmol) and TPC contents (115.9–119.2 mg ferulic acid/100 g and 1.02–1.40 mg GA/g) also differed for the cold-pressed black cumin oils analyzed by Symoniuk et al. [11] and Lutterodt et al. [9].

These differences between the AC and TPC results for black cumin oils observed by other authors can be explained by the use of various types of raw material, the country of origin, the cultivation conditions, the oil extraction techniques, and the applied analytical methods based on different reaction mechanisms.

### 3.3. *Oxidative Stability and Quality of Cold-Pressed Black Cumin Oils*

#### 3.3.1. Oxidative Stability

The oxidative stability of CPBCO samples was analyzed using the Rancimat method, and the obtained results were expressed via the induction period (IP). It is known that the IP is associated with the oil shelf life. The more extended the IP, the higher the stability against oil oxidation over time. The IP values for the studied CPBCO samples ranged between 8.31 and 37.34 h (Table 5).

**Table 5.** Oxidative stability and quality parameters of cold-pressed black cumin oils.

| Parameter | Mean Value * ± SD | | | | | | |
|---|---|---|---|---|---|---|---|
| | CPBCO1 | CPBCO2 | CPBCO3 | CPBCO4 | CPBCO5 | CPBCO6 | CPBCO7 |
| IP (h) | 22.05 ± 0.40 [e] | 37.34 ± 0.20 [g] | 10.14 ± 0.40 [c] | 19.7 ± 0.40 [d] | 24.92 ± 0.20 [f] | 8.31 ± 0.30 [a] | 9.60 ± 0.40 [b] |
| PV (meq $O_2$/kg) | 89.60 ± 0.01 [f] | 123.77 ± 0.56 [g] | 22.38 ± 0.01 [b] | 35.99 ± 0.01 [d] | 50.48 ± 0.02 [e] | 23.17 ± 0.02 [c] | 21.36 ± 0.05 [a] |
| AV (mg KOH/g) | 6.40 ± 0.06 [a] | 7.86 ± 0.02 [b] | 21.86 ± 0.08 [f] | 15.04 ± 0.08 [d] | 11.84 ± 0.09 [c] | 22.02 ± 0.08 [g] | 20.92 ± 0.10 [e] |
| WVC (%) | 0.05 ± 0.00 [a] | 0.05 ± 0.00 [a] | 0.05 ± 0.00 [a] | 0.26 ± 0.01 [c] | 0.08 ± 0.00 [b] | 0.05 ± 0.00 [a] | 0.06 ± 0.00 [a] |
| B(a)P (μg/kg) | 0.01 ± 0.00 [a] | 3.76 ± 0.05 [g] | 0.38 ± 0.01 [b] | 0.43 ± 0.01 [c] | 0.85 ± 0.00 [e] | 1.10 ± 0.02 [f] | 0.70 ± 0.03 [d] |
| Chry (μg/kg) | 1.89 ± 0.01 [e] | 2.79 ± 0.01 [f] | 0.35 ± 0.01 [b] | 0.76 ± 0.01 [c] | 1.79 ± 0.09 [d] | 0.06 ± 0.00 [a] | 2.80 ± 0.07 [f] |
| B(a)A (μg/kg) | 16.05 ± 0.21 [b] | 36.98 ± 1.63 [c] | 3.40 ± 0.01 [a] | 14.48 ± 0.01 [b] | 16.47 ± 0.42 [b] | 3.20 ± 0.18 [a] | 4.00 ± 0.34 [a] |
| B(b)F (μg/kg) | 0.03 ± 0.00 [a] | 3.14 ± 0.08 [e] | 0.69 ± 0.00 [c] | 0.41 ± 0.02 [b] | 0.63 ± 0.03 [c] | 1.60 ± 0.07 [d] | 0.60 ± 0.01 [c] |
| Σ4PAHs (μg/kg) | 17.98 | 46.68 | 4.48 | 16.09 | 19.73 | 7.90 | 6.60 |

Note: * $n$ = 3; different letters ([a–g]) within the same row indicate significant differences between oxidative stability and quality parameters of cold-pressed black cumin oils (one-way ANOVA and Duncan test, $p < 0.05$). Abbreviations: SD—standard deviation; IP—induction period; PV—peroxide value; AV—acid value; WVC—water and volatile matter content; B(a)P—benzo(a)pyrene; Chry—chrysene; B(a)A—benzo(a)anthracene; B(b)F—benzo(b)fluoranthene; Σ4PAHs—sum of four specific polycyclic aromatic hydrocarbons; CPBCO—cold-pressed black cumin oil.

As can be seen, the IP values of the examined oils differed significantly (Duncan test, $p < 0.05$, Table 5). The differences were probably related to the various cultivars used and quality levels of seeds, the different technological processes, and the country of *Nigella sativa* L. seed origin, as well as the amounts of natural antioxidants and profiles of fatty acids [5]. The highest significant oxidative stability was found for CPBCO2, and then decreased in the following order: CPBCO2 > CPBCO5 > CPBCO1 > CPBCO4 > CPBCO3 > CPBCO7 > CPBCO6 (Table 5). As expected, in the same order, the examined oils lost their radical scavenging activity, as measured using the DPPH and ABTS methods (Table 4). It is noteworthy that the low IP values for CPBCO6 (8.31 h), CPBCO7 (9.60 h), and CPBCO3 (10.14 h) indicated the low antioxidant potential of these oils, as analyzed using different analytical methods (Tables 4 and 5).

Similar IP values that ranged between 12.00–50.33 h at 100 °C and 76.37–157.58 h at 80 °C for black cumin oils were found by other authors [5,9,11]. The higher oxidative stability of black cumin oils (13.45–38.22 h) compared to other cold-pressed oils (IP = 3.67–13.98 h for linseed, camelina, evening primrose, hempseed, poppy, and rapeseed oils) can be explained by the different fatty acid compositions and strong antioxidants such as thymoquinone [9,11,38].

#### 3.3.2. Amounts of Primary Oxidation Products and Free Fatty Acids

The CPBCO samples were evaluated for severity in oxidation and hydrolysis using respectively the PV and AV, as shown in Table 5. The PV and AV are regular indices of quality control in oils. The Duncan test indicated that the overall intensity levels of the primary oxidation products (PV = 21.36–123.77 meq $O_2$/kg) were significantly different

($p < 0.05$) between the investigated oils. The amount of hydroperoxides in each CPBCO was higher than permitted by legal requirements (PV = 15 meq $O_2$/kg for cold-pressed oils) [39]. Unexpectedly, the formation of hydroperoxides in the tested CPBCOs increased almost linearly (correlation coefficient, r = 0.9064) with the increase in IP values. This can be explained by the fact that lipid oxidation is a multifactorial phenomenon that depends on a number of variables.

For comparison, two cold-pressed black cumin oils and oils extracted from two Nigella seeds having an Iranian and Tunisian origin with similar PV (1.03–1.56 meq $O_2$/kg and 4.35–5.65 meq $O_2$/kg, respectively) showed different IP (13.45–38.22 h and 12.00–50.33 h, respectively) [5,11]. In contrast, black cumin oils obtained using microwave-assisted extraction and Soxhlet extraction with comparable PV (21.45 and 25.23 meq $O_2$/kg) revealed similar IP (18.46 and 19.62 h) [31].

The tested CPBCO samples with high PV results (50.48–123.77 meq $O_2$/kg) had low AV values (6.40–11.84 mg KOH/g). In contrast, the oils with lower amounts of hydroperoxides (PV = 21.36–23.17 meq $O_2$/kg) and shorter IP results (8.31–10.14 h) demonstrated a higher hydrolysis degree (AV = 20.92–22.02 mg KOH/g). It is noteworthy that the AV results (6.40–22.02 mg KOH/g) of the studied CPBCOs exceeded the level recommended by the Codex Alimentarius Commission (AV = 4.0 mg KOH/g) for cold-pressed oils [39]. In our investigation, the CPBCO samples showed significantly higher acidity compared to the AV results (0.23 and 0.35 mg KOH/g) for cold-pressed black cumin oils obtained by Symoniuk et al. [11].

The obtained PV and AV results suggest that none of the assessed CPBCO samples were of acceptable oxidative or hydrolytic status, meaning they could potentially pose a health risk to consumers. In this regard, Gotoh and Wada [40] reported that oxidized fats and oils with PVs at a level of about 100 meq $O_2$/kg could be neurotoxic. Furthermore, the presence of these oxidized compounds affected the sensory quality of the analyzed oils.

The estimated oxidation and hydrolytic status of the investigated oils available for retail indicated the use of low-quality raw seeds and/or inappropriate oil extraction and preservation.

#### 3.3.3. Water and Volatile Matter Contents

The water and volatile compound (WVC) levels, some of basic quality parameters for vegetable oils, were in the range of 0.05–0.26% (Table 5). As can be seen, the WVC results (0.05–0.06%) for the five CPBCO samples available for retail did not differ significantly (Duncan test, $p > 0.05$, Table 5). According to the Codex Alimentarius [39], the volatile matter of cold-pressed oils should not exceed 0.20%. The amounts of water and volatiles in only one sample, CPBCO4, were somewhat higher (0.26%) than the prescribed limit in the Codex Alimentarius [39]. It is well known that the water excess in the oil can influence its stability through triacylglycerol hydrolysis. The higher amount of water in the oil may be the result of excessive seed humidity or technological processes.

For comparison, a lower water content (0.03%) in commercial cold-pressed black cumin oil was found by Mikołajczak et al. [41].

#### 3.3.4. Polycyclic Aromatic Hydrocarbons Content

All investigated samples were analyzed for carcinogenic compounds, namely PAHs, as it is known that black cumin seeds are often dried with a conventional method where the air used for drying is heated by charcoal or gasoline.

Unfortunately, the Σ4PAHs results for the four CPBCO samples ranged between 16.09 and 46.68 µg/kg (Table 5) and exceeded the legal limit of 10 µg/kg [42]. Moreover, the level of B(a)P in the CPBCO2 (3.76 µg/kg) was significantly higher than that recommended by the European Commission [42] for foodstuffs (2 µg/kg). Interestingly, the B(a)P values of all studied oils differed significantly ($p < 0.05$), while CPBCO3, CPBCO5, and CPBCO7 showed similar B(b)F contents (Duncan test, $p > 0.05$, Table 5). Badary et al. [43] reported that thymoquinone, which is present in all black cumin seeds and oils, can reduce B(a)P-induced forestomach carcinogenesis.

It is evident that the content of Σ4PAHs was the highest in CPBCO2 (46.68 μg/kg) due to the presence of the highest amounts of B(a)P (3.76 μg/kg), B(a)A (36.98 μg/kg) and B(b)F (3.14 μg/kg) in this oil. It is noteworthy that the Σ4PAHs value for CPBCO2 was approximately 10 times higher compared to CPBCO3 containing the lowest amounts of these four carcinogenic compounds (Σ4PAHs = 4.48 μg/kg). However, the Duncan test indicated that Chry concentrations in CPBCO2 and CPBCO7 were similar ($p > 0.05$, Table 5). Insignificant differences ($p > 0.05$) in results of B(a)A concentration were observed between oil samples CPBCO3, CPBCO6, and CPBCO7 (3.20–4.00 μg/kg), as well as CPBCO1, CPBCO4, and CPBCO5 (14.48–16.47 μg/kg), respectively.

The high levels of PAHs in all investigated oil samples and a lack of professional references for PAH analyses in black cumin oils indicate it is not a well-researched topic and should be explored for human and livestock health.

### 3.4. *Fatty Acid Compositions of Cold-Pressed Black Cumin Oils*

The fatty acid compositions of the investigated CPBCO samples available on the Polish market are presented in Table 6. It is noteworthy that most studied oils had the typical fatty acid profile for black cumin oil reported by other authors [5,6,8,9,11,30–32,37,38,41,44]. The percentages of fatty acids cannot be referenced to the Codex Alimentarius, the recognized standard for vegetable oils, because black cumin oil and seeds belong to the novel food category according to the European Commission Novel Food Catalogue; thus, these values are not present in this document. As can be seen, all CPBCO samples contained moderately low amounts of saturated fatty acids (SAFA = 12.14–16.76%) and moderate levels of monounsaturated fatty acids (MUFA = 24.49–31.71%), with the predominant component being oleic acid (C18:1, omega-9 = 23.89–31.01%). It is known that MUFA, among all unsaturated fatty acids, are the most resistant to lipid oxidation, and oleic acid plays a major role in the prevention of cardiovascular diseases. However, the polyunsaturated fatty acids (PUFA), ranging between 55.27 and 59.27%, predominantly linoleic acid (C18:2, omega-6 = 32.80–59.08%), are much more prone to lipid oxidation, but on the other hand are supportive of the human cardiovascular and immune systems [44].

All investigated samples contained summed unsaturated fatty acids in the range of 83.11–87.65%, allowing the recognition of black cumin oil as a healthy fat in general.

Surprisingly, the α-linolenic acid content (C18:3, omega-3 = 23.33%) and the C18:1 percentage (31.01%) in the CPBCO4 sample were significantly higher than in all other investigated samples and were not typical for the usual fatty acid profile of black cumin oil (Duncan test, $p < 0.05$, Table 6). However, the lowest concentrations of C18:2 (32.80%) and C16:0 (8.27%) were determined in this oil. This was probably due to the high-linolenic black cumin seed cultivar used for oil production or the adulteration of this oil with other omega-3 and omega-9 fatty acid-rich oils. Similar amounts of C16:0 (8.70%) and C18:2 (34.20%) were found in cumin seed oil by Ramadan et al. [44].

**Table 6.** Fatty acid compositions of cold-pressed black cumin oils.

| Fatty Acid | Content * ± SD (%) | | | | | | |
|---|---|---|---|---|---|---|---|
| | CPBCO1 | CPBCO2 | CPBCO3 | CPBCO4 | CPBCO5 | CPBCO6 | CPBCO7 |
| C 16:0 | 12.02 ± 0.17 [b,c] | 13.03 ± 0.12 [d] | 12.06 ± 0.09 [b,c] | 8.27 ± 0.05 [a] | 12.39 ± 0.17 [c] | 11.98 ± 0.12 [b] | 12.45 ± 0.18 [c] |
| C 18:0 | 3.10 ± 0.03 [a,b] | 3.01 ± 0.04 [a] | 3.53 ± 0.05 [c] | 3.41 ± 0.04 [c] | 3.13 ± 0.09 [a,b] | 3.50 ± 0.04 [c] | 3.40 ± 0.07 [c] |
| C 20:0 | 0.20 ± 0.01 [a] | 0.20 ± 0.00 [a] | 0.26 ± 0.01 [c] | 0.20 ± 0.01 [a] | 0.21 ± 0.00 [a] | 0.28 ± 0.01 [d] | 0.24 ± 0.00 [b,c] |
| C 22:0 | 0.14 ± 0.00 [c] | 0.03 ± 0.00 [a] | 0.09 ± 0.00 [b] | 0.17 ± 0.01 [c] | 0.05 ± 0.00 [a] | 0.16 ± 0.00 [c] | <LOD |
| ΣSAFA | 15.91 | 16.76 | 16.23 | 12.14 | 16.12 | 16.12 | 16.25 |
| C 16:1 | 0.21 ± 0.01 [b] | 0.24 ± 0.01 [c] | 0.18 ± 0.01 [a] | 0.17 ± 0.00 [a] | 0.20 ± 0.01 [b] | 0.18 ± 0.00 [a] | 0.20 ± 0.01 [b] |
| C 18:1 | 26.05 ± 0.84 [c] | 24.74 ± 0.11 [b] | 26.39 ± 0.12 [c] | 31.01 ± 0.08 [d] | 23.89 ± 0.09 [a] | 26.11 ± 0.81 [c] | 25.32 ± 0.89 [b] |
| C 20:1 | 0.41 ± 0.02 [b,c] | 0.44 ± 0.01 [c] | 0.38 ± 0.01 [a,b] | 0.30 ± 0.00 [a] | 0.40 ± 0.01 [b] | 0.34 ± 0.00 [a] | 2.89 ± 0.04 [d] |
| ΣMUFA | 26.62 | 25.37 | 26.95 | 31.71 | 24.49 | 26.64 | 28.41 |
| C 18:2 | 57.08 ± 0.82 [d] | 57.63 ± 1.47 [d] | 56.53 ± 2.10 [c] | 32.80 ± 1.05 [a] | 59.08 ± 2.61 [e] | 56.49 ± 2.01 [c] | 55.06 ± 1.78 [b] |
| C 18:3 | 0.34 ± 0.01 [b] | 0.19 ± 0.00 [a] | 0.25 ± 0.02 [a] | 23.33 ± 0.90 [d] | 0.26 ± 0.01 [a] | 0.75 ± 0.03 [c] | 0.29 ± 0.01 [a] |
| ΣPUFA | 57.20 | 57.74 | 56.69 | 55.94 | 59.27 | 57.03 | 55.27 |

Note: * $n = 3$; different letters ([a–e]) within the same row indicate significant differences between the percentages of fatty acids of cold-pressed black cumin oils (one-way ANOVA and Duncan test. $p < 0.05$). Abbreviations: SD—standard deviation; C 16:0—palmitic acid; C 18:0—stearic acid; C 20:0—arachidic acid; C 22:0—behenic acid; C 16:1—palmitoleic acid; C 18:1—oleic acid; C 20:1—eicosenoic acid; C 18:2—linoleic acid; C 18:3—α-linolenic acid; ΣSAFA—sum of saturated fatty acids; ΣMUFA—sum of monounsaturated fatty acids; ΣPUFA—sum of polyunsaturated fatty acids; CPBCO—cold-pressed black cumin oil.

### *3.5. Sensory Analysis of Cold-Pressed Black Cumin Oils*

To the best of our knowledge, the sensory characteristics and sensory quality of oils cold-pressed from black cumin seeds have not been reported. In the literature, we have only found overviews on the chemical composition, nutritional properties, or influence of black cumin on the sensory characteristics of mayonnaise [45]. Bendini et al. [13] and Wroniak et al. [14] employed the QDA methodology to describe only the sensory profiles of cold-pressed sunflower oils and cold-pressed rapeseed oils, respectively.

The sensory quality and attribute intensity data of the studied CPBCO samples are listed in Tables 1 and 2.

From the sensory quality scores, the analyzed samples can be divided into the acceptance sensory quality group (OSQ > 4.0 for CPCBO3, CPBCO6, and CPBCO7) and the unacceptance group (OSQ < 3.5 for CPBCO1, CPBCO2, CPBCO4, and CPBCO5). The PCA applied to all samples involved the mean attributes scores presented in Table 7, which were used to interpret and correlate data from 12 attributes and OSQ measured on the seven oils (Figure 1).

The first two principal components accounted for 88.69% (PC1 = 78.54% and PC2 = 10.15%, respectively) of the variability in the data set. The PCA sensory map (Figure 1) depicted that three samples (CPBCO3, CPBCO6, and CPBCO7) located on the right side of the PCA map were characterized by good sensory quality scores, as presented by the use of typical flavor descriptors (herb-like, flower-like, spicy-like and sweet-taste) for oils cold-pressed from black cumin seeds. However, some oil samples (CPBCO1, CPBCO2, CPBCO4, and CPBCO5) had poor sensory quality scores (score below 3.5 on a 5-pointing scale) with negative flavor descriptors (fuel-like, medical-like) and were situated on the left side of the PCA map. As shown in Figure 1, the CPBCO2 sample characterized by very high mouth feeling scores, mainly for astringency and pungency (about 9 points on the 10 cm scale), created an evidently distinct cluster. Furthermore, the CPBCO4 sample having a brighter (light-yellow) color and the highest bitterness perception indicated that it might be adulterated with cold-pressed linseed oil. According to Brühl et al. [46], freshly pressed linseed oil provides a delicate nutty flavor, and a lingering bitter off-taste develops upon storage at room temperature.

**Table 7.** Means for the scoring of the overall sensory quality and sensory attributes.

| Sensory Attribute | Mean Value * ± SD | | | | | | |
|---|---|---|---|---|---|---|---|
| | CPBCO1 | CPBCO2 | CPBCO3 | CPBCO4 | CPBCO5 | CPBCO6 | CPBCO7 |
| OSQ | 2.0 ± 0.0 [a] | 2.0 ± 0.1 [a] | 4.0 ± 0.1 [c] | 3.0 ± 0.1 [b] | 2.0 ± 0.0 [a] | 5.0 ± 0.2 [d] | 5.0 ± 0.1 [d] |
| OFI | 9.0 ± 0.3 [c] | 9.0 ± 0.1 [c] | 6.0 ± 0.2 [a] | 7.2 ± 0.3 [b] | 7.1 ± 0.2 [b] | 6.5 ± 0.1 [a,b] | 6.3 ± 0.2 [a] |
| Color intensity | 9.1 ± 0.2 [e] | 8.2 ± 0.2 [d] | 6.0 ± 0.2 [a] | 7.0 ± 0.1 [b] | 7.4 ± 0.3 [b,c] | 6.5 ± 0.2 [a,b] | 6.5 ± 0.1 [a,b] |
| Flavor | | | | | | | |
| Herb-like | 1.3 ± 0.0 [a] | 1.1 ± 0.0 [a] | 7.8 ± 0.1 [d,e] | 5.0 ± 0.2 [c] | 2.5 ± 0.0 [b] | 8.3 ± 0.3 [e] | 7.2 ± 0.2 [d] |
| Flower-like | 0.0 ± 0.0 [a] | 1.0 ± 0.0 [b] | 5.6 ± 0.1 [d] | 3.0 ± 0.1 [c] | 2.5 ± 0.0 [c] | 8.3 ± 0.1 [f] | 7.2 ± 0.1 [e] |
| Medical-like | 8.0 ± 0.3 [e] | 7.5 ± 0.1 [d] | 3.4 ± 0.1 [b] | 4.0 ± 0.2 [c] | 7.8 ± 0.1 [d,e] | 3.0 ± 0.1 [b] | 1.7 ± 0.0 [a] |
| Fuel-like | 6.2 ± 0.2 [b] | 7.9 ± 0.1 [c] | 0.0 ± 0.0 [a] | 0.0 ± 0.0 [a] | 7.2 ± 0.3 [c] | 0.0 ± 0.0 [a] | 0.0 ± 0.0 [a] |
| Spicy-like | 1.9 ± 0.0 [a] | 2.9 ± 0.1 [b] | 4.5 ± 0.2 [c] | 2.6 ± 0.1 [b] | 2.1 ± 0.1 [a,b] | 4.7 ± 0.1 [c,d] | 5.1 ± 0.2 [d] |
| Taste | | | | | | | |
| Sweet | 1.1 ± 0.0 [a] | 0.9 ± 0.0 [a] | 3.7 ± 0.1 [c] | 1.5 ± 0.0 [a,b] | 1.2 ± 0.0 [a] | 4.3 ± 0.1 [d] | 3.1 ± 0.1 [c] |
| Bitter | 6.5 ± 0.2 [c] | 4.5 ± 0.1 [b] | 4.9 ± 0.1 [b] | 7.1 ± 0.3 [d] | 4.9 ± 0.1 [b] | 3.9 ± 0.1 [a] | 3.9 ± 0.2 [a] |
| Mouthfeeling | | | | | | | |
| Astringency | 8.5 ± 0.4 [d] | 9.0 ± 0.4 [d] | 4.6 ± 0.1 [a] | 7.0 ± 0.3 [c] | 7.1 ± 0.2 [c] | 5.5 ± 0.2 [b] | 5.0 ± 0.1 [a,b] |
| Pungency | 7.1 ± 0.1 [b] | 9.1 ± 0.1 [c] | 6.2 ± 0.2 [a] | 7.3 ± 0.1 [b] | 6.0 ± 0.3 [a] | 7.0 ± 0.1 [b] | 6.7 ± 0.3 [a,b] |
| Painty | 7.0 ± 0.3 [e] | 6.0 ± 0.1 [d] | 2.0 ± 0.0 [a] | 4.2 ± 0.1 [c] | 7.2 ± 0.2 [e] | 3.0 ± 0.1 [b] | 3.0 ± 0.0 [b] |

Note: * $n$ = 20 (10 assessors × 2 repetitions); different letters ([a–f]) within the same row indicate significant differences between attributes of the studied cold-pressed black cumin oils (one-way ANOVA and Duncan test, $p < 0.05$). Abbreviations: SD—standard deviation; OSQ—overall sensory quality; OFI—overall flavor intensity; CPBCO—cold-pressed black cumin oil.

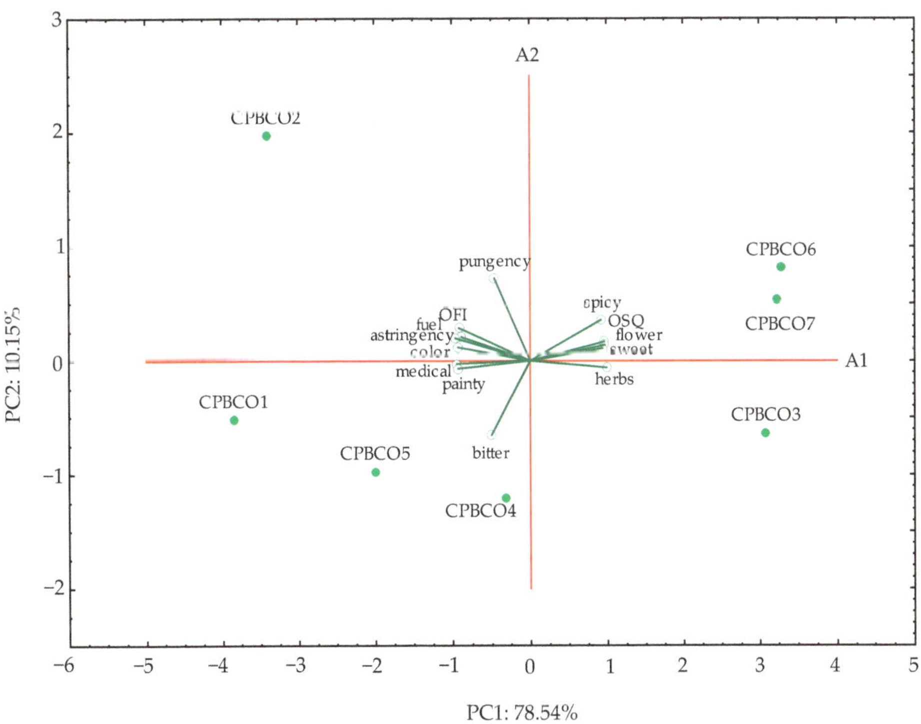


**Figure 1.** Biplot for PC1 and PC2 scores generated by the PCA of 12 attributes and OSQ for cold-pressed black cumin oils.

Nevertheless, high positive correlations were found between the OSQ and flavor attributes such as herb-like (r = 0.9462), flower-like (r = 0.9668), spicy-like (r = 0.9440), and sweet taste (r = 0.9273). However, negative correlations were observed between the OSQ and attributes such as medical-like (r = −0.9464) and fuel-like (r = −0.8642), and mouth feeling attributes such as astringency (r = −0.8724) and painty (r = −0.8880). As can be seen in Table 7, all oil samples presented high scores in terms of the OFI and were positively correlated with the color intensity (r = 0.9587). Generally, in all studied samples, the pungency, astringency, and bitter taste intensities were perceived in medium–high scores. An intense taste with the presence of bitter notes associated with pungent and astringent mouth feeling might be due to bioactive compounds such as polyphenols [47]. Different factors might contribute to off-flavors in cold-pressed oils, such as incorrect storage conditions causing the oxidative degradation of the oil, the presence of impurities in the raw material, or even the use of improper processing technology.

### 3.6. Principal Component Analysis on the Entire Set of Chemical and Sensory Data

The PCA was applied to compare the multidimensional chemical and sensory quality levels of the seven CPBCO samples. The PCA model was applied to all data to determine the most important variables that explain the relationships between the investigated oils and to identify any group patterns. The PCA model retained two principal components (PC1 and PC2), which gave eigenvalues greater than 1.00 (8.63 and 4.50, respectively) and explained 87.52% of the total variability. Therefore, only the first two PCs were used to understand the similarities or dissimilarities of the CPBCO samples available for retail, as illustrated in Figure 2. The PC1 was inversely correlated with IP (−0.9012), PV (−0.9331), ∑4PAHs (−0.8727), PUFA (−0.7004), and all AC (−0.7389––0.9534) variables, whereas PC2 was highly correlated with the moisture content and individual antioxidants (0.9027, 0.9097, and 0.7962 for WVC, TTC, and TSC, respectively).

It is noteworthy that CPBCO3, CPBCO4, CPBCO6, and CPBCO7, with high OSQ and AV scores, as well as low antioxidant properties were located to the right in the score biplot and had positive values for PC1, while the three oils CPBCO1, CPBCO2, and CPBCO5, with the highest oxidative stability (IP), hydroperoxide content (PV), and antioxidant potential (DPPH, ABTS, CUPRAC, and FC values), were situated to the left in the diagram and had negative values for PC1. However, the samples CPBCO3, CPBCO5, CPBCO6, and CPBCO7, with moderate concentrations of total phytosterols (TSC = 201–222 mg/100 g) and tocopherols (TTC = 3.9–7.2 mg/100 g), were located under the A1 axis.

It should be noted that the studied oils fell into three distinct groups (Figure 2).

The CPBCO4 with the longest distance from the other oils revealed the highest contents of tocopherols (TTC), phytosterols (TSC), MUFAs, and water and volatiles (WVC). Three oils (CPBCO3, CPBCO6 and CPBCO 7) with the highest sensory quality (OSQ) and free fatty acid (AV) contents but the lowest amounts of peroxides (PV), ∑4PAHs, and antioxidant potential as determined by different analytical assays (DPPH, ABTS, CUPRAC, and FC) created an evidently distinct cluster. Moreover, three oil samples (CPBCO1, CPBCO2, and CPBCO5) with the highest oxidative stability (IP), antioxidant properties, PV results, and ∑4PAHs, as well as the lowest AV and sensory quality, were separated from the other investigated oils. Additionally, the oils with the highest concentrations of the primary oxidation products were characterized by poor sensory quality.

Regarding the influence of impurities on the sensory profile, Wroniak et al. [14] reported a negative correlation between the level of contaminants in cold-pressed rapeseed oil and the sensory quality with the presence of off-flavors.

The positive and negative correlations between the chemical and sensory characteristics of the seven commercial CPBCO samples are presented as a correlation matrix in Figure 3.

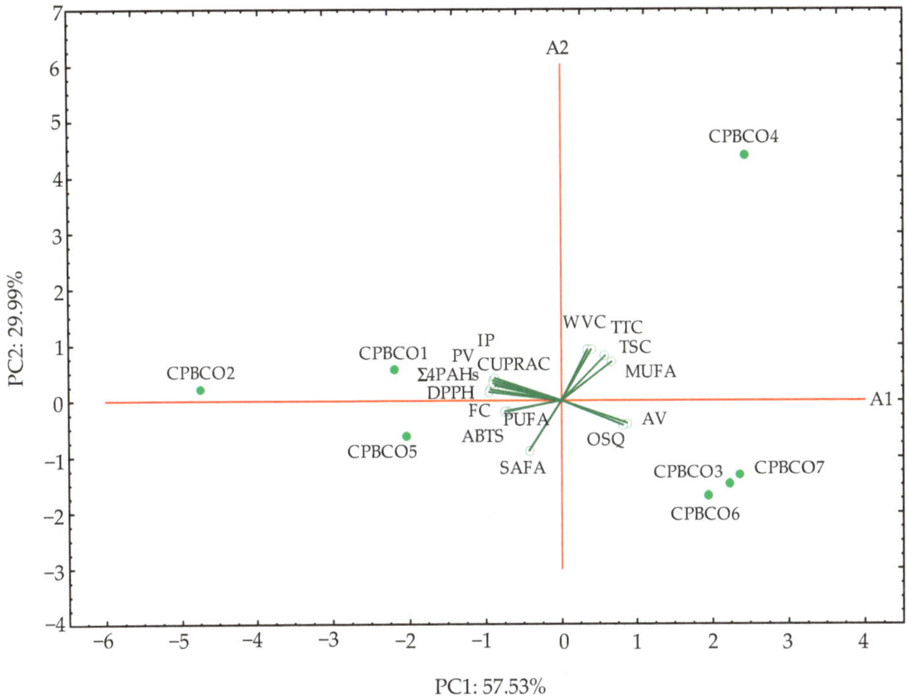


**Figure 2.** Biplot of scores and loadings of data obtained for chemical and overall sensory characteristics of the seven cold-pressed black cumin oils.

Unfortunately, the OSQ of the discussed oils was significantly negatively correlated with their antioxidant potential (r = −0.8107–−0.9325, $p$ = 0.0022–0.027), oxidative stability (−0.8746, $p$ = 0.010), and peroxide content (r = −0.7818, $p$ = 0.038), but positively associated with the level of free fatty acids (r = 0.9314, $p$ = 0.0023). Moreover, there were significant positive correlations between the AC results obtained using different analytical methods (r = 0.8522–0.9732, $p$ = 0.00022–0.014), AC and IP (r = 0.8842–0.9267, $p$ = 0.0027–0.0082), AC and PV (r = 0.8542–0.9734, $p$ = 0.00022–0.014), AC and Σ4PAHs (r = 0.8014–0.8940, $p$ = 0.0066–0.030), Σ4PAHs and IP (r = 0.9597, $p$ = 0.00061), and TTC and TSC (r = 0.9719, $p$ = 0.00025). However, significant negative correlations between AC and AV (r = −0.9200–−0.9521, $p$ = 0.00094–0.0033), SAFA and MUFA (r = −0.8569, $p$ = 0.014), MUFA and PUFA (r = −0.7939, $p$ = 0.033), SAFA and TTC (r = −0.9939, $p$ = 0.000006), and SAFA and TSC (r = −0.9823, $p$ = 0.00008) for CPBCOs were observed.

Unexpectedly, TTC and TSC did not correlate with AC, as determined by four analytical methods (r = −0.4238–0.0060, $p$ = 0.3433–0.9897 for TTC and AC; r = −0.2239–−0.5376, $p$ = 0.2133–0.6293 for TSC and AC). These low negative correlation coefficients suggest that the antioxidant properties of the oils may be due to more hydrophilic compounds such as polyphenols in the prepared extracts or to the synergistic interactions of tocopherols and sterols with other antioxidants present in the CPBCO samples. Therefore, the TTC and TSC in the studied oils are not reliable indicators of their overall AC. The applied DPPH, ABTS, CUPRAC, and FC methods for the determination of the AC were not equivalent by virtue of their varying sensitivities to different antioxidants found in CPBCOs. The effects of the different antioxidants did not appear to be additive but were most likely governed by synergistic and antagonistic interactions among them.

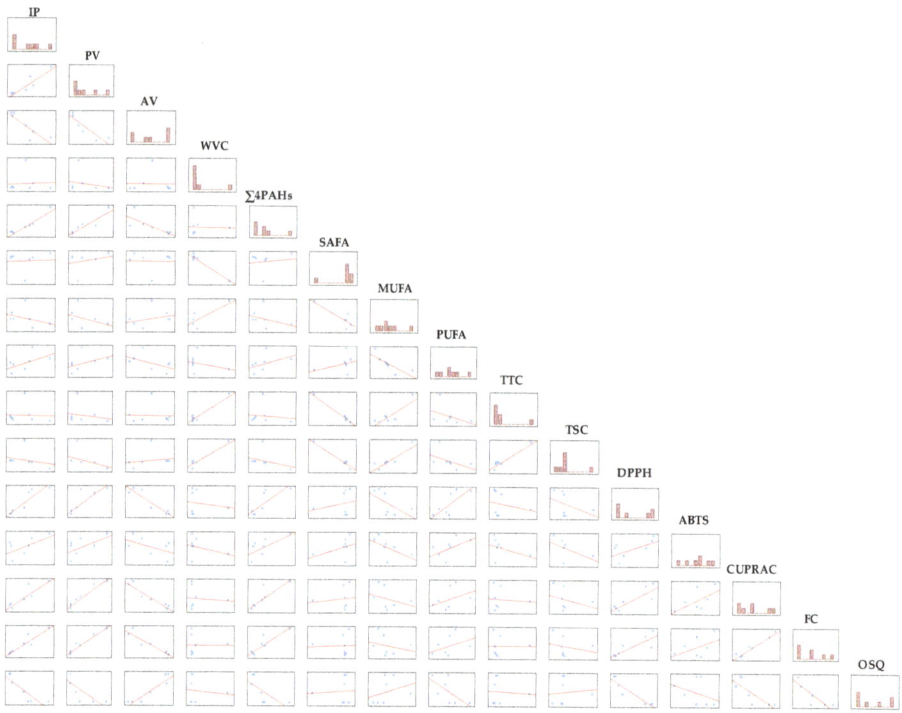


**Figure 3.** Correlation matrix between the chemical parameters and overall sensory quality levels of cold-pressed black cumin oils.

## 4. Conclusions

This study was one of the first attempts to thoroughly investigate the similarities between the sensory attributes, the composition of the CPBCO samples, and their oxidative stability.

The CPBCOs (except CPBCO4) available on the Polish market had similar chemical and nutritional values. The amounts of antioxidants and fatty acid compositions of the studied samples can be considered typical for black cumin oil. The studied oils were rich sources of unsaturated fatty acids and consisted of large amounts of linoleic acid (18:2, omega-6, >50%) and oleic acid (18:1, omega-9, >20%), while a low level of SAFA (about 16%) was found. The CPBCO4 sample was probably adulterated or unintentionally contaminated with cold-pressed linseed oil. Additionally, the sensory attributes of typical black cumin oil (including bitter taste) can be similar to linseed oil's characteristics.

The antioxidant properties determined as the AC, TTC, and TSC and the oxidative stability of the investigated oils differed significantly ($p < 0.05$) due to differences in the raw material varieties, country of cultivation, or oil age. However, the high antioxidant potential enhanced the oxidative stability of the CPBCOs and can provide potential health benefits to consumers. On the other hand, some analyzed chemical parameters indicating the oil oxidative and hydrolytic status (PV, AV) and cancerogenic contaminant content (∑4PAHs) were above the legal limits recommended by the EU food regulations and the Codex Alimentarius standards.

Moreover, the sensory quality levels of three oil samples with the perception of off-flavor attributes were unacceptable for fresh cold-pressed oils (OSQ = 2.0). All of these defects can occur due to poor raw material quality, improper storage of the seeds, or the industrial process. For the first time, a sensory evaluation of CPBCOs was conducted by applying the QDA methodology. The sensory glossary developed in this study can be helpful for the assessment of CPBCO quality.

**Author Contributions:** Conceptualization, M.M., B.S., D.R.-K. and A.S.-C.; methodology, M.M., B.S., D.R.-K. and A.S.-C.; software, A.S.-C.; validation, A.S.-C., M.M., B.S. and D.R.-K.; formal analysis, M.M., B.S. and D.R.-K.; investigation, M.M., B.S., D.R.-K. and A.S.-C.; resources, M.M., B.S., D.R.-K. and A.S.-C.; data curation, A.S.-C., D.R.-K., M.M. and B.S.; writing—original draft preparation, M.M., B.S. and D.R.-K.; writing—review and editing, A.S.-C.; visualization, M.M., B.S., D.R.-K. and A.S.-C.; supervision, A.S.-C.; project administration, A.S.-C., D.R.-K., M.M. and B.S.; funding acquisition, M.M. and A.S.-C. All authors have read and agreed to the published version of the manuscript.

**Funding:** This research received no external funding.

**Institutional Review Board Statement:** Not applicable.

**Informed Consent Statement:** Not applicable.

**Data Availability Statement:** All of the data is contained within the article.

**Conflicts of Interest:** The authors declare no conflict of interest. Two authors are employees of Bunge Polska Sp. z o.o. These studies were carried out as part of the "Implementation Doctorate" Programme for favourable cooperation between university and business.

## Abbreviations

| | |
|---|---|
| ABTS | 2,2′-azino-bis(3-ethylbenzothiazoline-6-sulfonic acid) |
| AC | antioxidant capacity |
| AV | acid value |
| ANOVA | analysis of variance |
| AOCS | American Oil Chemists' Society |
| B(a)A | benzo(a)anthracene |
| B(a)P | benzo(a)pyrene |
| B(b)F | benzo(b)fluoranthene |
| Chry | chrysene |
| CPBCO | cold-pressed black cumin oil |
| CUPRAC | Cupric-reducing antioxidant capacity |
| DGF | German Society for Fat Science |
| DPPH | 2,2-diphenyl-1-picrylhydrazyl |
| EU | European Union |
| FAC | fatty acid composition |
| FAME | fatty acid methyl esters |
| FC | Folin–Ciocalteu |
| FFA | free fatty acids |
| FID | flame ionization detector |
| FRAP | Ferric-reducing antioxidant power |
| GA | gallic acid |
| GC-FID | gas chromatograph equipped with flame ionization detector |
| HAT | hydrogen atom transfer |
| HPLC | high performance liquid chromatography |
| HPLC-FLD | high performance liquid chromatography with the fluorescence detector |
| $IC_{50}$ | half-maximal inhibitory concentration |
| IP | induction period |
| ISO | International Organization for Standardization |
| LDL | low-density lipoprotein |
| LOD | limit of detection |
| LOQ | limit of quantification |
| MUFA | monounsaturated fatty acids |
| OFI | overall flavor intensity |
| OSQ | overall sensory quality |

| | |
|---|---|
| PAH | polycyclic aromatic hydrocarbon |
| Σ4PAHs | sum of four specific polycyclic aromatic hydrocarbons |
| PC | principal component |
| PCA | principal component analysis |
| PUFA | polyunsaturated fatty acids |
| PV | peroxide value |
| QDA | quantitative descriptive analysis |
| SAFA | saturated fatty acids |
| SET | single electron transfer |
| SD | standard deviation |
| TE | Trolox equivalents |
| TPC | total phenolic content |
| TTC | total tocopherol content |
| TSC | total sterol content |
| WVC | water and volatile matter content |

## References


1. Dedebas, T.; Ekici, L.; Sagdic, O. Chemical characteristics and storage stabilities of different cold-pressed seed oils. *J. Food Process. Preserv.* **2021**, *45*, e15107. [CrossRef]
2. Ketenoglu, O.; Kiralan, S.S.; Kiralan, M.; Ozkan, G.; Ramadan, M.F. Cold Pressed Black Cumin (*Nigella sativa* L.) Seed Oil. In *Cold Pressed Oils*; Ramadan, M.F., Ed.; Academic Press: Cambridge, MA, USA; Elsevier: Amsterdam, The Netherlands, 2020; pp. 53–64. [CrossRef]
3. Kamal-Eldin, A. Nigella (Black Cumin) Seed Oil. In *Gourmet and Health-Promoting Specialty Oils*, 3rd ed.; Moreau, R.A., Kamal-Eldin, A., Eds.; AOCS Press: Urbana, IL, USA; Elsevier: Amsterdam, The Netherlands, 2009; pp. 299–311. [CrossRef]
4. Kabir, Y.; Shirakawa, H.; Komai, M. Nutritional composition of indigenous cultivar of black cumin seeds from Bangladesh. *Prog. Nutr.* **2019**, *21*, 428–434. [CrossRef]
5. Cheikh-Rouhou, S.; Besbes, S.; Hentati, B.; Blecker, C.; Deroanne, C.; Attia, H. *Nigella sativa* L.: Chemical composition and physicochemical characteristics of lipid fraction. *Food Chem.* **2007**, *101*, 673–681. [CrossRef]
6. Ramadan, M.F. Healthy blends of high linoleic sunflower oil with selected cold pressed oils: Functionality, stability and antioxidative characteristics. *Ind. Crops Prod.* **2013**, *43*, 65–72. [CrossRef]
7. Tulukcu, E. A comparative study on fatty acid composition of black cumin obtained from different regions of Turkey, Iran and Syria. *Afr. J. Agric. Res.* **2011**, *6*, 892–895.
8. Ramadan, M.F.; Mörsel, J. Oxidative stability of black cumin (*Nigella sativa* L.), Coriander (*Coriandrum sativum* L.) and Niger (*Guizotia Abyssinica Cass.*) crude seed oils upon stripping. *Eur. J. Lipid Sci. Technol.* **2004**, *106*, 35–43. [CrossRef]
9. Lutterodt, H.; Luther, M.; Slavin, M.; Yin, J.-J.; Parry, J.; Gao, J.-M.; Yu, L. Fatty acid profile, thymoquinone content, oxidative stability, and antioxidant properties of cold-pressed black cumin seed oils. *LWT-Food Sci. Technol.* **2010**, *43*, 1409–1413. [CrossRef]
10. Burits, M.; Bucar, F. Antioxidant activity of *Nigella sativa* essential oil. *Phytother. Res.* **2000**, *14*, 323–328. [CrossRef]
11. Symoniuk, E.; Ratusz, K.; Ostrowska-Ligęza, E.; Krygier, K. Impact of selected chemical characteristics of cold-pressed oils on their oxidative stability determined using the Rancimat and pressure differential scanning calorimetry method. *Food Anal. Methods* **2018**, *11*, 1095–1104. [CrossRef]
12. Novel Food Catalogue. Available online: https://ec.europa.eu/food/safety/novel-food/novel-food-catalogue_en (accessed on 25 April 2022).
13. Bendini, A.; Barbieri, S.; Valli, E.; Buchecker, K.; Canavari, M.; Toschi, T.G. Quality evaluation of cold pressed sunflower oils by sensory and chemical analysis. *Eur. J. Lipid Sci. Technol.* **2011**, *113*, 1375–1384. [CrossRef]
14. Wroniak, M.; Rękas, A.; Ratusz, K. Influence of impurities in raw material on sensory and physicochemical properties of cold-pressed rapeseed oil produced from conventionally and ecologically grown seeds. *Acta Sci. Pol. Technol. Aliment.* **2016**, *15*, 289–297. [CrossRef] [PubMed]
15. Tauferova, A.; Dordevic, D.; Jancikova, S.; Tremlova, B.; Kulawik, P. Fortified cold-pressed oils: The effect on sensory quality and functional properties. *Separations* **2021**, *8*, 55. [CrossRef]
16. *ISO 9936:2016*; Animal and Vegetable Fats and Oils-Determination of Tocopherol and Tocotrienol Contents by High Performance Liquid Chromatography. ISO: Geneva, Switzerland, 2016.
17. *ISO 1228-1:2014*; Animal and Vegetable Fats and Oils-Determination of Individual and Total Sterols Contents-Gas Chromatographic Method-Part 1. ISO: Geneva, Switzerland, 2014.
18. Szydłowska-Czerniak, A.; Tułodziecka, A.; Momot, M.; Stawicka, B. Physicochemical, antioxidative, and sensory properties of refined rapeseed oils. *J. Am. Oil Chem. Soc.* **2019**, *96*, 405–419. [CrossRef]
19. AOCS. Official Method Cd 12b-92: Oil stability index. In *Official Method and Recommended Practices of the American Oil Chemist's Society*; AOCS Publishing: Champaing, IL, USA, 2017.

20. *ISO 27107:2010*; Animal and Vegetable Fats and Oils-Determination of Peroxide Value-Potentiometric End-Point Determination. ISO: Geneva, Switzerland, 2010.
21. *ISO 660:1996*; Animal and Vegetable Fats and Oils-Determination of Acid Value and Acidity. ISO: Geneva, Switzerland, 1996.
22. *ISO 662:1996*; Animal and Vegetable Fats and Oils-Determination of Moisture and Volatile Matter Content. ISO: Geneva, Switzerland, 1996.
23. *ISO 5508:1996*; Animal and Vegetable Fats and Oils-Analysis by Gas Chromatography of Methyl Esters of Fatty Acids. ISO: Geneva, Switzerland, 1996.
24. *ISO 5509:2000*; Animal and Vegetable Fats and Oils-Preparation of Methyl Esters of Fatty Acids. ISO: Geneva, Switzerland, 2000.
25. *ASTM E1627-94:2004*; Standard Practice for Sensory Evaluation of Edible Oils and Fats. ASTM: West Conshohocken, PA, USA, 2004.
26. *DGF C-II 1, 09:2009*; Quality Evaluation of Cold Pressed Oil. Section C-Fats. DGF: Frankfurt, Germany, 2009.
27. *ISO 11035:1994*; Sensory Analysis-Identification and Selection of Descriptors for Establishing a Sensory Profile by a Multi-Dimensional Approach. ISO: Geneva, Switzerland, 1994.
28. *ISO 8586-1:1996*; Sensory Analysis-General Guidance for the Selection, Training and Monitoring of Assessors-Part 1: Selected Assessors. ISO: Geneva, Switzerland, 1996.
29. *ISO 8586-2:1996*; Sensory Analysis-General Guidance for the Selection, Training and Monitoring of Assessors-Part 2: Expert Sensory Assessors. ISO: Geneva, Switzerland, 1996.
30. Kostadinović Veličkovska, S.; Brühl, L.; Mitrev, S.; Mirhosseini, H.; Matthäus, B. Quality evaluation of cold-pressed edible oils from Macedonia. *Eur. J. Lipid Sci. Technol.* **2015**, *117*, 2023–2035. [CrossRef]
31. Kiralan, M.; Özkan, G.; Bayrak, A.; Ramadan, M.F. Physicochemical properties and stability of black cumin (*Nigella sativa*) seed oil as affected by different extraction methods. *Ind. Crops Prod.* **2014**, *57*, 52–58. [CrossRef]
32. Hassanein, M.M.; El-Shami, S.M.; El-Mallah, M.H. Investigation of lipids profiles of nigella, lupin and artichoke seed oils to be used as healthy oils. *J. Oleo Sci.* **2011**, *60*, 99–107. [CrossRef]
33. Grajzer, M.; Szmalcel, K.; Kuźmiński, Ł.; Witkowski, M.; Kulma, A.; Prescha, A. Characteristics and antioxidant potential of cold-pressed oils—possible strategies to improve oil stability. *Foods* **2020**, *9*, 1630. [CrossRef] [PubMed]
34. Choe, E.; Min, D.B. Mechanisms of antioxidants in the oxidation of foods. *Compr. Rev. Food Sci. Food Saf.* **2009**, *8*, 345–358. [CrossRef]
35. Gulcin, İ. Antioxidants and antioxidant methods: An updated overview. *Arch. Toxicol.* **2020**, *94*, 651–715. [CrossRef] [PubMed]
36. Mohammed, N.K.; Abd Manap, M.Y.; Tan, C.P.; Muhialdin, B.J.; Alhelli, A.M.; Meor Hussin, A.S. The effects of different extraction methods on antioxidant properties, chemical composition, and thermal behavior of black seed (*Nigella sativa* L.) oil. *J. Evid. Based Complement. Altern. Med.* **2016**, *2016*, 6273817. [CrossRef]
37. Sielicka, M.; Małecka, M.; Purłan, M. Comparison of the antioxidant capacity of lipid-soluble compounds in selected cold-pressed oils using photochemiluminescence assay (PCL) and DPPH method. *Eur. J. Lipid Sci. Technol.* **2014**, *116*, 388–394. [CrossRef]
38. Mazaheri, Y.; Torbati, M.; Azadmard-Damirchi, S.; Savage, G.P. A comprehensive review of the physicochemical, quality and nutritional properties of *Nigella sativa* oil. *Food Rev. Int.* **2019**, *35*, 342–362. [CrossRef]
39. Codex Alimentarius. Codex Standard for Named Vegetable Oils. Codex-Stan 210-1999. Adopted in 1999. Revised in 2001, 2003, 2009, 2017, 2019. Amended in 2005, 2011, 2013, 2015, 2019, 2021. Available online: https://www.fao.org/fao-who-codexalimentarius/codex-texts/list-standards/en/ (accessed on 25 April 2022).
40. Gotoh, N.; Wada, S. The importance of peroxide value in assessing food quality and food safety. *J. Am. Oil Chem. Soc.* **2006**, *83*, 473–474. [CrossRef]
41. Mikołajczak, N.; Tańska, M.; Ogrodowska, D.; Czaplicki, S. Efficacy of canolol and guaiacol in the protection of cold-pressed oils being a dietary source linoleic acid against oxidative deterioration. *Food Chem.* **2022**, *393*, 133390. [CrossRef]
42. Commission Regulation (EU) No 835/2011 of 19 August 2011 Amending Regulation (EC) No 1881/2006 as Regards Maximum Levels for Polycyclic Aromatic Hydrocarbons in Foodstuffs. 2011. Available online: https://eur-lex.europa.eu/LexUriServ/LexUriServ.do?uri=OJ:L:2011:215:0004:0008:EN:PDF (accessed on 25 April 2022).
43. Badary, O.A.; Abd-Ellah, M.F.; El-Mahdy, M.A.; Salama, S.A.; Hamada, F.M. Anticlastogenic activity of thymoquinone against benzo(a)pyrene in mice. *Food Chem. Toxicol.* **2007**, *45*, 88–92. [CrossRef] [PubMed]
44. Ramadan, M.F.; Asker, M.M.S.; Tadros, M. Antiradical and antimicrobial properties of cold-pressed black cumin and cumin oils. *Eur. Food Res. Technol.* **2012**, *234*, 833–844. [CrossRef]
45. Ozdemir, N.; Kantekin-Erdogan, M.N.; Tat, T.; Tekin, A. Effect of black cumin oil on the oxidative stability and sensory characteristics of mayonnaise. *J. Food Sci. Technol.* **2018**, *55*, 1562–1568. [CrossRef] [PubMed]
46. Brühl, L.; Matthäus, B.; Fehling, E.; Wiege, B.; Lehmann, B.; Luftmann, H.; Bergander, K.; Quiroga, K.; Scheipers, A.; Frank, O.; et al. Identification of bitter off-taste compounds in the stored cold pressed linseed oil. *J. Agric. Food Chem.* **2007**, *55*, 7864–7868. [CrossRef] [PubMed]
47. Gutiérrez-Rosales, F.; Ríos, J.J.; Gómez-Rey, M.L. Main polyphenols in the bitter taste of virgin olive oil. structural confirmation by on-line high-performance liquid chromatography electrospray ionization mass spectrometry. *J. Agric. Food Chem.* **2003**, *51*, 6021–6025. [CrossRef] [PubMed]

*antioxidants*

MDPI

*Article*

# Antioxidant Properties of Thymoquinone, Thymohydroquinone and Black Cumin (*Nigella sativa* L.) Seed Oil: Scavenging of Superoxide Radical Studied Using Cyclic Voltammetry, DFT and Single Crystal X-ray Diffraction

**Raiyan Sakib, Francesco Caruso, Sandjida Aktar, Stuart Belli, Sarjit Kaur, Melissa Hernandez and Miriam Rossi ***

Department of Chemistry, Vassar College, Poughkeepsie, NY 12604, USA
* Correspondence: rossi@vassar.edu

**Abstract:** Black cumin seeds and seed oil have long been used in traditional foods and medicine in South Asian, Middle Eastern and Mediterranean countries and are valuable flavor ingredients. An important ingredient of black cumin is the small molecule thymoquinone (TQ), which manifests low toxicity and potential therapeutic activity against a wide number of diseases including diabetes, cancer and neurodegenerative disorders. In this study, the antioxidant activities of black seed oil, TQ and a related molecule found in black cumin, thymohydroquinone (THQ), were measured using a direct electrochemical method to experimentally evaluate their superoxide scavenging action. TQ and the black seed oil showed good superoxide scavenging ability, while THQ did not. Density Functional Theory (DFT) computational methods were applied to arrive at a chemical mechanism describing these results, and confirmed the experimental Rotating Ring Disk Electrode (RRDE) findings that superoxide oxidation to $O_2$ by TQ is feasible, in contrast with THQ, which does not scavenge superoxide. Additionally, a thorough inquiry into the unusual cyclic voltammetry pattern exhibited by TQ was studied and was associated with formation of a 1:1 TQ-superoxide radical species, $[TQ\text{-}O_2]^{-}\bullet$. DFT calculations reveal this radical species to be involved in the $\pi$-$\pi$ mechanism describing TQ reactivity with superoxide. The crystal structures of TQ and THQ were analyzed, and the experimental data reveal the presence of stacking intermolecular interactions that can be associated with formation of the radical species, $[TQ\text{-}O_2]^{-}\bullet$. All three of these methods were essential for us to arrive at a chemical mechanism that explains TQ antioxidant activity, that incorporates intermolecular features found in the crystal structure and which correlates with the measured superoxide scavenging activity.

**Keywords:** thymoquinone; thymohydroquinone; black seed oil; cyclovoltammetry; superoxide; DFT

## 1. Introduction

The annual herb *Nigella sativa* L. is native to the Mediterranean region, northern Africa and south Asia. Historical evidence of its use by ancient civilizations in those areas exists and seeds of the plant (black cumin) were even found in King Tutankhamen's tomb [1]. Today, populations in those geographical locations still use the black seeds and the related oil, containing the phytochemicals thymoquinone (TQ) and thymohydroquinone (THQ), in traditional foods as valuable flavor ingredients and for medicinal purposes, ingested with food or honey [1,2]. The main active phytochemicals in the seeds and oil include several monoterpenes, TQ (by far, the highest concentration), THQ, p-cymene, carvacrol and thymol [3]. Additionally, black cumin seed oil is rich in nutritionally beneficial fatty acids such as linoleic acid (57.71%) and oleic acid (24.46%) [2]. As with many agricultural materials, black seeds and the related cold-pressed oil show compositional variation depending on the area of cultivation [4].

Importantly, black seeds and their oil have demonstrated widespread action against numerous diseases. Because the low toxicity and potential therapeutic activity of TQ is notable, including diseases such as diabetes, cancer and neurodegenerative disorders, it has been the subject of many review articles [5–7]. Recently, TQ therapeutic activity has extended to include its use as a viable treatment against Covid-19 [8]. The presence of TQ and THQ in other plant species besides *N. Sativa*, including the *Thymus vulgaris* L. and *Origanum* species, is confirmed [5,9]. Studies on the biosynthesis of TQ and THQ in Lamiaceae established that the monoterpenes, thymol and carvacrol, are used to produce THQ by cytochrome P-450 enzymes, CYP76S40 or CYP736A300, which then is converted to TQ [10].

An objective of this study was to obtain chemical data to help clarify the relationship between the antioxidant and free radical scavenging properties of these substances and to gain insight into the chemical mechanisms underlying the therapeutic benefits of these phytochemical ingredients. We investigated the crystal and molecular structures of the small molecules TQ and THQ, as this method also reveals information regarding intermolecular interactions among neighboring molecules. We measured the antioxidant activity of black seed oil, TQ and THQ using the RRDE and one-electrode cyclic voltammetry electrochemical methods to obtain information about scavenging properties. Last, we used DFT methods to describe electronic properties since these are related to a molecule's ability to donate and/or receive electron density and its antioxidant capability. These computational methods were essential for us to arrive at a superoxide scavenging chemical mechanism that incorporates intermolecular features found in the crystal structure and which correlates with the measured superoxide scavenging activity.

## 2. Materials and Methods

### 2.1. Reagents

Thymoquinone (Toronto Research Chemicals, Toronto, ON, CA), thymohydroquinone (Cayman Chemical, Ann Arbor, MI, USA) and Cold Pressed Black Seed Oil (Maju Superfoods, San Diego, CA, USA). For electrochemical studies, tetrabutylammonium bromide (TBAB; TCI Chemicals, Portland, OR, USA) and 99.9% anhydrous Dimethyl Sulfoxide (DMSO; Sigma-Aldrich, Inc., St. Louis, MO, USA).

### 2.2. Equipment

#### 2.2.1. Electrochemistry

A Pine Research WaveDriver 20 bipotentiostat with the Modulated Speed Electrode Rotator was used to perform the hydrodynamic voltammetry at a rotating ring-disk electrode (RRDE) as well as cyclic voltammetry (CV). The working electrode is the AFE6R2 gold disk and gold ring rotator tip (Pine Research, Durham, NC, USA) combined with a coiled platinum wire counter electrode and a reference electrode consisting of an AgCl coated silver wire immersed in 0.1 M tetrabutylammonium bromide (TBAB) in dry DMSO in a fritted glass tube. The electrodes were placed in a five-neck electrochemical cell together with means for either bubbling or blanketing the solution with gas. Voltammograms were collected using Aftermath software provided by Pine Research. Careful cleaning of the electrodes was performed by polishing with 0.05 µm alumina-particle suspension (Allied High Tech Products, Inc., Rancho Dominguez, CA, USA) on a moistened polishing microcloth to eliminate potential film formation [11]. Cyclic voltammetry was performed using the same electrodes and cell.

#### 2.2.2. X-ray Diffraction

An APEX2 DUO platform X-ray diffractometer from Bruker Advanced X-ray Solutions (Madison, WI, USA) was used to obtain X-ray data measurements on suitable crystals at 125 K. Temperature was maintained using a cold liquid nitrogen stream from Oxford Cryosystems (UK).

### 2.3. Electrochemical Studies

#### 2.3.1. Hydrodynamic Voltammetry (RRDE)

Stock clear light-yellow 0.03 M solution of TQ (0.0492 g in 10 mL) and clear colorless 0.03 M solution THQ (0.0498 g in 10 mL), in anhydrous DMSO (99.9% purity), were used in trials, whereas black seed oil was directly introduced in the electrochemical cell.

For the experiment, a solution of 0.1 M TBAB electrolyte in anhydrous DMSO was bubbled for 5 min with a dry $O_2/N_2$ (35%/65%) gas mixture to establish the dissolved oxygen level in the electrochemical cell of 50 mL. The Au/Au disc electrode was then rotated at 1000 rpm and potential sweep applied to the disk from 0.2 V to −1.2 V and then back to 0.2 V while the ring was held constant at 0.0 V; the disk voltage sweep rate was set to 25 mV/s. The molecular oxygen reduction peak (reaction 1) is observed around −0.6 V, at the disk electrode; the oxidation current (reaction 2) occurs at the ring electrode. An initial blank, in the absence of an antioxidant, was run on this solution and the ratio of the ring/disk current was calculated as the "efficiency". This blank efficiency was found to be about 20%. Next, an antioxidant aliquot was added, the solution bubbled with the gas mixture for 5 min, the voltammogram was rerecorded, and efficiency obtained. In this way, the rate at which increasing concentrations of antioxidant scavenge the generated superoxide radicals during the electrochemical reaction is determined as each additional antioxidant aliquot is added. Results from each run were collected on Aftermath software and represented as voltammograms showing current vs. potential graphs that were later analyzed using Microsoft Excel. The aliquots used are indicated in related RRDE graphs. Ultimately, the slope of the overall decrease in efficiency with addition of antioxidant serves as a quantitative measure of the antioxidant activity of each compound. Any decrease in the collection efficiency is expected to be due to the amount of superoxide removed by the antioxidant.

In an RRDE voltammetry experiment, the generation of the superoxide radicals occurs at the disk electrode while the oxidation of the residual superoxide radicals (that have not been scavenged by the antioxidant) occurs at the ring electrode.

Reaction 1: Reduction of molecular oxygen at the disk electrode

$$\text{Disk Reaction} \qquad O_2 + e^- \rightarrow O_2\bullet^- \tag{1}$$

Reverse Reaction 2: Oxidation of superoxide radicals at the ring electrode

$$\text{Ring Reaction} \qquad O_2\bullet^- \rightarrow O_2 + e^- \tag{2}$$

#### 2.3.2. Cyclic Voltammetry

On completion of the RRDE study of TQ, CV was performed both in the presence and absence of oxygen dissolved in the DMSO solvent in order to further study the reactions occurring at the electrode surfaces.

### 2.4. Computational Study

Calculations were performed using software programs from Biovia (San Diego, CA, USA). Density Functional Theory (DFT) included in $DMol^3$ was applied to calculate energy, geometry, and frequencies implemented in Materials Studio 7.0 (PC platform) [12]. We employed the double numerical polarized (DNP) basis set that included all the occupied atomic orbitals plus a second set of valence atomic orbitals, and polarized d-valence orbitals [13]; the correlation generalized gradient approximation (GGA) was applied including Becke exchange [14] plus BLYP-D correlation including Grimme's correction when van der Waals interactions were involved [15]. All electrons were treated explicitly and the real space cutoff of 5 Å was imposed for numerical integration of the Hamiltonian matrix elements. The self-consistent field convergence criterion was set to the root mean square change in the electronic density to be less than $10^{-6}$ electron/Å$^3$. The convergence criteria applied during geometry optimization were $2.72 \times 10^{-4}$ eV for energy and 0.054 eV/Å for force. Calculations included the effect of DMSO solvent using the continuous model of Dmol[3] [16], to allow correlation with the experimental features from cyclovoltammetry results.

### 2.5. Diffraction Study

Suitable large yellow single crystals of TQ were grown from a 2:1 ($v/v$) methanol:water solution. Colorless crystals of THQ were recrystallized from 1:1 ($v/v$) ethanol:water solution. The crystal structures were solved and refined using full-matrix least-squares on $F^2$ with the Bruker incorporated ShelX programs [17]. We input the X-ray data into the MERCURY program from Cambridge Structural Database (CSD) to produce images of the molecules and crystal packing [18]. Crystal data of THQ and TQ have been deposited at the CSD and are available at https://www.ccdc.cam.ac.uk/structures/? (accessed on 1 January 2023) using Identifier CCDC number 2239493-2239494 and are available upon request [19].

## 3. Results and Discussion

### 3.1. Diffraction Study

Crystal data for THQ and TQ are given in Table 1. Both THQ and TQ crystallized with two molecules in the asymmetric unit, displayed in Figure 1. Atomic distances and angles in both structures agree with expected values.

**Table 1.** Crystal data and structure refinement for thymohydroquinone (THQ), thymoquinone (TQ).

| Identification code | THQ | TQ |
|---|---|---|
| Empirical formula | $C_{10}H_{14}O_2$ | $C_{10}H_{12}O_2$ |
| Formula weight | 166.221 | 164.206 |
| Temperature/K | 125.15 | 125.15 |
| Crystal system | monoclinic | triclinic |
| Space group | Pc | P-1 |
| a/Å | 10.7442(11) | 6.6965(7) |
| b/Å | 10.020(1) | 10.4246(10) |
| c/Å | 9.0363(9) | 13.4314(13) |
| $\alpha$/° | 90 | 98.553(5) |
| $\beta$/° | 110.201(2) | 102.981(5) |
| $\gamma$/° | 90 | 92.841(5) |
| Volume/Å$^3$ | 912.98(16) | 900.15(16) |
| Z | 4 | 4 |
| $\rho_{calc}$ g/cm$^3$ | 1.209 | 1.212 |
| $\mu$/mm$^{-1}$ | 0.083 | 0.673 |
| F(000) | 360.2 | 353.2 |
| Crystal size/mm$^3$ | 0.1 × 0.08 × 0.06 | 0.29 × 0.25 × 0.01 |
| Radiation | Mo K$\alpha$ ($\lambda$ = 0.71073) | Cu K$\alpha$ ($\lambda$ = 1.54178) |
| 2$\Theta$ range for data collection/° | 4.04 to 50.92 | 6.84 to 143.16 |
| Index ranges | $-12 \leq h \leq 12, -12 \leq k \leq 12, -10 \leq l \leq 10$ | $-8 \leq h \leq 8, -12 \leq k \leq 12, -16 \leq l \leq 16$ |
| Reflections collected | 15,163 | 17,559 |
| Independent reflections | 3375 [$R_{int}$ = 0.0678, $R_{sigma}$ = 0.0610] | 3363 [$R_{int}$ = 0.0435, $R_{sigma}$ = 0.0340] |
| Data/restraints/parameters | 3375/2/332 | 3363/0/269 |
| Goodness-of-fit on $F^2$ | 1.070 | 1.166 |
| Final R indexes [I >= 2$\sigma$ (I)] | $R_1$ = 0.0407, $wR_2$ = 0.0628 | $R_1$ = 0.0493, $wR_2$ = 0.1631 |
| Final R indexes [all data] | $R_1$ = 0.0762, $wR_2$ = 0.0734 | $R_1$ = 0.0719, $wR_2$ = 0.2379 |
| Largest diff. peak/hole/e Å$^{-3}$ | 0.17/−0.16 | 0.40/−0.27 |
| Flack parameter | 0.2(3) | |

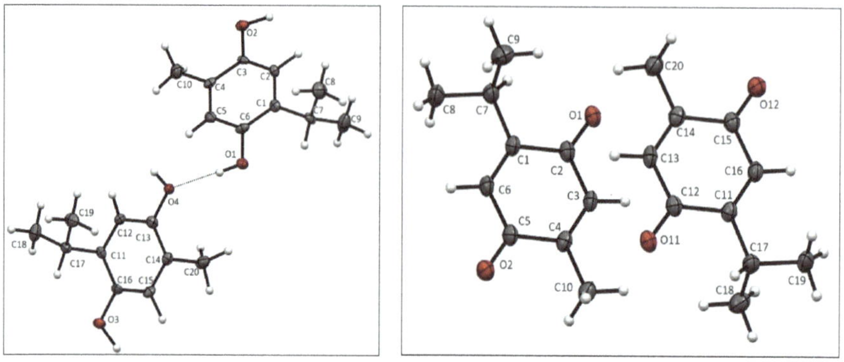


**Figure 1.** Asymmetric unit of THQ, (**left**), and TQ, (**right**), with atom labeling.

### 3.1.1. Thymoquinone

The two molecules in the asymmetric unit differ mainly in the rotation of the isopropyl group, as seen in Figure S1.

Alternating inversion center related molecules of TQ are stacked in a displaced manner primarily along the *a* axis, allowing the hydrophobic isopropyl and methyl groups to be close to each other. The displaced stacking among these planes is close to 3.4 Å. Similarly, TQ molecules in the same plane are arranged with CH . . . O close contacts of about 3.4 Å. These are displayed in Figure 2 and illustrate that only van der Waals forces are seen in the crystal structure.

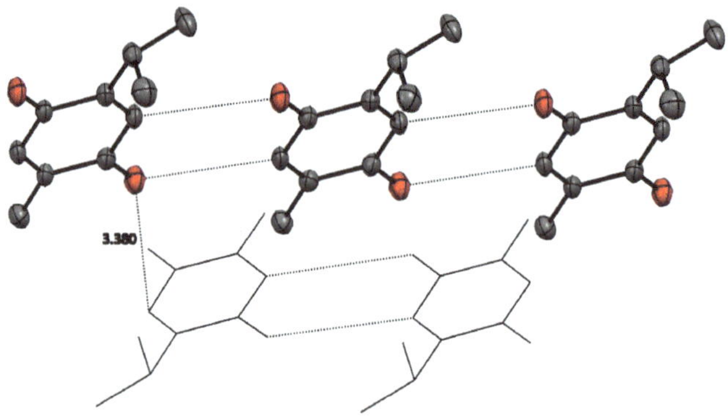


**Figure 2.** TQ molecules have planar quinone rings which have close contacts with other TQ molecules in the same plane of about 3.4 Å. Stacking between these planes is also about 3.4 Å. Hydrogen atoms removed for clarity.

A lower-resolution powder diffraction structure of thymoquinone had been earlier reported with CSD refcode NIDKER [20]. This structure determination was not able to detect the hydrogen atoms in the molecule.

### 3.1.2. Thymohydroquinone

The space group for THQ was ultimately chosen to be the non-centric space group Pc. The centrosymmetric $P2_1/c$ was attempted, but refinement in that space group gave an unacceptable solution with much higher R-value and abnormally high anisotropic displacement parameters. As expected, THQ manifests extensive and strong hydrogen-bonding intermolecular interactions, which are shown in Figure 3, and whose geometric details are reported in Table 2. Each hydroxyl group (O1H, O2H in Molecule 1 and O3H and O4H of Molecule 2 in the asymmetric unit) has two strong intermolecular hydrogen bonds creating an infinite hydrogen bond network throughout the crystal structure. No stacking interactions are evident.

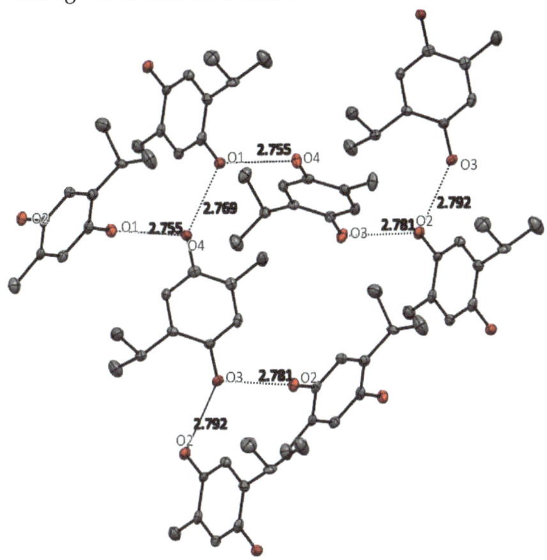


**Figure 3.** Hydrogen bonding among the THQ molecules in the crystal. Hydrogen atoms are not shown for clarity.

**Table 2.** Hydrogen Bonds from crystal structure of THQ.

| D | H | A | d(D-H)/Å | d(H-A)/Å | d(D-A)/Å | D-H-A/° |
|---|---|---|---|---|---|---|
| O1 | H1 | O4 | 0.85(4) | 1.93(5) | 2.769(5) | 170(5) |
| O4 | H4 | O1 [1] | 0.80(4) | 1.96(4) | 2.755(5) | 172(4) |
| O3 | H3 | O2 [2] | 1.05(8) | 1.74(8) | 2.781(5) | 172(7) |
| O2 | H2 | O3 [3] | 1.10(8) | 1.71(8) | 2.791(5) | 164(7) |

[1] +x, 1 − y, −1/2 + z; [2] 1 + x, 1 − y, 1/2 + z; [3] −1 + x, 1 + y, +z.

### 3.2. DFT Study

#### 3.2.1. Thymoquinone

After X-ray atomic coordinates of TQ were input in Materials Studio (Dmol$^3$) quantum mechanical program, geometry optimization was applied and the resulting minimum energy structure is shown in Figure 4A. Single bonds in the quinone ring are in the range 1.47–21.501 Å, obviously longer than the two double bonds, 1.356 Å (compared with X-ray values of 1.473–1.493 Å (single) and 1.337–1.342 Å (double)).

Next, we studied the TQ structural modifications after a superoxide radical was π-π placed at van der Waals separation (3.50 Å) between quinone and superoxide centroids. DFT geometry minimization, Figure 4B, shows the superoxide being more distant, 3.579 Å apart, and with the initial superoxide O-O bond length distance of 1.373 Å shortening to 1.283 Å, which rather corresponds to the $O_2$ double bond distance. Meanwhile, the TQ ring bond lengths become modified, for instance, C=C double bonds become longer, 1.372 Å (from initial values shown in Figure 4A 1.356 Å), while the quinone single C-C bonds become 0.02–0.03 Å shorter. Shorter single bonds and longer double bonds can be associated with ring aromatization, suggesting the unpaired electron of superoxide is donated to the quinone ring. This result has already been observed by us in related polyphenols scavenging superoxide [21]. The electron captured by the ring also has an effect on the C=O bond lengths as they become elongated by 0.03 Å. We call this process (a).

From Figure 4 we conclude that superoxide oxidation (and $O_2$ formation) by TQ is feasible.

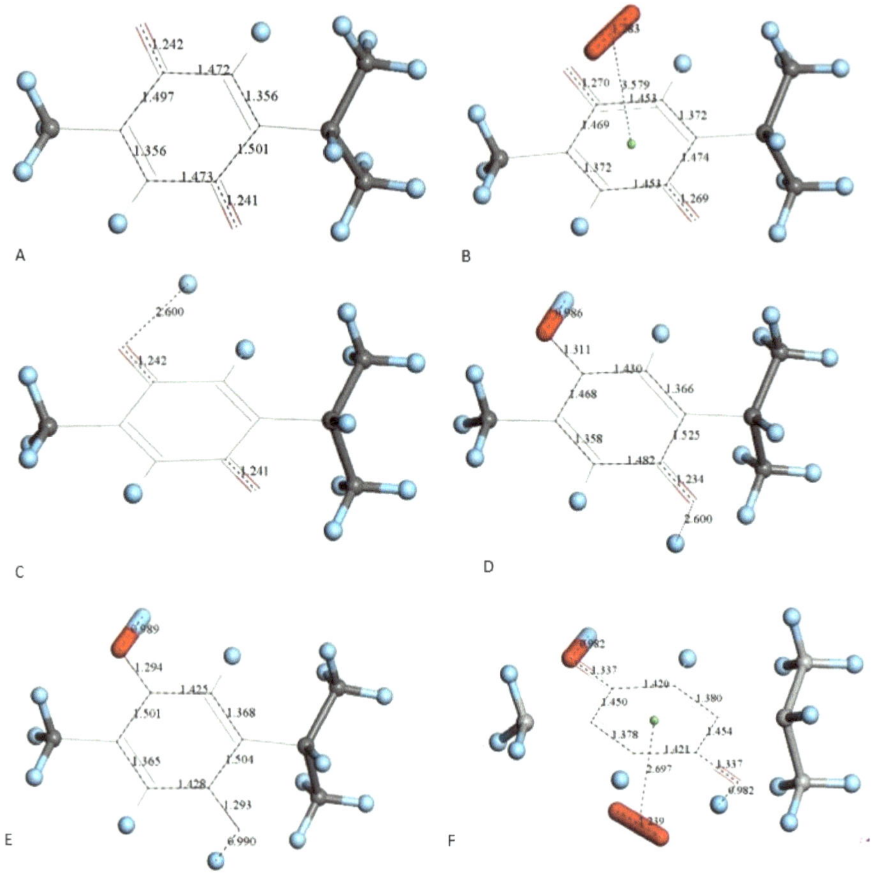


**Figure 4.** TQ superoxide scavenging DFT calculations. H atoms are light blue to avoid white background misperception. (**A**) Minimized TQ molecule obtained after input of corresponding X-ray coordinates. (**B**) Superoxide was $\pi$-$\pi$ placed (3.50 Å) above ring shown in A, and upon DFT minimization both moieties became more distant, 3.579 Å, while ring C-C bond lengths were modified due to electron transferred from superoxide to the ring and O-O bond length shortens to 1.283 Å. (**C**) Structure shown in A has a proton van der Waals separated (2.60 Å) from one O(carbonyl). (**D**) DFT outcome of structure shown in C, plus an additional proton 2.60 Å separated from the second O(carbonyl). (**E**) DFT outcome of structure shown in D whose charge is 2+, showing two C-OH moieties. (**F**) A superoxide radical was $\pi$-$\pi$ placed in E, and DFT shows formation of the radical [thymohydroquinone-$\eta$-$O_2$]$\bullet^+$, with separation between centroids of 2.697 Å.

However, since TQ has no hydroxyls available, an alternative process (b) can be envisioned involving interaction between the quinone and available protons. This route can induce reduction of a C=O carbonyl to a C-OH moiety and has been described for the quinone natural product embelin [22]. The initial state of this process is shown in Figure 4C, where a proton is placed at van der Waals separation from the O(carbonyl), 2.60 Å (1.20 Å + 1.40 Å, for H and O van der Waals radii, respectively), Figure 4A. After DFT minimization, Figure 4D shows the proton captured by the TQ, O-H bond length = 0.986 Å, and lengthening of the corresponding C-O bond, 1.311 Å, compared to initial 1.241 Å in Figure 4A, thus confirming C-OH formation. Indeed, Figure 4D shows the input for the next DFT calculation, i.e., the approach of an additional proton to the second O(carbonyl) in TQ, located initially at 2.60 Å, and which results in the energy minimum moiety shown in Figure 4E, a 2+ charged polyphenol (from 2 captured protons), and having 2 hydroxyls, $[THQ]^{2+}$. As observed in Figure 4B, superoxide is able to transfer its unpaired electron to the TQ ring, and we can expect that the dicationic polyphenol in Figure 4E will permit an even easier electron transfer to the ring. Indeed, this happens and the DFT outcome after energy minimization of the initial structure, a superoxide $\pi$-$\pi$ posed at 3.50 Å (Figure 4E), is shown in Figure 4F, having shorter separation between stacked centroids, 2.697 Å. Upon an additional $\pi$-$\pi$ attack by a second superoxide on the opposite side of the polyphenol ring, this radical also binds to the ring with a longer separation, 2.987 Å, Figure 5. This resulting thymohydroquinone-η-2$O_2$ complex is a neutral species with aromatic character, as shown by the ring C-C bonds in the range 1.394–1.418 Å; this is process (b). Attempts to extract one H(hydroxyl) from the thymohydroquinone-η-2$O_2$ structure shown in Figure 5, by an additional superoxide via σ attack, were not successful, as the possible expected radical product {[thymohydrosemiquinone-η-2$O_2$]• $HO_2^-$} is +4.3 Kcal/mol higher than the reagent, thus indicating no further reactivity of the thymohydroquinone-η-2$O_2$ complex.

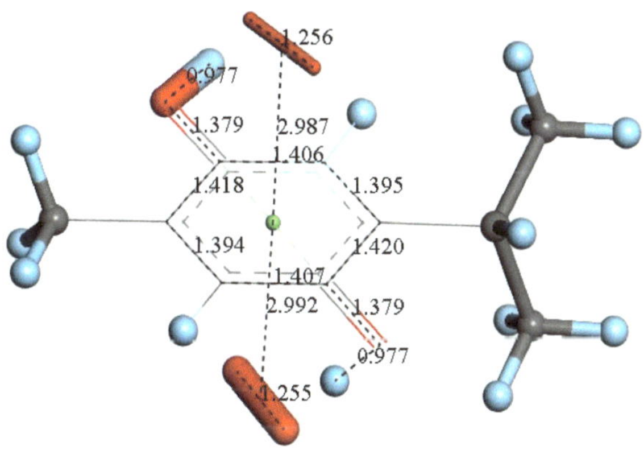


**Figure 5.** An additional superoxide (large stick style) was $\pi$-$\pi$ placed below the structure shown in Figure 4F and its DFT minimization shows the neutral final complex [thymohydroquinone-η-$O_2$], obtained for process (b).

As recently described, molecular oxygen is an important product when polyphenols act as mimics of superoxide dismutase enzymes, following reaction (3) [23].

$$2\,O_2\bullet^- + 2H^+ \rightarrow O_2 + H_2O_2 \quad (3)$$

The reacting proton component of reaction (3) is associated with a polyphenol hydroxyl H atom transfer (HAT, σ attack, [23]) and is more often described in polyphenol antioxidant studies than the $\pi$-$\pi$ attack. According to our DFT results, in TQ scavenging of superoxide, the reaction (3) reactants are involved, since two protons (Figure 4C,D) are captured as well as two superoxide radicals (Figures 4F and 5), but the products of reaction (3), $O_2$ and $H_2O_2$, are not produced by process (b). However, the independent process (a), shown in Figure 4B, suggests $O_2$ formation. In any case, $H_2O_2$ is not produced, and so TQ does not mimic SOD action, unlike behavior recently described for some polyphenols [23]. We conclude that TQ scavenging of superoxide can be described only in terms of stoichiometric reactions.

### 3.2.2. Thymohydroquinone

THQ X-ray atomic coordinates were treated in a similar way with $Dmol^3$. The DFT energy minimum structure Figure 6A shows aromatic ring distances (1.40–1.41 Å), which can be compared to the corresponding X-ray values (1.37–1.41 Å). A superoxide σ-oriented towards one H(hydroxyl), initially at van der Waals separation 2.60 Å, results in a structure having shorter separation, 1.583 Å, Figure 6B. This structure was compared with the one obtained after posing a $HO_2^-$ anion near the corresponding thymohydrosemiquinone radical (a calculation describing the potential expected product), which resulted in a slightly shorter H–O separation, 1.564 Å, Figure 6C. A comparison between the results of both minimizations showed ΔG of 0.3 kcal/mol, indicating no product feasibility. The π-π attack was also analyzed after placing superoxide on top of Figure 6A, and DFT minimization showed both reagents rejected, Figure 4D. We conclude that THQ does not scavenge superoxide.

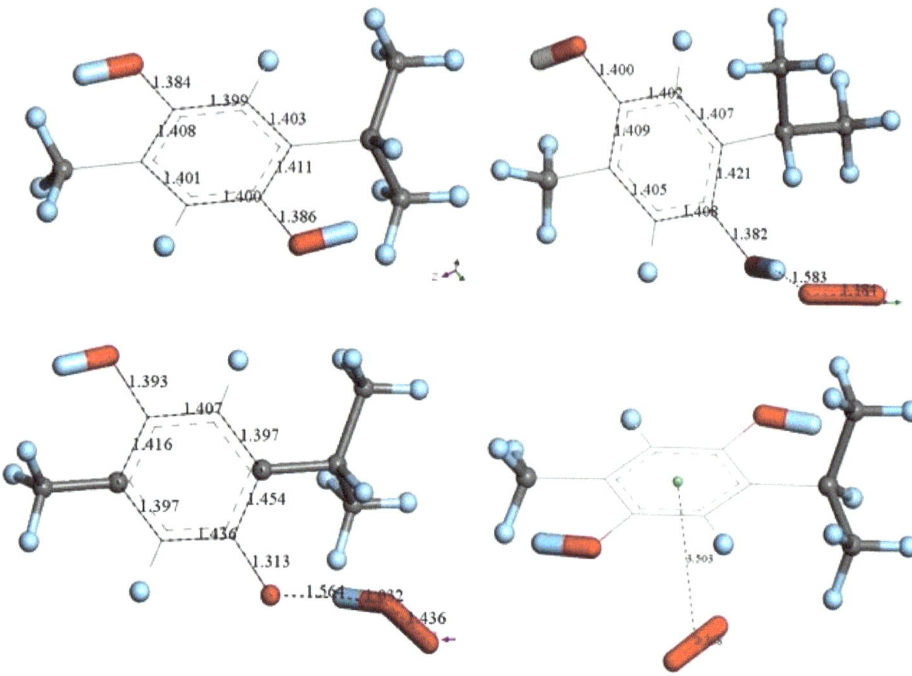


**Figure 6.** THQ DFT calculations. (**A**) Geometry minimization of thymohydroquinone X-ray coordinates. (**B**) Minimization obtained after placing a van der Waals separated superoxide from one H(hydroxyl). (**C**) Minimization obtained after placing a van der Waals separated $HO_2^-$ from thymohydroquinone radical. (**D**) Minimization obtained after placing a π-π van der Waals separated superoxide from THQ, 3.50 Å. Initial superoxide bond length of 1.373 Å had only a slight variation, 1.368 Å. This indicates rejection between both reagents.

### 3.3. *Electrochemistry*

#### 3.3.1. Thymoquinone

The antioxidant activity of TQ was studied using an electrochemical technique developed by Belli et al. [24]. Figure 7 shows RRDE graphs of TQ for all aliquots. Figure 8 shows the corresponding collection efficiency, while Figure 9 shows the collection efficiency of the initial five TQ additions. In addition, CV was performed on a $O_2$ saturated solution (bubbled 5 min) containing a single TQ aliquot (Figure 10), and then after purging with dry $N_2$ gas, Figure 11. Both CVs show two complete scans.

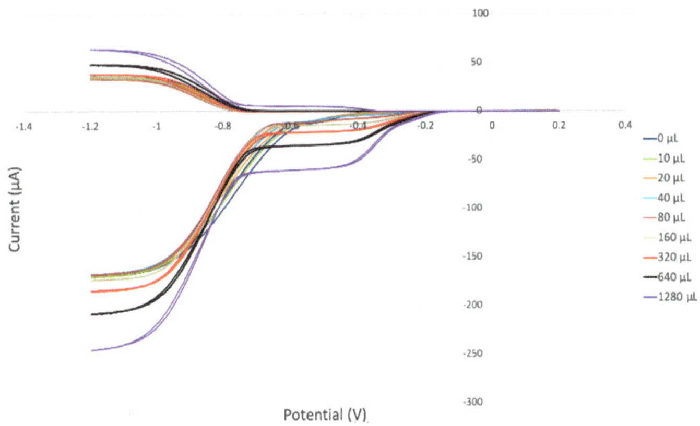


**Figure 7.** TQ RRDE cyclovoltammetry.

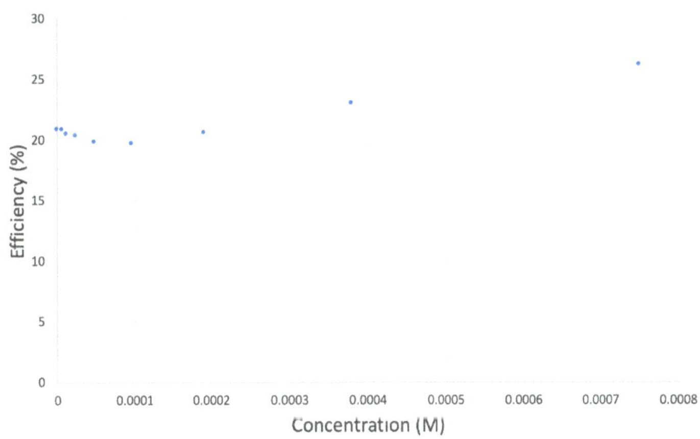


**Figure 8.** TQ efficiency.

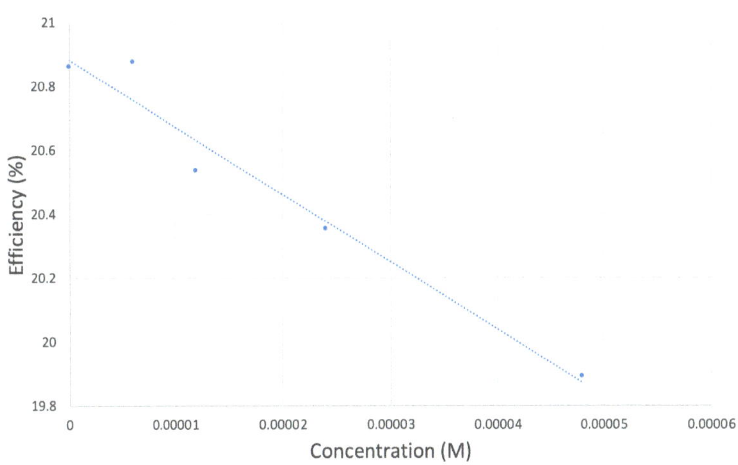


**Figure 9.** TQ collection efficiency for blank plus first four aliquots, whose line equation is $y = -21104x + 20.9$ ($R^2 = 0.9965$).

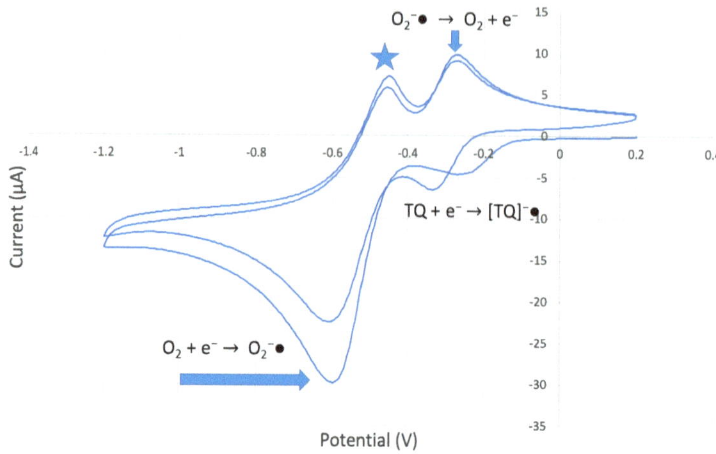


**Figure 10.** CV of TQ 0.00074 M solution (1280 μL aliquot) saturated with the $O_2/N_2$ mixture. The star and arrows are described in the text.

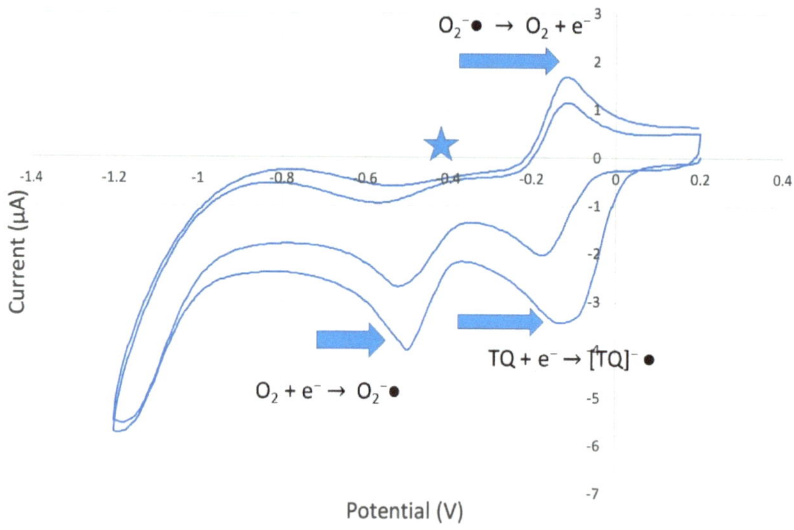


**Figure 11.** CV of TQ 0.00074 M solution (1280 μL aliquot) purged with $N_2$ gas showing evidence of incomplete purging of the $O_2$.

The collection efficiency for the TQ RRDE experiment (Figure 8) is anomalous, showing an initial decrease in efficiency through the first four TQ additions (as normally observed), but then the efficiency increased with each further addition of TQ aliquot, which has never been observed in previous studies [22,24–27]. To clarify this behavior of TQ, cyclic voltammetry of an RRDE solution yielded the voltammogram in Figure 10. Under these conditions, the reduction of oxygen to generate superoxide, reaction (1), −0.60 V, as well as the reverse reaction (2) (oxidation of superoxide), −0.30 V, was previously examined by analyzing a blank solution (without antioxidant), Figure 12 [24]. We analyzed the peaks seen in Figure 10, starting with the first reduction peak (non-reversible) at about −0.30 to −0.35 V and assigned it to incorporation of an electron into TQ (TQ + e- → [TQ]$^{\bullet}$). A reverse reaction was not observed, as after reversing the potential (more positive) than 0.20 V, no related oxidation peak was seen (not shown).

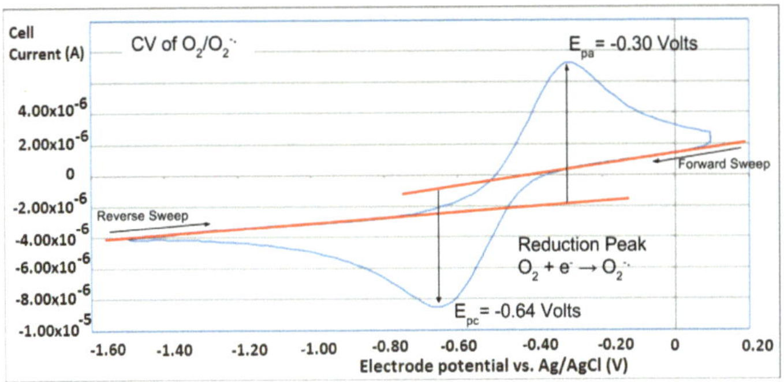


**Figure 12.** A cyclovoltammogram of a blank experiment using a classical one working electrode, is taken from [24], where meaning of red lines are described.

To assign the first oxidation peak, labelled ★ in Figure 10 at about −0.45 V, we used DFT to analyze the reactivity of $[TQ]^{-}\bullet$ (seen at peak −0.30 to −0.35 V), and saw that this species is able to react with bubbled oxygen to form the radical $[TQ\text{-}O_2]^{-}\bullet$, Figure 13, a reaction not influenced by the further negative electrochemical potential. This capture of $O_2$ by a radical species is uncommon, and the resulting species, $[TQ\text{-}O_2]^{-}\bullet$, is therefore associated with its oxidation ★ peak in Figure 10. By purging the electrochemical cell with bubbled nitrogen, in Figure 11, the ★ peak was strongly decreased, consistent with decreasing $O_2$, although not completely eliminated because of the presence of some residual non-purged oxygen. Thus, the anomalous increase in efficiency, Figure 8, is not associated with antioxidant features of TQ, and contrary to results seen for previously studied scavengers [22,24–27]. Rather, this increase in efficiency is due to increasing $[TQ\text{-}O_2]^{-}\bullet$ concentration, formed as TQ aliquots are added and react with bubbled $O_2$.

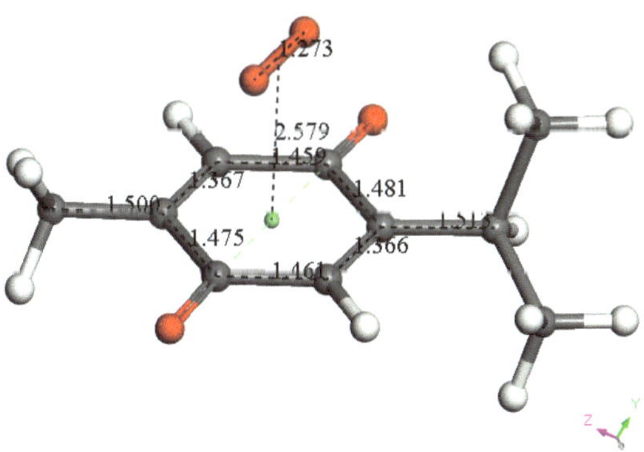


**Figure 13.** DFT minimum energy reaction product, obtained after geometry minimization of $\pi$-$\pi$ approach of a van der Waals separated, 3.50 Å, molecular oxygen and the TQ radical $[TQ]\bullet^{-}$. This radical product had a shortened $\pi$-$\pi$ interaction of 2.579 Å and is associated with ★ peak oxidation in Figure 9, to give TQ plus $O_2$.

However, it is interesting to compare the peaks associated with reactions (1) and (2) in Figure 10, as reaction (2) peak height (indicated by short vertical arrow) is shorter than that of reaction (1) (represented by longer horizontal arrow). For comparison, Figure 12 shows a typical blank, also using one working electrode, where both peak heights are seen to be equal. Thus, in Figure 10, the difference in peak height between the amount of superoxide present at the oxidation peak and the amount of superoxide at the reduction peak is due to TQ scavenging of superoxide. These features have been characterized in a previous study using both electrochemical methods, RRDE and CV [24]. The interference of radical $[TQ\text{-}O_2]^{-}\bullet$ seems not to be important at low concentrations of TQ, as seen in the first five data points in Figure 8. These initial values are plotted in Figure 9 to obtain the efficiency at low concentrations; the related slope (slope = $-2.11 \times 10^4$) for the linear expression $y = -21104x + 20.9$ ($R^2 = 0.9965$), is a good indicator of superoxide scavenging ability of TQ. We conclude that RRDE experiments on TQ show a mixed effect: on one hand, at low concentrations, TQ is able to scavenge superoxide, and shows the expected decreasing efficiency while at higher TQ concentrations, an unexpected increase in efficiency can be associated with formation of the $[TQ\text{-}O_2]^{-}\bullet$ radical. This radical species is oxidized after reversing the potential at about −0.45 V in the electrochemistry experiment using one working electrode, or alternatively, is detected at the ring electrode, whose setting is 0.0 V in the RRDE experiment.

Table 3 shows electrochemical RRDE data of other polyphenols analyzed previously. The slope of TQ, $-2.11 \times 10^4$ falls between chrysin and eriodyctiol, but is closer in value to the latter.

**Table 3.** RRDE efficiency slopes of various polyphenols.

| BHT. | Chrysin | Eriodictyol | DHDM | Butein | Clovamide | Quercetin | Galangin |
|---|---|---|---|---|---|---|---|
| $-0.16 \times 10^4$ [22] | $-1.10 \times 10^4$ [24] | $-2.20 \times 10^4$ [24] | $-8.0 \times 10^4$ [21] | $-11.2 \times 10^4$ [21] | $-12.0 \times 10^4$ [25] | $-15.5 \times 10^4$ [23,24] | $-19.0 \times 10^4$ [26] |

Our results support earlier investigations that describe TQ superoxide scavenging ability [28] as well as a thorough inquiry on the unusual cyclic voltammetry pattern shown by TQ [29].

### 3.3.2. Thymohydroquinone

THQ was also explored with the RRDE method, and results are shown in Figure 14. There was no variation in disk and ring current, and so THQ does not scavenge superoxide, in agreement with the results of the DFT study described earlier, and so the data show no change from the blank run, both below and above the potential axis, indicating unmodified reaction (1) and (2), respectively. In other words, THQ has no effect on the RRDE outcome.

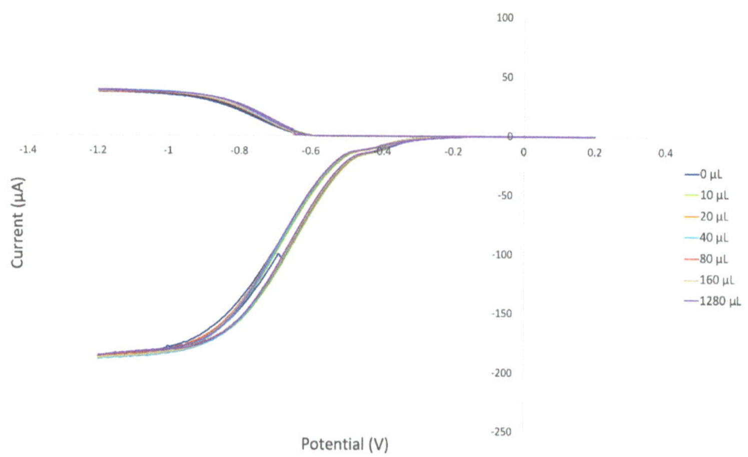


**Figure 14.** THQ is not able to scavenge superoxide, as no variation is found with increasing concentration of THQ.

3.3.3. Black Seed Oil

Black seed oil was also analyzed, and its RRDE cyclovoltammetry data are shown in Figure 15. Figure 16 shows decreasing collection efficiency overall, with the greatest decrease at low concentrations. Figure 17 shows the collection efficiency of the first four aliquots, having good linear behavior (y = −0.0781x + 20.08, $R^2$ = 0.9763), demonstrating a good scavenging ability by the black seed oil.

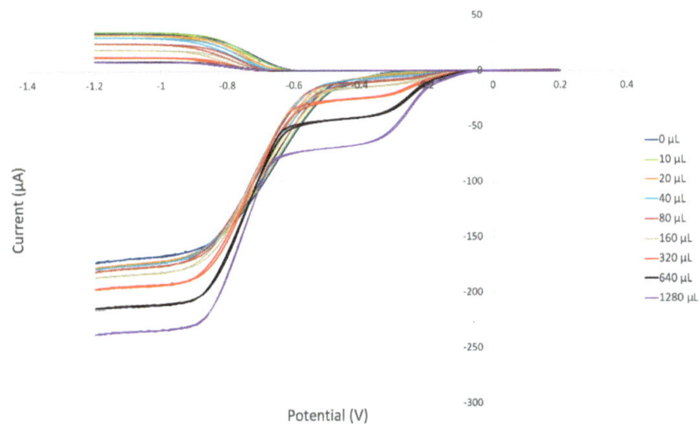


**Figure 15.** Black seed oil RRDE electrochemistry.

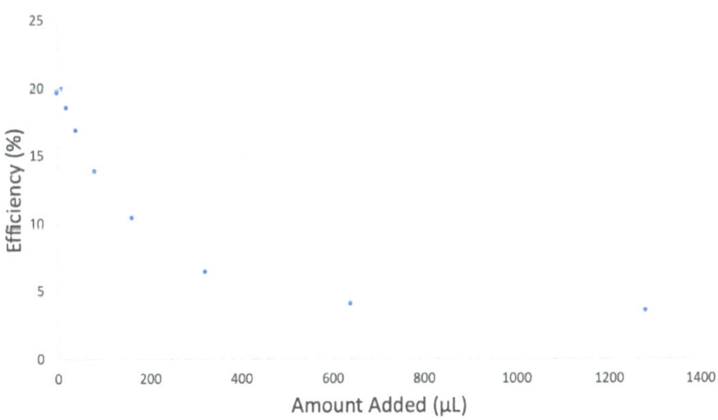


**Figure 16.** Black seed oil collection efficiency.

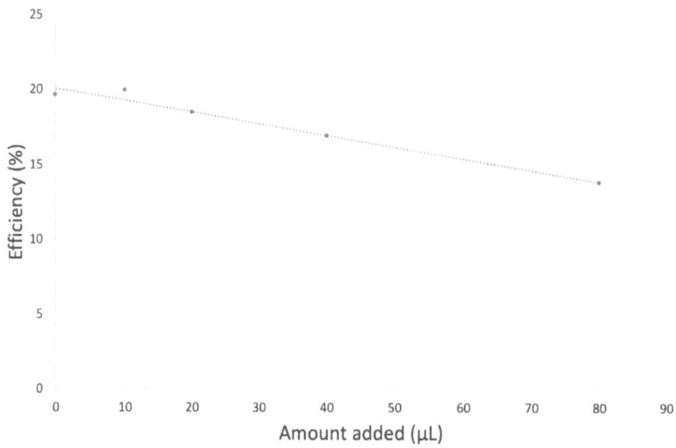


**Figure 17.** Black seed oil collection efficiency of blank plus first four aliquots, y = −0.0781x + 20.08, $R^2$ = 0.9763.

Earlier RRDE studies by our group on another cold-pressed vegetable oil, extra virgin olive oil [30], allow us to compare the slope of the RRDE efficiency of black seed oil (−0.078) and that of extra virgin olive oil (−0.0838). The latter is characterized by a linear trend (y = −0.0838x +19.73, $R^2$ = 0.99348) for all 11 runs, (aliquot range 0–100 µL), and shows a slightly better scavenging of superoxide by olive oil.

## 4. Conclusions

The molecular mechanisms that underlie the therapeutic effects of TQ are not completely understood, and literature reports show that TQ likely interacts with a number of receptors. Its low toxicity and the fact that TQ as well as the oil and seeds from which it is isolated have been used for thousands of years, make studying its therapeutic capacity appealing [6].

In this work, we have measured the superoxide scavenging ability of cold-pressed black cumin oil and two of its ingredients, TQ and THQ. Our results show that the oil and TQ are strong scavengers of the superoxide radical, whereas THQ has no effect on the scavenging, despite the presence of two para-hydroxyl groups. Structural X-ray diffraction data support subsequent DFT calculations that show a path of energetically feasible molecular transformations which corroborate cyclovoltammetry experimental features, i.e., the scavenging of superoxide by TQ, and no scavenging by THQ. The suggested stoichiometric reactivity of TQ when reacting with superoxide involves 1) direct $\pi$-$\pi$ attack, involving superoxide oxidation to produce $O_2$ as a product, and 2) superoxide $\sigma$ proton attack on a TQ O(carbonyl), which gets reduced to C-OH, and is not sensitive to further attack by superoxide.

This theoretical assessment is confirmed using two cyclovoltammetry techniques, cyclic voltammetry and hydrodynamic voltammetry at an RRDE. Lastly, our work shows the efficacy of combining multidisciplinary approaches, including experimental X-ray crystallography and voltammetry as well as computational DFT methods, to provide understanding of the mechanism of action of small molecule antioxidants. This research supports further exploration of other natural products having important biomedical activities.

**Supplementary Materials:** The following supporting information can be downloaded at: https://www.mdpi.com/article/10.3390/antiox12030607/s1, Figure S1: Overlay of two molecules in the asymmetric unit.

**Author Contributions:** Conceptualization, F.C. and M.R.; methodology, F.C., M.R. and S.B.; software, F.C.; validation, M.R., F.C. and S.B.; formal analysis, F.C., M.R., S.B., R.S. and S.K.; investigation, F.C.,

M.R., R.S., M.H. and S.A.; data curation, F.C., S.B., R.S. and M.R.; writing—original draft preparation, F.C.; writing—review and editing, F.C. and M.R.; supervision, M.R. All authors have read and agreed to the published version of the manuscript.

**Funding:** M.R. gratefully acknowledges NSF Grant CHE 0521237 for X-ray diffractometer.

**Institutional Review Board Statement:** Not applicable.

**Data Availability Statement:** Crystal data of TQ and THQ have been deposited at the Cambridge Structural Database (CSD) and are available at https://www.ccdc.cam.ac.uk/structures/? using Identifier CCDC numbers 2239493-2239494 (accessed on 1 January 2023).

**Acknowledgments:** We thank Karly Andreassen and Vassar College CIS for computational resources and support with Dassault Systems Biovia software.

**Conflicts of Interest:** The authors declare no conflict of interest.

## Abbreviations

| | |
|---|---|
| TQ, | Thymoquinone; |
| THQ, | Thymohydroquinone; |
| DFT, | Density |
| Functional | Theory; |

| | |
|---|---|
| TBAB, | tetrabutylammonium |
| bromide; | RRDE, |
| Rotating | Ring |
| Disk | Electrode. |

## References

1. Padhye, S.; Banerjee, S.; Ahmad, A.; Mohammad, R.; Sarkar, F.H. From here to eternity—The secret of Pharaohs: Therapeutic potential of black cumin seeds and beyond. *Cancer Ther.* **2008**, *6*, 495–510. [PubMed]
2. Albakry, Z.; Karrar, E.; Ahmed, I.A.M.; Oz, E.; Proestos, C.; El Sheikha, A.F.; Oz, F.; Wu, G.; Wang, X. Nutritional Composition and Volatile Compounds of Black Cumin (*Nigella sativa* L.) Seed, Fatty Acid Composition and Tocopherols, Polyphenols, and Antioxidant Activity of Its Essential Oil. *Horticulturae* **2022**, *8*, 575. [CrossRef]
3. Singh, S.; Das, S.S.; Singh, G.; Schuff, C.; de Lampasona, M.P.; Catalán, C.A. Composition, in vitro antioxidant and antimicrobial activities of essential oil and oleoresins obtained from black cumin seeds (*Nigella sativa* L.). *Biomed. Res. Int.* **2014**, *2014*, 918209. [CrossRef] [PubMed]
4. Kabir, Y.; Akasaka-Hashimoto, Y.; Kubota, K.; Komai, M. Volatile compounds of black cumin (*Nigella sativa* L.) seeds cultivated in Bangladesh and India. *Heliyon* **2020**, *6*, e05343. [CrossRef] [PubMed]
5. Goyal, S.N.; Prajapati, C.P.; Gore, P.R.; Patil, C.R.; Mahajan, U.B.; Sharma, C.; Talla, S.P.; Ojha, S.K. Therapeutic potential and pharmaceutical development of thymoquinone: A multitargeted molecule of natural origin. *Front. Pharmacol.* **2017**, *8*, 656. [CrossRef]
6. Darakhshan, S.; Bidmeshki Pour, A.; Hosseinzadeh Colagar, A.; Sisakhtnezhad, S. Thymoquinone and its therapeutic potentials. *Pharmacol. Res.* **2015**, *95–96*, 138–158. [CrossRef]
7. Sarkar, C.; Jamaddar, S.; Islam, T.; Mondal, M.; Islam, M.T.; Mubarak, M.S. Therapeutic perspectives of the black cumin component thymoquinone: A review. *Food Funct.* **2021**, *12*, 6167–6213. [CrossRef]
8. Badary, O.A.; Hamza, M.S.; Tikamdas, R. Thymoquinone: A Promising Natural Compound with Potential Benefits for COVID-19 Prevention and Cure. *Drug Design. Develop. Ther.* **2021**, *15*, 1819–1833. [CrossRef]
9. Taborsky, J.; Kunt, M.; Kloucek, P.; Lachman, J.; Zeleny, V.; Kokoska, L. Identification of potential sources of thymoquinone and related compounds in Asteraceae, Cupressaceae, Lamiaceae, and Ranunculaceae families. *Open Chem.* **2012**, *10*, 1899–1906. [CrossRef]
10. Krause, S.T.; Liao, P.; Crocoll, C.; Boachon, B.; Förster, C.; Leidecker, F.; Wiese, N.; Zhao, D.; Wood, J.C.; Buell, C.R.; et al. Biosynthesis of thymol, carvacrol, and thymohydroquinone in Lamiaceae proceeds via cytochrome P450s and a short-chain dehydrogenase. *Proc. Natl. Acad. Sci. USA* **2021**, *118*, e2110092118. [CrossRef]
11. Mayrhofer, K.; Strmcnik, D.; Blizanac, B.; Stamenkovic, V.; Arenz, M.; Markovic, N. Measurement of oxygen reduction activities via the rotating disc electrode method: From Pt model surfaces to carbon-supported high surface area catalysts. *Electrochim. Acta* **2008**, *53*, 3181–3188. [CrossRef]
12. Delley, B. From molecules to solids with the DMol$^3$ approach. *J. Chem. Phys.* **2000**, *113*, 7756–7764. [CrossRef]

13. Perdew, J.P.; Chevary, J.A.; Vosko, S.H.; Jackson, K.A.; Pederson, M.R.; Singh, D.J.; Fiolhais, C. Atoms, molecules, solids, and surfaces: Applications of the generalized gradient approximation for exchange and correlation. *Phys. Rev. B Cond. Mat. Mater. Phys.* **1992**, *46*, 6671–6687. [CrossRef] [PubMed]
14. Becke, A.D. Density-functional exchange-energy approximation with correct asymptotic behavior. *Phys. Rev. A* **1988**, *38*, 3098–3100. [CrossRef]
15. Grimme, S. Semiempirical GGA-type density functional constructed with a long-range dispersion correction. *J. Comput. Chem.* **2006**, *27*, 1787–1799. [CrossRef]
16. Lee, C.; Yang, W.; Parr, R.G. Development of the Colle-Salvetti correlation-energy formula into a functional of the electron density. *Phys. Rev. B* **1988**, *37*, 785–789. [CrossRef]
17. Sheldrick, G.M. A short history of SHELX. *Acta Cryst.* **2008**, *A64*, 112–122. [CrossRef]
18. The Cambridge Crystallographic Data Centre (CCDC) Mercury Visualization Software. Available online: https://www.ccdc.cam.ac.uk/solutions/software/mercury/ (accessed on 27 February 2023).
19. The Cambridge Crystallographic Data Centre (CCDC), Access Structures. Available online: https://www.ccdc.cam.ac.uk/structures/? (accessed on 1 January 2023).
20. Pagola, S.; Benavente, A.; Raschi, A.; Romano, E.; Molina, M.A.; Stephens, P.W. Crystal structure determination of thymoquinone by high-resolution X-ray powder diffraction. *AAPS Pharm. Sci. Tech.* **2004**, *5*, e28. [CrossRef]
21. Okoye, I.; Yu, S.; Caruso, F.; Rossi, M. X-ray Structure Determination, Antioxidant Voltammetry Studies of Butein and 2′,4′-Dihydroxy-3,4-dimethoxychalcone. Computational Studies of 4 Structurally Related 2′,4′-diOH Chalcones to Examine Their Antimalarial Activity by Binding to Falcipain-2. *Molecules* **2021**, *26*, 6511. [CrossRef]
22. Caruso, F.; Rossi, M.; Kaur, S.; Garcia-Villar, E.; Molasky, N.; Belli, S.; Sitek, J.D.; Gionfra, F.; Pedersen, J.Z.; Incerpi, S. Antioxidant Properties of Embelin in Cell Culture. Electrochemistry and Theoretical Mechanism of Scavenging. Potential Scavenging of Superoxide Radical through the Cell Membrane. *Antioxidants* **2020**, *9*, 382. [CrossRef] [PubMed]
23. Caruso, F.; Incerpi, S.; Pedersen, J.; Belli, S.; Kaur, S.; Rossi, M. Aromatic Polyphenol π-π Interactions with Superoxide Radicals Contribute to Radical Scavenging and Can Make Polyphenols Mimic Superoxide Dismutase Activity. *Curr. Issues Mol. Biol.* **2022**, *44*, 5209–5220. [CrossRef] [PubMed]
24. Belli, S.; Rossi, M.; Molasky, N.; Middleton, L.; Caldwell, C.; Bartow-McKenney, C.; Duong, M.; Chiu, J.; Gibbs, E.; Caldwell, A.; et al. Effective and Novel Application of Hydrodynamic Voltammetry to the Study of Superoxide Radical Scavenging by Natural Phenolic Antioxidants. *Antioxidants* **2019**, *8*, 14. [CrossRef] [PubMed]
25. Ye, N.; Belli, S.; Caruso, F.; Roy, G.; Rossi, M. Antioxidant studies by hydrodynamic voltammetry and DFT, quantitative analyses by HPLC-DAD of clovamide, a natural phenolic compound found in *Theobroma cacao* L. beans. *Food Chem.* **2021**, *341 Pt 2*, 128260. [CrossRef] [PubMed]
26. Caruso, F.; Berinato, M.; Hernandez, M.; Belli, S.; Smart, C.; Rossi, M. Antioxidant properties of bee propolis and an important component, galangin, described by X-ray crystal structure, DFT-D and hydrodynamic voltammetry. *PLoS ONE* **2022**, *17*, e0267624. [CrossRef]
27. Rossi, M.; Wen, K.; Caruso, F.; Belli, S. Emodin scavenging of superoxide radical. X-ray crystal structure, hydrodynamic voltammetry and theoretical studies. *Antioxidants* **2020**, *9*, 194. [CrossRef]
28. Badary, O.A.; Taha, R.A.; Gamal el-Din, A.M.; Abdel-Wahab, M.H. Thymoquinone is a potent superoxide anion scavenger. *Drug Chem. Toxicol.* **2003**, *26*, 87–98. [CrossRef]
29. Petrucci, R.; Marrosu, G.; Astolfi, P.; Lupidi, G.; Greci, L. Cyclic voltammetry, spectroelectrochemistry and electron spin resonance as combined tools to study thymoquinone in aprotic medium. *Electrochim. Acta* **2012**, *60*, 230–238. [CrossRef]
30. Rossi, M.; Caruso, F.; Kwok, L.; Lee, G.; Caruso, A.; Gionfra, F.; Candelotti, E.; Belli, S.L.; Molasky, N.; Raley-Susman, K.M.; et al. Protection by extra virgin olive oil against oxidative stress in vitro and in vivo. Chemical and biological studies on the health benefits due to a major component of the Mediterranean diet. *PLoS ONE* **2017**, *12*, e0189341. [CrossRef]

*antioxidants*

*Article*

# Effect of Flavorization on Virgin Olive Oil Oxidation and Volatile Profile

**Enrique Jacobo Díaz-Montaña , Ramón Aparicio-Ruiz and María T. Morales ***

Department of Analytical Chemistry, Faculty of Pharmacy, University of Seville, Calle Profesor García González, 41012 Seville, Spain
* Correspondence: tmorales@us.es

**Abstract:** The volatile compounds of virgin olive oil (VOO) have an important role from a sensory point of view as they are responsible for the aroma of the oil. Once the oil is obtained, auto-oxidation is the main process contributing to its deterioration, modifying the volatiles profile and aroma. The addition of aromatic herbs to VOO is a traditional technique to change the flavor and to preserve the oil. The aim of this study was to evaluate the effect on the volatile profile and sensory properties of flavoring VOO with rosemary and basil herbs and its impact on the evolution of the oxidative process during a six-month shelf-life study at 15.7 ± 3.6 °C and exposed to 500 ± 100 lx of light for 12 h each day. The determination of quality parameters, volatiles concentrations and VOO sensory properties and their comparison with the flavored VOO samples showed that the addition of basil or rosemary herbs, in addition to retarding the oxidation of the oil, allowed the discrimination of the flavored samples due to the migration of compounds from herbs to the oil. The aroma of basil olive oil (BOO) samples was mainly due to β-pinene, ocimene and 1,8-cineol compounds while for rosemary olive oil (ROO) samples, their aroma was mainly due to the concentrations of camphene, β-myrcene, α-terpinolene, limonene and 1,8-cineol. From the antioxidant standpoint, the effect of the herbs was more noticeable from the third month onwards.

**Keywords:** volatile compounds; virgin olive oil; flavored olive oil; rosemary; basil; oxidation

## 1. Introduction

Olive oil is the main fat and one of the dietary staples in countries of the Mediterranean basin. In recent years, the consumption of olive oil has increased worldwide due to its well-known health benefits which are mainly due to the presence of antioxidant compounds, among others. Virgin olive oil (VOO) is obtained from the fruit of the olive tree (*Olea europaea* L.) when it is fresh, sound and has its optimal maturity [1]. VOO, and extra virgin olive oil (EVOO) in particular, differs from other edible oils in its nutritional, health and organoleptic properties [1–3]; it is one of the most-studied antioxidant food sources [4]. The health properties of VOO are attributed to a high content of oleic acid (monounsaturated fatty acid) and the presence of a series of minor components such as triterpenic alcohols, phenols, tocopherols, etc. [1,5]. In recent years, more health properties of olive oil have been reported, including oxidative stress prevention, anti-inflammatory properties, lowering the risk of hypertension and antitumoral properties, among many others. These properties have allowed olive oil to stand out from other fats and edible oils [2].

VOO is also appreciated for its sensory characteristics; it has a pleasant aroma and delicious taste, which contribute to appetite, satiety and consumer preference [6]. Olive oil's flavor is due to the presence of phenolic compounds which are responsible for its bitterness and pungency, and the volatile compounds, which are responsible for the aroma [4,7]. The biogenesis of VOO volatiles is mainly due to the oxidation of unsaturated fatty acids by the endogenous enzyme lipoxygenase, which generates volatile compounds using linoleic and linolenic acids as precursors [8]. C6 aldehydes and alcohols, and their corresponding esters, are mainly responsible for positive sensory perception, whereas chemical oxidation, hydrolysis and exogenous enzymes (usually from microorganisms) produce the volatiles associated with sensory defects [9]. The quality deterioration of oil is linked to food safety since oxidation products are formed when oil ages. The oxidative degradation of olive oil causes an increase in free radicals and the formation of off-flavors [10], producing the appearance of a rancid sensory defect [11] and a decrease in the antioxidant content [12]. The oxidation process is highly influenced by the processing and storage conditions of the oil, which affects not only its sensory and healthy qualities [13], but also its economic value and consumer acceptability [6].

The addition of aromatic herbs, bulbs or fruits—such as oregano, basil, rosemary, thyme, garlic and lemon—to olive oil results in aromatized or flavored olive oils (FOO) [14–20], thereby modifying their shelf life [12], and perhaps, masking undesirable sensory perceptions. The flavoring process of VOOs is varied and can include the infusion of spices into the oil, ultrasound-assisted maceration and joint malaxation of olive paste and herbs, among others [14,18–21]. Some studies have characterized the flavoring process through the detection of new sensory characteristics. Additionally, olive oil flavoring has also been studied because it enhances its health properties including antimicrobial, anticancer or antioxidant properties [22–24]. Thus, the VOO flavoring not only modifies the initial VOO sensory characteristics and health properties, but also can change its shelf life and quality parameters as consequence of the migration/transference of antioxidants and/or pro-oxidant compounds from the herbs to the oils [25,26].

Research is usually focused on characterizing flavored VOOs and only a few studies have been conducted on how these FOOs evolve over time [25,27–29]. Hence, the aims of this work were twofold: (i) to study and compare the changes produced on the profiles of volatile compounds in non-flavored olive oils and virgin oils flavored by the contact method with rosemary (*Rosmarinus officinalis* L.) and basil (*Ocimum basilicum* L.) during six months of storage at 15.7 ± 3.6 °C and exposed to 500 ± 100 lx of light for 12 h each day; and, (ii) to characterize the new volatiles appearing in flavored olive oils (FOO) and to study their quantitative changes and influence on the shelf life.

## 2. Materials and Methods

### 2.1. Chemicals

The reagents employed were of pure analytical grade. Acetic acid (99%), ethanol (96%) β-caryophyllene (80%), β-myrcene (90%), camphene (95%), α-pinene (98%), borneol (97%), 2-methylbutanal (95%), naphthalene (98%), p-cymene (99%), α-linalool (97%), β-farnesene (90%), β-pinene (99%), limonene (97%), 1,8-cineole (99%), octadecane (99%), γ-terpinene (97%), β-farnesene (90%), naphthalene (99%) and isobutyl acetate (<99.9%), used as internal standards, were purchased from Sigma-Aldrich (St. Louis, MO, USA).

Ethyl acetate (99.5%), butanoic acid (99%), propanoic acid (99.5%), pentanoic acid (99%), butan-1-ol (99%), butan-2-ol (99.5%), heptan-2-ol (99%), heptan-2-one (98%), (E)-2-heptenal (98%), hexanal (98%), hexan-1-ol (98%), (E)-2-hexenal (97%), 2-methyl-butan-1-ol (99.5%), 3-methyl-butan-1-ol (99.5%), 4-methyl-pentan-2-one (99%), nonanal (95%), octanal (98%), pentanal (98%), octane (99%), 1-penten-3-one (98%) and ethyl propanoate (99%) were purchased from Fluka (Zwijndrecht, The Netherlands).

Ethyl butanoate (98%), heptanal (95%), (E)-3-hexen-1-ol (98%), ethyl 2-methylbutanoate (99%), 6-methyl-5-hepten-2-one (99%), octan-3-one (97%), 1-octen-3-one (50%) and pentan-3-one (99.5%) were purchased from Panreac (Barcelona, Spain).

### 2.2. Samples and Storing Conditions

An EVOO (var. Manzanilla) from the 2020/21 harvest was purchased from a local market (Sevilla, Andalusia, Spain). The oil was filtered and its regulated quality parameters were determined [30]. Next, the sample (VOO 0) was divided into 18 aliquots stored in 1 L glass containers, 6 of them contained only EVOO (VOO), 6 contained EVOO dressed with dried rosemary (ROO, Rosemary Olive Oil) leaves at 5% *w*/*w* and another set of 6 samples contained EVOO dressed with dried basil (BOO, Basil Olive Oil) leaves at 5% *w*/*w* during the whole storage conditions, respectively. The aromatic herbs, collected in September of 2021 from the countryside of Seville (Andalusia, Spain) (37.320471, −5.979523), were treated to separate the leaves and discarded wooden parts, drying only the leaves at 60 °C for 10 h [31].

The 18 samples were stored for six months at 15.7 ± 3.6 °C and exposed to 500 ± 100 lx of light for 12 h each day. Once per month the volatile compounds, physical–chemical parameters and sensory characteristics were determined in triplicate by sampling one bottle from each group (VOO, ROO, BOO).

### 2.3. *Quality Parameters*

Free acidity (FA), peroxide value (PV) and specific extinction values (K232, K270 and ΔK) were determined according to The International Olive Council (IOC) trade standards [32].

### 2.4. *Sensory Analysis*

VOO-0 was evaluated according to [33] to assign the initial category of the purchased olive oil [30]. Evaluation was based on the determination of the median of the fruity attributes and the detection, or not, of sensory defects [33]. In addition, a sensory evaluation of all the samples was carried out by Quantitative Descriptive Analysis (QDA) to assess possible sensory changes occurring during storage [34]. For this purpose, a protocol was developed for the study including an evaluation sheet (Figure S1), which was created considering the different sensory attributes, and an unstructured 5 cm scale was used to evaluate the intensity of the attributes. The sensory evaluation was carried out by 8 assessors, all of whom were trained panelists, to check the consistency of the results.

### 2.5. *Volatile Compounds Determination*

Solid Phase Micro Extraction-Gas Chromatography-Mass spectrometry (SPME-GC-MS, Varian 3900 GC, Bruker, Palo Alto, CA, USA; MSD 5975, Agilent Tech, Santa Clara, CA, USA) and Dynamic Head Space-Gas Chromatography with Flame Ionization Detector (DHS-GC-FID, Varian 3900 GC, Bruker, Palo Alto, CA, USA) were employed to determine the volatile compounds.

#### 2.5.1. SPME-GC-MS

For SPME-GC-MS, 2 g of oil or herb samples was placed in a 30 mL glass vial tightly capped with silicone/polytetrafluoroethylene (PTFE) septum. To reach oil–air equilibrium, the samples were left for 10 min at 40 °C, after which a 1 cm SPME fiber with a 50/30 µm film thickness with divinylbenzene/carboxen/polydimethylsiloxane (DVB/CAR/PDMS) as a stationary phase, purchased from Supelco (Bellefonte, PA, USA), was exposed to the headspace for 40 min. The fiber was previously conditioned following the supplier instructions.

Desorption of volatiles was carried out in the hot injection port by heating the fiber for 5 min at 280 °C. The volatile compounds were injected into a TR-WAX capillary column (60 m × 0.25 mm i.d., 0.25 µm; Teknokroma, Barcelona, Spain), in splitless mode, of a Varian 3900 gas chromatograph (Bruker, Palo Alto, CA, USA) coupled to a mass spectrometer, using hydrogen at 0.9 mL/min as a carrier gas. The oven was kept in isothermal mode for 10 min at 35 °C and then increased by 3 °C/min until 200 °C. The GC–MS interface was heated to 280 °C with the actual temperature reaching 180 °C in MS source and 150 °C in MS-quadrupole [35]. The energy used for electron impact was set at 70 eV, and data were collected in the range of 40–300 atomic mass units (amu). The signal was recorded, and the integration was carried out with Enhanced Chemstation MSD ChemStation E.02.02.1431 (Agilent Tech, Santa Clara, CA, USA). For the volatile identification the library, Wiley 7 was used (John Wiley & Sons Limited, Hoboken, NJ, USA), and later confirmed by pure standards, when available.

2.5.2. DHS-GC-FID

The dynamic headspace employed 1.5 g of samples. Once the samples were placed into a 20 mL glass vial and tightly capped with silicone/polytetrafluoroethylene (PTFE) septum, they were pre-heated for 18 min at 40 °C and mixed for 15 min using a HT3 Dynamic Headspace System (Teledyne Tekmar, Mason, OH, USA). After mixing, helium employed as a portable gas, using a flow rate of 5 mL/min, moved the volatile compounds to the adsorbent trap (Tenax TA$^{TM}$, Supelco, Bellefonte, PA, USA).

The volatiles were thermally desorbed from the trap (5 min at 260 °C) and then injected with a split mode 7:1. As in the case of SPME, the injection was carried out in a TR-WAX capillary column (60 m × 0.25 mm i.d., 0.25 µm; Teknokroma, Barcelona, Spain) of a Varian 3900 gas chromatograph (Bruker, Palo Alto, CA, USA) with a flame ionization detector (FID), using hydrogen with a flow rate of 1.5 mL/min. The oven conditions were 10 min at 35 °C and then programmed to rise to 200 °C at 3 °C/min. The temperature of the flame ionization detector (FID) was set at 280 °C, and the signal was recorded with Star Chromatography Workstation, System Control version 6 (Bruker, Palo Alto, CA, USA).

Quantification was carried out using the internal standard methodological calibration procedure. The internal standard used was isobutyl acetate.

*2.6. Statistical Analysis*

The whole dataset was imported into Excel from the instrument and the Statistica 8 package (StatSoft, Tulsa, OK, USA) was used to carry out the data treatment and the statistical analyses. A t-test was used, which allowed the selection of compounds that showed significant differences ($p < 0.05$) in their relative concentration among the groups. Cluster analysis of the VOO samples, using the information of the total set of volatiles, was performed by a complete standardization of the data and was implemented with the City-Block (Manhattan) distance measure and the Ward's method linkage amalgamation rule. Multidimensional Scaling (MDS)—an unsupervised statistical procedure—was implemented to explore the level of similarity of individual samples of the flavored and unflavored oils during the six-month storage experiment. The MDS algorithm is based on the results of the Cluster Analysis [36].

## 3. Results and Discussion

Rosemary and basil are usually used for VOO flavoring and to take advantage of its antioxidant properties, which could slow down the auto-oxidative process of VOO [37]. The volatile compounds of VOO vary according to different factors, such as fruit ripeness, pedoclimatic conditions, cultivation, time and conditions of storage, among others [38]. Particularly, virgin olive oil var. Manzanilla is characterized as having a different amount of alcohol dehydrogenase than other varieties, which leads to an accumulation of aldehydes. This variety is also characterized by having a low taste intensity related to bitterness and pungency, as well as a low aroma intensity of the fruity attribute, which make flavoring this oil a good option [39,40].

*3.1. Quality Parameters*

The initial sample (VOO-0), filtered and without storing, was classified as extra virgin olive oil once its physico-chemical and organoleptic-regulated parameters were determined [30]. VOO samples did not overstep the limits of any of the physico-chemical quality parameters during storage, but they showed a linear increase during the studied period. Conversely, the values of these parameters for the flavored samples (FOO) remained below their limits for peroxide value and K232, but they exceeded the limits for the free acidity and K270 parameters (Table S1) [30]. The free acidity, which is mainly increased due to the hydrolysis of triacylglycerides to free fatty acids, was exceeded at the fifth month due to a higher water activity [31] and the migration of organic acids from herbs to the oil but not due to a higher oxidation. The limit for K270 parameters was exceeded by ROO samples from the third month onwards, which is not an anomalous result as it depends on the phenolic content in the oil and the bottling [12].

### *3.2. Virgin Olive Oil Shelf Life*

The changes of the VOO samples throughout storage were first assessed, and later, the comparison with the FOO samples was carried out. More than 180 volatile compounds have been reported in VOO [38], and it is not only those with a higher concentration than their odor threshold that contribute to the aroma. In order to reduce the complexity of the system, only concentrations above the odor threshold were considered [38,41]. This study focused on several of the latter volatiles related to the aroma, and some of them, such as 3-pentanone or E-2-hexenal, are related to positive sensory perceptions. Others, such as E-2-heptenal or nonanal, are related to sensory defects. Table 1 shows the volatile compounds identified in the initial sample (VOO-0), their concentration (mg/Kg), odor threshold (mg/Kg) and associated sensory descriptors. The concentration of volatiles attributed to sensory defects in the initial sample were below their odor thresholds [35,42], therefore, they did not contribute to the sample aroma, which agrees with its classification as extra virgin olive oil.

**Table 1.** Kovats index, retention time (min), concentration (mg/Kg), odor threshold (mg/Kg) and sensory descriptors of the volatile compounds of the sample VOO-0.

| Volatile Compound | Kovats I | Rt [a] (min) | Conc. [b] (mg/Kg) | O. Thres. [c] (mg/Kg) | Sensory Descriptor [38] |
|---|---|---|---|---|---|
| Octane | 800 | 4.126 | 0.71 ± 0.05 | 0.94 | Alkane, solvent |
| Ethyl acetate | 892 | 6.398 | 6.09 ± 0.43 | 0.94 | Sweet, aromatic |
| Ethanol | 932 | 7.457 | 29.51 ± 2.09 | 30.00 | Apple, sweet, alcohol |
| 3-Pentanone | 964 | 8.271 | 0.36 ± 0.03 | 70.00 | Sweet, fruity |
| Isobutyl acetate (Internal standard) | - | 9.852 | - | - | - |
| 1-Penten-3-one | 1016 | 11.548 | 0.16 ± 0.01 | 0.70 | Spicy, mustard |
| Pentanal | 969 | 12.873 | 27.30 ± 1.93 | 0.24 | Oily, wood, bitter, almond |
| Hexanal | 1074 | 14.479 | 0.87 ± 0.06 | 0.08 | Green, green apple, grass, oily, fatty |
| Heptanal | 1174 | 20.502 | 1.11 ± 0.08 | - | Citrus, fatty |
| 1-Butanol | 1145 | 22.261 | 0.38 ± 0.08 | 0.40 | Sweet, oily, drug |
| E-2-hexen-1-al | 1216 | 23.440 | 2.59 ± 0.05 | 0.42 | Fruity, green, bitter almond |
| 2-Heptanone | 1170 | 24.159 | 0.27 ± 0.10 | 0.30 | Musty, soapy, cinnamon |
| 2-Methyl-1-butanol | 1210 | 24.949 | 2.50 ± 0.36 | - | Ethereal, alcoholic, fatty |
| 3-Octanone | 1244 | 26.177 | 0.040 ± 0.001 | - | Grass, musty, green, butter |
| E-2-heptenal | 1282 | 27.552 | 4.14 ± 0.53 | 4.20 | Soapy, fatty, almond, spicy |
| 2-Heptanol | 1288 | 28.299 | 0.01 ± 0.06 | 0.01 | Mushroom, dirt, sweet |
| 6-Methyl-5-hepten-2-one | 1317 | 28.643 | 0.15 ± 0.01 | 1.00 | Fruity, green, herb |
| E-3-Hexen-1-ol | 1366 | 30.060 | 3.30 ± 0.23 | 1.00 | Green grass |
| Nonanal | 1385 | 31.467 | Not detected | 0.15 | Rancid, citrus, soapy |
| 1-Octen-3-ol | 1390 | 33.115 | 0.54 ± 0.03 | 1.00 | Musty, mushroom |
| Acetic acid | 1452 | 35.110 | 0.48 ± 0.05 | 0.50 | Bitter, winery, vinegar |
| Propionic acid | 1527 | 38.316 | 0.61 ± 0.04 | 0.72 | sweaty, pungent |
| Butanoic acid | 1628 | 40.340 | 0.56 ± 0.09 | 0.65 | Rancid, sweaty |
| Pentanoic acid | 1720 | 44.162 | 0.33 ± 0.02 | 0.60 | Spicy, rancid |
| Hexanoic acid | 1829 | 49.000 | Not detected | 0.70 | Rancid, cheese |
| Heptanoic acid | 1990 | 53.450 | Not detected | 0.10 | Rancid, fatty |
| Octanoic acid | 2083 | 55.103 | Not detected | 3.00 | Rancid, cheese, oily, fatty |

Note: [a], Retention time; [b], Concentration; [c], Odor threshold.

The profiles of the volatiles changed throughout the storage, due to the modifications produced by the oxidation process. Initially VOO samples were rich in biogenically-formed C5 and C6 compounds—including 3-pentanone or E-2-hexenal—that can contribute to positive sensory perceptions [38]; the samples did not contain (or contained only at low concentrations) organic acids. After six months of storage, VOO samples presented a high number of compounds characterized by negative sensory descriptors such as acetic, propanoic and hexanoic acids, as well as other C7 and C9 compounds [38] such as heptanal or nonanal. The compounds generated throughout the storage were mainly due to the oxidation of unsaturated fatty acids. The study of the generation of sensory defects is crucial because when a VOO reaches high intensities of sensory defects, it is classified as lampante virgin olive oil and must undergo a refining process before it can be consumed [25], which contributes to economic losses.

All the studied compounds showed significant differences (*t*-test) ($p < 0.05$) during the storage, except for 1-butanol. Particularly, in VOO samples, the difference started becoming significant from the third month onwards (*t*-test). The behavior of each compound varied during the storage; compounds related to sensory defects increased while compounds related to positive perceptions decreased. Table 2 displays the information of several representative volatile compounds, of which the concentrations greatly differed during the storage, most of them being characterized as volatiles produced by the auto-oxidation of unsaturated fatty acids.

**Table 2.** Concentration (mg/Kg; mean ± SD) of volatile compounds of the VOO samples during the six months of storage.

| Compound | VOO-1 | VOO-2 | VOO-3 | VOO-4 | VOO-5 | VOO-6 |
|---|---|---|---|---|---|---|
| Pentanal | 38.25 ± 2.71 [a] | 50.74 ± 3.59 [b] | 53.43 ± 3.78 [b] | 65.96 ± 4.66 [c] | 67.15 ± 4.75 [c] | 68.35 ± 4.83 [c] |
| Nonanal | 0.18 ± 0.04 [a] | 0.59 ± 0.09 [b] | 0.62 ± 0.16 [b] | 0.82 ± 0.17 [b] | 1.00 ± 0.21 [c] | 1.34 ± 0.26 [c] |
| 2-Heptanol | 0.87 ± 0.06 [a] | 1.06 ± 0.08 [a] | 1.19 ± 0.08 [b] | 1.29 ± 0.09 [b] | 1.36 ± 0.9 [b] | 1.43 ± 0.10 [b] |
| 1-Octen-3-ol | 0.52 ± 0.04 [a] | 0.52 ± 0.04 [a] | 0.53 ± 0.04 [a] | 0.55 ± 0.04 [a] | 0.61 ± 0.04 [a] | 0.80 ± 0.06 [b] |
| Butanoic acid | 1.14 ± 0.08 [a] | 8.26 ± 0.58 [b] | 10.52 ± 0.74 [c] | 21.80 ± 1.54 [d] | 43.87 ± 3.10 [e] | 61.66 ± 4.36 [f] |
| Hexanoic acid | 12.90 ±0.91 [a] | 19.74 ± 1.40 [b] | 38.24 ± 2.70 [c] | 59.57 ± 4.91 [d] | 72.94 ± 5.16 [d] | 76.31 ±5.40 [d] |
| Ethyl acetate | 4.96 ± 0.35 [a] | 3.36 ± 0.24 [b] | 3.14 ± 0.22 [b] | 2.82 ± 0.20 [b] | 2.35 ± 0.17 [c] | 2.13 ± 0.15 [c] |
| Ethanol | 26.38 ± 1.87 [a] | 24.44 ± 1.73 [a] | 23.76 ± 1.68 [a] | 20.81 ± 1.47 [b] | 20.83 ± 1.47 [b] | 20.85 ± 1.46 [b] |
| 1-penten-3-one | 0.08 ± 0.01 [a] | 0.07 ± 0.01 [a] | 0.04 ± 0.01 [b] | 0.03 ± 0.00 [c] | 0.02 ± 0.00 [d] | 0.01 ± 0.00 [e] |

Note: Values with different letters ([a,b,c,d,e,f]) within the same row are significantly different ($p < 0.05$). Experiments were set up in two repetitions and samples for each repetition were analyzed in triplicate ($n = 6$).

Pentanal is a volatile compound which is normally associated with a green fruity perception but it can also be related to a rancid off-flavor in olive oil undergoing oxidation [12]. Pentanal and hexanoic acid are good examples of a progressive increment during the study (Figure 1). Both compounds increased but with different slopes—higher for hexanoic acid, in part produced from hexanal oxidation, and lower for pentanal.

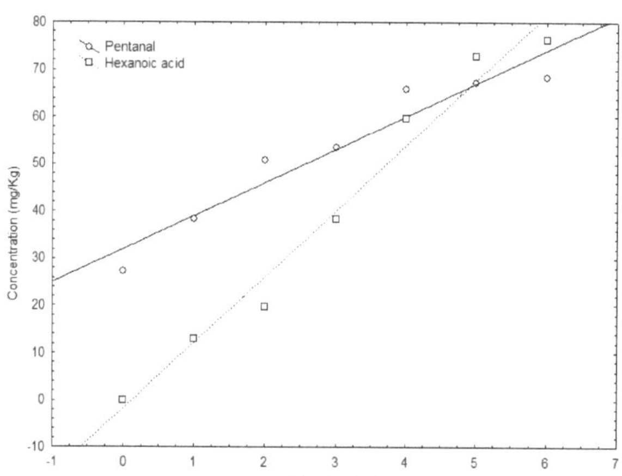


**Figure 1.** Evolution of pentanal and hexanoic acid concentration (mg/Kg) during the storage in VOO samples.

In addition to the compounds described in Table 2, there are other compounds (Table 1) that are also important because their sensory impact and their relationship with virgin olive oil quality as to be considered. Thus, 2-heptanone and 3-octanone, both related to sensory defects, increased their concentrations from 1.43 ± 0.01 to 11.01 ± 0.78 mg/Kg and from 0.04 ± 0.00 to 1.64 ± 0.12 mg/Kg, respectively. Other studied organic acids, such as acetic, propanoic and pentanoic acids, also increased between 5 and 30 times their initial concentrations. These results disagree with Cecchi et al. [11], who reported that the concentration of organic acids never exceeded 0.4 mg/Kg and the amount of acetic acid did not show a significant increase ($p < 0.05$).

The concentrations of all the volatile compounds were used to perform a cluster analysis, which allowed the classification of the samples according to the storage time. Figure 2 shows that the initial sample (VOO-0) and the samples from the first three months were clustered separately from the rest of samples, which formed other clusters where the most oxidized samples were included. The first cluster, however, was split into two groups: one group joins the initial and the first month samples, showing that both samples had fewer differences; and another group joins the samples corresponding to the second and third months, showing that they had differences in their volatile content due to the incipient progression of the oxidative process.

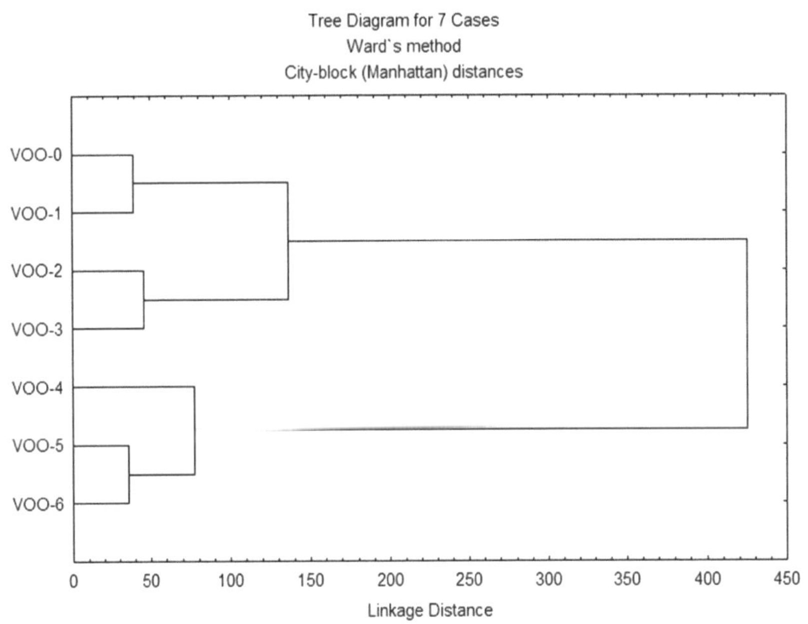


**Figure 2.** Cluster analysis of the volatile compounds of the VOO samples.

In the second cluster, the samples corresponding to the fourth month appeared separately from the other two more altered samples. The separation between the sample of the third month and the samples of the fourth, fifth and sixth months is in accordance with the information obtained from the *t*-test.

### 3.3. Comparison between Unflavored and Flavored Samples

The results of the study of the volatile compounds in VOO samples were compared with those obtained in FOO samples. The addition of rosemary and basil altered the content of some volatiles due to their antioxidant activity and the migration of volatile compounds. These changes varied not only depending on whether the sample was flavored or not, but also depending on the type of flavoring agent. The volatile composition of both vegetal materials was first identified to evaluate the potential migration of volatiles from the herbs to the oil. Table 3 shows the volatile compounds identified in the two vegetal materials: 23 in rosemary and 26 in basil. The presence of volatiles of the herbs in the FOO samples allowed the confirmation of the migration of several of them, but not all the basil and rosemary volatiles migrated from the herb to the oil; some volatiles remained in the herbs and others were already present in VOO, not all of which influenced the aroma.

**Table 3.** Volatile compounds identified in rosemary and basil herbs by SPME-GC-MS.

| **Rosemary Compounds** | | | | | | | | |
|---|---|---|---|---|---|---|---|---|
| **Volatile Compound *** | **Kovats I **** | **Molecular Formula** | **Molecular Weight** | **Fragments (m/z)** | | | | |
| Octane | 800 | $C_8H_{18}$ | 114.23 | 43 | 85 | 57 | 71 | 56 |
| Pentanal | 935 | $C_5H_{10}O$ | 86.13 | 44 | 29 | 58 | 41 | 27 |
| α-pinene | 1035 | $C_{10}H_{16}$ | 136.23 | 93 | 92 | 79 | 43 | 94 |
| Camphene | 1080 | $C_{10}H_{16}$ | 136.23 | 93 | 121 | 41 | 79 | 39 |
| m-menthene | 1150 | $C_{10}H_{18}$ | 138.25 | 95 | 138 | 67 | | |
| p-menthene | 1152 | $C_{10}H_{18}$ | 138.25 | 139.1 | 97.1 | 95.1 | 113.1 | 83.1 |
| β-myrcene | 1158 | $C_{10}H_{16}$ | 136.23 | 93 | 69 | 94 | 79 | 77 |
| α-terpinene | 1184 | $C_{10}H_{16}$ | 136.23 | 121 | 93 | 136 | 91 | 77 |
| Limonene | 1208 | $C_{10}H_{16}$ | 136.23 | 93 | 68 | 136 | 121 | 67 |
| 1,8-cineole | 1222 | $C_{10}H_{18}O$ | 154.25 | 43 | 81 | 108 | | |
| γ-terpinene | 1262 | $C_{10}H_{16}$ | 136.23 | 93 | 91 | 136 | 77 | 121 |
| m-cymene | 1267 | $C_{10}H_{14}$ | 134.22 | 93 | 135 | 121 | 133 | 119 |
| 3-octanone | 1285 | $C_8H_{16}O$ | 128.21 | 43 | 57 | 29 | 72 | 99 |
| α-terpinolene | 1297 | $C_{10}H_{16}$ | 136.23 | 93 | 121 | 136 | 91 | 79 |
| 3-octanol | 1390 | $C_8H_{18}O$ | 130.23 | 59 | 83 | 55 | 41 | 101 |
| Tetrahydrolinalool | 1397 | $C_{10}H_{22}O$ | 158.28 | 73 | 69 | 55 | 43 | 129 |
| Acetic acid | 1452 | $C_2H_4O_2$ | 60.05 | 43 | 45 | 60 | 15 | 42 |
| L-camphor | 1498 | $C_{10}H_{16}O$ | 152.23 | 152 | 108 | 81 | 95 | 109 |
| Bornyl acetate | 1610 | $C_{12}H_{20}O_2$ | 196.29 | 95 | 43 | 93 | | |
| β-caryophyllene | 1618 | $C_{15}H_{24}$ | 204.35 | 91.1 | 93.1 | 133.1 | 79.1 | 41.1 |
| Borneol | 1642 | $C_{10}H_{18}O$ | 154.25 | 95 | 93 | 96 | 94 | 97 |
| 2,6-dimethyloctane | 1727 | $C_{10}H_{22}$ | 142.28 | 57 | 43 | 71 | | |
| Octadecane | 1800 | $C_{18}H_{38}$ | 254.50 | 71 | 85 | 99 | 70 | 113 |
| **Basil compounds** | | | | | | | | |
| **Volatile compound *** | **Kovats I **** | **Molecular formula** | **Molecular weight** | **Fragments (m/z)** | | | | |
| 2-methylbutanal | 864 | $C_5H_{10}O$ | 86.13 | 41 | 29 | 57 | | |
| 3-methylbutanal | 912 | $C_5H_{10}O$ | 86.13 | 44 | 41 | 43 | 58 | 27 |
| Pentanal | 935 | $C_5H_{10}O$ | 86.13 | 44 | 29 | 58 | 41 | 27 |
| 2,6-dimethyl-2-octene | 936 | $C_{10}H_{20}$ | 140.27 | 69 | 70 | 41 | | |
| Hexanal | 1088 | $C_6H_{12}O$ | 100.16 | 44 | 41 | 56 | 43 | 27 |
| β-pinene | 1118 | $C_{10}H_{16}$ | 136.23 | 121.1 | 136.1 | 135.1 | | |
| Limonene | 1208 | $C_{10}H_{16}$ | 136.23 | 93 | 68 | 136 | 121 | 67 |
| 1,8-cineole | 1222 | $C_{10}H_{18}O$ | 154.25 | 43 | 81 | 108 | | |
| β-ocimene | 1245 | $C_{10}H_{16}$ | 136.23 | 93.1 | 91.1 | 79.1 | 92.1 | 77.1 |
| γ-terpinene | 1262 | $C_{10}H_{16}$ | 136.23 | 93 | 91 | 136 | 77 | 121 |
| p-cymene | 1280 | $C_{10}H_{14}$ | 134.22 | 119 | 134 | 91 | 120 | 117 |
| 6-methyl-5-hepten-2-one | 1340 | $C_8H_{14}O$ | 126.20 | 43 | 41 | 55 | 69 | 39 |
| Tetrahydrolinalool | 1397 | $C_{10}H_{22}O$ | 158.28 | 73 | 69 | 55 | 43 | 129 |
| Acetic acid | 1452 | $C_2H_4O_2$ | 60.05 | 43 | 45 | 60 | 15 | 42 |
| Copaene | 1510 | $C_{15}H_{24}$ | 204.35 | 105 | 91 | 41 | 119 | 93 |
| Dihydrolinalool | 1525 | $C_{10}H_{20}O$ | 156.26 | 109 | 69 | 73 | 41 | 43 |
| α-linalool | 1544 | $C_{10}H_{18}O$ | 154.25 | 71 | 93 | 55 | 69 | 80 |
| β-farnesene | 1674 | $C_{15}H_{24}$ | 204.35 | 41 | 69 | 93 | 81 | 67 |
| Aristolene | 1697 | $C_{15}H_{24}$ | 204.35 | 105 | 161 | 91 | | |
| Pentanoic acid | 1698 | $C_5H_{10}O_2$ | 102.13 | 60 | 27 | 29 | 41 | 39 |
| Naphthalene | 1718 | $C_{10}H_8$ | 128.17 | 128 | 127 | 51 | 64 | 126 |
| 2,6-dimethyloctane | 1727 | $C_{10}H_{22}$ | 142.28 | 57 | 43 | 71 | | |
| α-curcumene | 1777 | $C_{15}H_{22}$ | 202.33 | 119 | 132 | 41 | | |
| α-bergamotene | 1779 | $C_{15}H_{24}$ | 204.35 | 93 | 119 | 41 | 69 | 55 |
| β-ionone | 1955 | $C_{13}H_{20}O$ | 192.30 | 177 | 43 | 41 | 178 | 135 |
| Ethyl vanillin | 2565 | $C_9H_{10}O_3$ | 166.17 | 137 | 166 | 138 | 81 | 27 |

* Identification of the rosemary and basil volatile compounds by mass spectrometry using the WILEY 7 (John Wiley & Sons Limited, NJ) library, and later confirmed by pure standards analysis. ** Kovats Index reported in literature [38].

Throughout the storage, several volatiles showed the same trend in FOO and VOO, such as pentanal, nonanal or ethyl acetate, but the rate of generation or degradation due to the oxidation varied between the different samples. As previously described, pentanal increased in VOO from the first month (38.25 ± 2.71 mg/Kg) to the sixth month (68.35 ± 4.83 mg/Kg), while in ROO, it increased from 33.90 ± 2.40 mg/Kg to 84.98 ± 6.00 mg/Kg, and in BOO it increased from 36.34 ± 2.57 mg/Kg to 70.86 ± 5.01 mg/Kg. The concentrations in the first months of storage were similar and increased to higher values; in the case of FOO samples, this was likely due to the migration of pentanal from the herbs to the oil and not only due to the auto-oxidation process.

The concentration of nonanal, in concordance with [43], increased during the storage even in flavored samples, but with different trend. Nonanal increased its concentration 10.91% in ROO and 34.93% in BOO, but these increments were lower than in VOO. Thus, nonanal—an aldehyde produced by the auto-oxidation of monounsaturated oleic acid—is a good example of the protective role of the herbs. While the concentration of nonanal increased significantly ($p < 0.05$) in VOO samples, it did not vary significantly ($p < 0.05$) in FOO samples; the concentration slope was higher in VOO than in FOO samples (Figure 3). In addition, the increase in the nonanal concentration was significant in VOO samples from the second month onwards, while the increase was delayed until the fourth month in FOO samples, especially in the case of ROO.

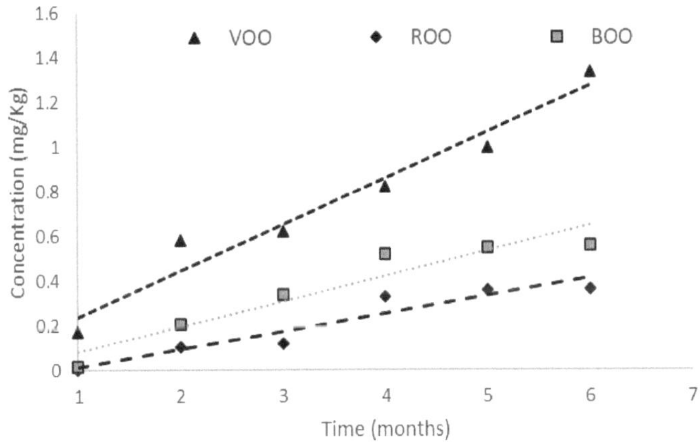


**Figure 3.** Variation in the concentration (mg/Kg) of nonanal in VOO, ROO and BOO samples throughout the storage.

Other compounds that showed the same behavior in the three types of samples (VOO, ROO, BOO) were 2 heptanol and ethyl acetate. Thus, 2-heptanol showed an increasing trend, reaching the highest value in VOO in the sixth month. The concentration of ethyl acetate, however, decreased, with BOO samples having the highest amount (4.52 ± 0.32 mg/Kg); the difference between ROO and BOO could be due to a greater antioxidant protection of BOO.

On the other hand, compounds such as ethanol, 1-penten-3-ol (associated to positive sensory descriptors) and 1-octen-3-ol (responsible for the musty/mushroom sensory perception) [9] showed different behavior at least in one of the FOO samples. Ethanol showed a completely different behavior between FOO and VOO samples; as an average of FOO samples, its initial concentration was 19.17 ± 4.83 mg/Kg and increased to 24.32 ± 0.51 mg/Kg, but without significant differences ($p < 0.05$).

Volatile 6-methyl-5-hepten-2-one showed a different behavior in each sample. In VOO, the concentration increased significantly ($p < 0.05$) due to the oxidation process [43], while the variations were not significant ($p < 0.05$) in FOO samples. In ROO samples, the concentration of 6-methyl-5-hepten-2-one showed a downward trend, whereas it did not change in BOO samples due to the decrease in its initial concentration which was compensated for by its transfer from the herb to the oil.

In relation to organic acids, the FOO samples showed large differences with VOO samples. Pentanoic and hexanoic acids maintained the same concentration as in the initial sample (VOO-0), indicating that there was a slowing down of the oxidative process that generated them due to the migration of antioxidant compounds. The concentration of propanoic and butanoic acids increased throughout the storage, but it remained below the concentration of the VOO samples. Additionally, a softer slope was observed (Figure 4), which implies that the generation of these acids was slower than in the unflavored VOO due to the protective role of the herbs against the oxidation process. Figure 4 also shows that ROO samples reached a higher concentration of butanoic acid than BOO samples, which supports the idea that basil has a greater protective role.

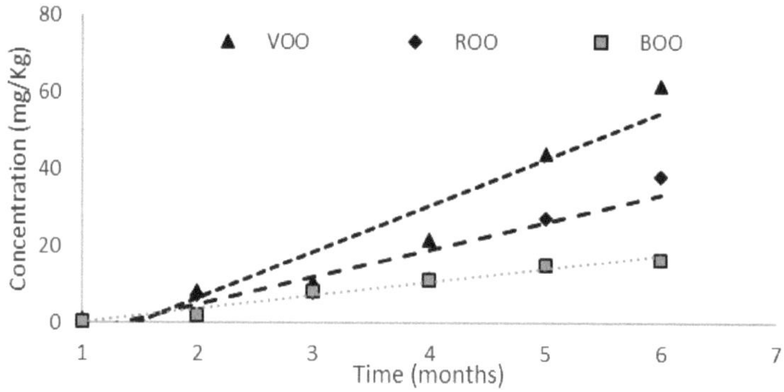


**Figure 4.** Concentration (mg/Kg) of butanoic acid during the storage in the three samples.

Multi-dimensional scaling was applied to all the samples using the volatiles previously described as variables (Figure 5). The results showed that the volatile compounds presented different concentrations depending on the storage time and the kind of flavoring. These differences were mainly due to the migration and effect of antioxidant compounds from the herbs to the virgin olive oil [37]. As mentioned above, it was from the third month onwards when significant differences began to appear, making it possible to separate older samples from the initial ones (Figure 5). The final two months of the ROO samples were also farther from the rest, but not as much as VOO-4 to VOO-6, while the BOO samples maintained similar concentrations, supporting once again the greater antioxidant capacity of basil compared to rosemary.

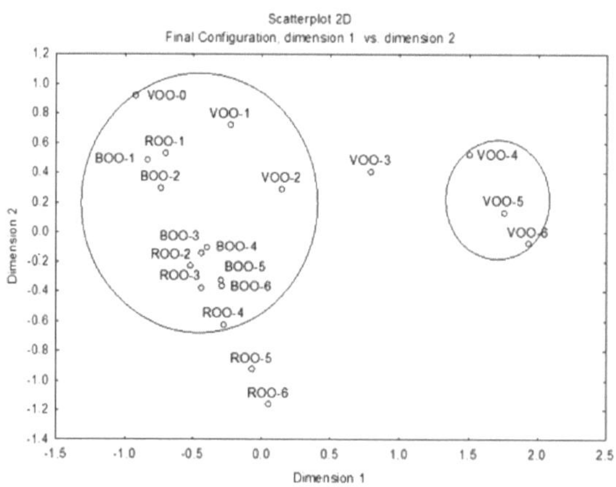


**Figure 5.** Multi-dimensional scaling (MDS) analysis applied to all the samples. VOO, virgin olive oil; ROO, virgin olive oil flavored with rosemary herbs; BOO, virgin olive oil flavored with basil herbs.

### 3.4. New Volatile Compounds

In addition to volatile compounds naturally existing in VOO and terpenes, rosemary and basil herbs have their own volatiles, for example, high concentrations of hydrocarbons and terpenes (Table 3). From the first month, FOO samples presented a different profile, justified by the transfer of these new compounds from the herbs to the oil. Some of the transferred compounds were present in both FOO (ROO, BOO), while others were unique of each one of them (Table 3). The presence of new volatiles in the oils is explained by their chemical properties and their capacity to move through biological membranes. The FOO samples did not show any decrease in these new volatiles throughout the study. On the contrary, most of the volatiles gradually increased their concentrations during storage due to the migration of the compounds from the herbs to the oil.

The common compounds identified in both herbs were γ-terpinene, limonene, 1,8-cineol and tetrahydrolinalool; both FOO samples showed an increase in their concentrations except for tetrahydrolinalool, the concentration of which remained unaltered during the storage. Depending on the flavoring, higher levels of terpenic compounds, such as limonene, were found [17]. The BOO samples showed significant ($p < 0.05$) differences between the first and sixth months of storage for γ-terpinene, which increased from $0.02 \pm 0.001$ mg/Kg to $0.04 \pm 0.01$ mg/Kg with significant differences ($p < 0.05$). Additionally, 1,8-cineol, which was not detected in the first month, increased to $0.04 \pm 0.01$ mg/Kg. The ROO samples, however, only showed significant ($p < 0.05$) differences for 1,8-cineol, the concentration of which increased from $0.01 \pm 0.001$ mg/Kg to $0.09 \pm 0.02$ mg/Kg. The increment in 1,8-cineol concentrations is important due to its impact on the sensory perception of the oil [44]. The concentrations of 1,8-cineol showed significant ($p < 0.05$) differences between the first and the sixth months in the ROO samples, but also between ROO and BOO samples.

In addition to the volatiles that are common to both herbs, there were others that were unique to each herb and migrated to the oil as well. The exclusive volatiles of rosemary herbs [45,46] that were quantified were m-mentene, α-terpene, β-caryophyllene, β-myrcene, camphene, borneol and α-pinene. The rosemary compounds whose concentrations were higher in ROO were m-mentene (6-fold increase) followed by α-terpinolene and camphene (5-fold increase). On the other hand, α-pinene, which has a fresh smell, could not be quantified until the fifth month, after a long time contact time between the herb and oil. The concentration of β-myrcene remained constant ($0.22 \pm 0.03$ mg/Kg) throughout the storage. Table 4 reports that all the volatiles, except for β-myrcene, showed significant differences throughout storage due to the migration of the new compounds from the herb to the oil. The presence of all these new compounds is not only reflected in a very different volatile profile, but also in the aroma of the olive oil. Some of the compounds shown in Table 4 have been also described by other authors [38] while others, such as m-mentene, have not been previously described.

**Table 4.** Concentration ± SD (mg/Kg) of volatile compounds from rosemary herb quantified in flavored virgin olive oil (ROO) during the six months of storage.

| Time (Month) | M-Mentene | A-Terpinolene | B-Caryophyllene | B-Myrcene | Camphene | A-Pinene | Borneol |
|---|---|---|---|---|---|---|---|
| 1 | $0.24 \pm 0.01$ ^a | $0.01 \pm 0.00$ ^a | $0.01 \pm 0.00$ ^a | $0.18 \pm 0.00$ ^a | $0.04 \pm 0.00$ ^a | n.d. * | $0.01 \pm 0.00$ ^a |
| 2 | $0.21 \pm 0.04$ ^a | $0.03 \pm 0.00$ ^b | $0.01 \pm 0.00$ ^a | $0.21 \pm 0.02$ ^a | $0.04 \pm 0.00$ ^a | n.d. | $0.02 \pm 0.00$ ^a |
| 3 | $0.84 \pm 0.14$ ^b | $0.03 \pm 0.00$ ^b | $0.01 \pm 0.00$ ^a | $0.21 \pm 0.01$ ^a | $0.14 \pm 0.01$ ^b | n.d. | $0.02 \pm 0.00$ ^a |
| 4 | $1.16 \pm 0.08$ ^b | $0.04 \pm 0.01$ ^b | $0.01 \pm 0.00$ ^a | $0.23 \pm 0.02$ ^a | $0.15 \pm 0.01$ ^c | n.d. | $0.02 \pm 0.01$ ^a |
| 5 | $1.38 \pm 0.15$ ^b | $0.05 \pm 0.01$ ^b | $0.02 \pm 0.01$ ^b | $0.24 \pm 0.02$ ^a | $0.20 \pm 0.06$ ^c | $0.20 \pm 0.02$ ^a | $0.02 \pm 0.00$ ^a |
| 6 | $1.43 \pm 0.01$ ^c | $0.05 \pm 0.00$ ^c | $0.02 \pm 0.00$ ^b | $0.25 \pm 0.01$ ^a | $0.32 \pm 0.06$ ^c | $0.27 \pm 0.02$ ^b | $0.02 \pm 0.00$ ^a |

Note: Values with different letters (a,b,c) within the same row are significantly different ($p < 0.05$). Experiments were set up in two repetitions and samples for each repetition were analyzed in triplicate ($n = 6$). * Not detected.

The volatiles that migrated from basil herbs to flavored olive oil (BOO) were different from flavored olive oil with rosemary herbs (ROO), except for the four compounds mentioned above. This resulted in different volatile and sensory profiles of the olive oils. Table 5 shows the main compounds that migrated to BOO samples from the basil herb. As in the ROO samples, none of the compounds decreased throughout the storage. However, β-farnesene was only detected in the samples from last two months, but in low concentrations, which may be due to the fact that this compound is mainly found in the root and in a lower concentration in the aerial part of the herb [47].

**Table 5.** Concentration ± SD (mg/Kg) of volatile compounds from basil herb quantified in flavored virgin olive oil (BOO) during the six months of storage.

| Time (Month) | 2-Methylbutanal | Naphthalene | P-Cymene | B-Ocimene | Dihydrolinalool | A-Linalool | B-Farnesene | B-Pinene |
|---|---|---|---|---|---|---|---|---|
| 1 | 0.25 ± 0.07 $^{a}$ | 0.04 ± 0.01 $^{a}$ | 0.01 ± 0.00 $^{a}$ | 0.01 ± 0.00 $^{a}$ | n.d.* | 0.01 ± 0.00 $^{a}$ | 0.00 ± 0.00 $^{a}$ | n.d. |
| 2 | 0.33 ± 0.06 $^{a}$ | 0.07 ± 0.01 $^{a}$ | 0.02 ± 0.00 $^{b}$ | 0.01 ± 0.00 $^{a}$ | 0.01 ± 0.00 $^{a}$ | 0.01 ± 0.00 $^{b}$ | 0.00 ± 0.00 $^{a}$ | 0.01 ± 0.00 $^{a}$ |
| 3 | 0.40 ± 0.02 $^{a}$ | 0.08 ± 0.03 $^{a}$ | 0.02 ± 0.00 $^{c}$ | 0.02 ± 0.00 $^{b}$ | 0.01 ± 0.00 $^{a}$ | 0.02 ± 0.01 $^{b}$ | 0.00 ± 0.00 $^{a}$ | 0.03 ± 0.00 $^{a}$ |
| 4 | 0.40 ± 0.02 $^{a}$ | 0.11 ± 0.03 $^{a}$ | 0.02 ± 0.00 $^{d}$ | 0.04 ± 0.01 $^{c}$ | 0.02 ± 0.00 $^{a}$ | 0.02 ± 0.00 $^{c}$ | 0.00 ± 0.00 $^{a}$ | 0.04 ± 0.01 $^{a}$ |
| 5 | 0.42 ± 0.04 $^{a}$ | 0.10 ± 0.01 $^{a}$ | 0.04 ± 0.00 $^{e}$ | 0.04 ± 0.00 $^{c}$ | 0.02 ± 0.00 $^{a}$ | 0.02 ± 0.00 $^{c}$ | 0.01 ± 0.00 $^{b}$ | 0.14 ± 0.02 $^{b}$ |
| 6 | 0.43 ± 0.04 $^{a}$ | 0.13 ±0.02 $^{b}$ | 0.04 ± 0.01 $^{e}$ | 0.04 ± 0.01 $^{c}$ | 0.02 ± 0.00 $^{a}$ | 0.03 ± 0.01 $^{d}$ | 0.01 ± 0.00 $^{b}$ | 0.16 ± 0.03 $^{b}$ |

Note: Values with different letters ($^{a,b,c,d,e}$) within the same row are significantly different ($p < 0.05$). Experiments were set up in two repetitions and samples for each repetition were analyzed in triplicate ($n = 6$). * Not detected.

The other volatiles transferred from herbs increased their concentrations in BOO and ROO samples during the storage. It is remarkable that, except for 2-methylbutanal and dihydrolinalool, all the volatiles showed significant differences throughout the storage (Table 5).

### 3.5. Sensory Evaluation

The new volatile profiles influenced the sensory assessment of the resulting flavored oils (ROO, BOO). However, as already described, not all the volatile compounds transferred from the herbs to flavored oils contribute to the aroma. Thus, a sensory evaluation, using a descriptive–quantitative analysis, was carried to characterize the aroma of all the VOO and FOO samples.

Figure 6 shows the initial sensory profile of the EVOO sample (VOO-0). The sensory attributes green, olive fruit (green), tomato, apple and sweet quantitatively described the sensory profile of the non-stored VOO-0 sample as a typical profile of an EVOO without sensory defects.

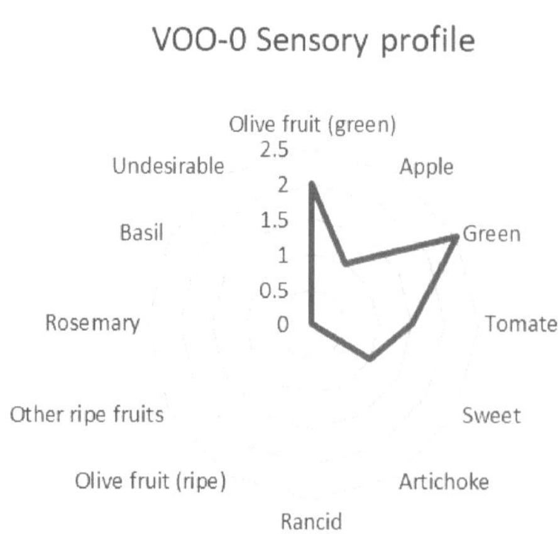


**Figure 6.** Spider chart of the sensory attributes evaluated in the non-stored extra virgin olive oil sample (VOO-0).

Table 6 shows the qualitative sensory modifications suffered by the VOO and FOO throughout the storage; the quantitative data are shown in Figures S2–S4. In the case of VOO, the initial profile was characterized by a gradual loss of the green and fruity nuances until the fourth month. At that time, the rancid sensory defect appeared for the first time and the intensity of the perception increased from then on, justified by higher concentrations of aldehydes and acids. The sensory profiles were different in FOO samples, though the presence of the herbs was detected from the first month. From the third month onwards, the sensory properties changed—decreasing the intensity of the perceptions in both herbs and becoming unpleasant in the case of BOO samples, which is in concordance with the physical—chemical analysis and the volatiles profiles. FOO samples also indicated that the presence of herbs in the oils from three to five months resulted in a masking effect of other perceptions. In the final month, assessors detected the aroma of dry herbs in the case of ROO and rancid began to be present in BOO, in accordance with the concentrations of volatiles in the flavored samples.

**Table 6.** Qualitative sensory evaluation of the aroma of VOO and flavored (BOO, ROO) samples during the storage experiment.

| Time (Month) | VOO | BOO | ROO |
|---|---|---|---|
| 1 | Fruity, green | Slightly basil | Rosemary |
| 2 | Slightly green | Basil | Intense rosemary |
| 3 | Slightly green, heated oil | Basil, unpleasant | Rosemary |
| 4 | Slightly green, slightly rancid | Slightly basil, unpleasant | Slightly rosemary |
| 5 | Slightly fatty, rancid | Slightly basil and altered | Slightly rosemary, dry herb |
| 6 | Pungent, rancid | Slightly basil, rancid | Rosemary, dry herb |

Considering the migration of volatiles from the herbs to the oil, the aroma of BOO samples was mainly due to β-pinene, ocimene and 1,8-cineol compounds. In the case of ROO samples, their aroma is due to the concentrations of camphene, β-myrcene, α-terpinolene, limonene and 1,8-cineol. Thus, the sensory assessment shows that the volatile compounds of the herbs modified the initial organoleptic characteristics of the virgin olive oil and mask other sensory perceptions, which needs to be further studied.

## 4. Conclusions

The application of DHS-GC in the study of the evolution of the volatile profiles of VOO and FOO for six months allowed the quantification of 27 volatile compounds that are responsible for their aroma. The volatiles were classified into two types: those responsible for the pleasant aroma, whose concentration decreased during storage; and others resulting from auto-oxidative or photo-oxidative processes that are responsible for the unpleasant aroma. The addition of rosemary and basil herbs to a VOO produced significant differences in volatile concentrations and a protective (or masking) effect was detected in rosemary and basil flavored oils. The volatile profile of rosemary and basil, identified by GC-MS, was mainly characterized by hydrocarbons and terpenes. Although some of them have already been quantified in the VOO, volatiles allowed distinguishing between flavored and non-flavored samples. Furthermore, the addition of basil or rosemary herbs to VOO enhances the initial VOO shelf life and helps to mask undesirable sensory perceptions. Further studies on the effect of rosemary and basil on virgin olive oil stored in darkness should be carried out and compared with the results obtained in this work, in addition to using different olive oil varieties.

The use of flavoring agents with a strong aroma that can mask the defects of VOOs can make fraud detection a difficult task, therefore, a future analytical challenge is the ability to guarantee an oil's authenticity. Despite this problem, what remains to be addressed is the lack of legislation at the level of the European Union on the marketing of this type of flavored olive oil, which not only makes it difficult to market, but also creates helplessness in the protection and expectations for of consumers for this product.

**Supplementary Materials:** The following supporting information can be downloaded at: https://www.mdpi.com/article/10.3390/antiox12020242/s1, Figure S1: Profile sheet employed for the evaluation of the olive oil samples (VOO, ROO, BOO).; Table S1: Evolution of the main physico-chemical parameters in the three types of olive oils (VOO, ROO, BOO).; Figure S2: Spider chart evolution of the sensory attributes evaluated during six month stored of rosemary olive oil.; Figure S3: Spider chart evolution of the sensory attributes evaluated during six month stored of basil olive oil.; Figure S4: Spider chart evolution of the sensory attributes evaluated during six month stored of virgin olive oil.

**Author Contributions:** Conceptualization, M.T.M.; methodology, M.T.M. and R.A.-R.; investigation, E.J.D.-M., M.T.M. and R.A.-R.; software, validation, and formal analysis, M.T.M., R.A.-R. and E.J.D.-M.; resources, M.T.M. and R.A.-R.; data curation, R.A.-R. and E.J.D.-M.; writing—original draft preparation, E.J.D.-M.; writing—review and editing, R.A.-R. and M.T.M.; visualization and supervision, M.T.M. and R.A.-R.; project administration, M.T.M.; funding acquisition, M.T.M. and R.A.-R. All authors have read and agreed to the published version of the manuscript.

**Funding:** This research was funded by Ministerio de Ciencia e Innovación-Spanish Research State Agency (research project RTI2018-101546-B-C22) and Proyecto FEDER US-1381617.

**Institutional Review Board Statement:** Not applicable.

**Informed Consent Statement:** Not applicable.

**Data Availability Statement:** The data that support the findings of this study are contained within the article. More information is available from the corresponding author, M.T.M.

**Acknowledgments:** E.J. Díaz-Montaña gives thanks for support from a Postgraduate Program of PIF fellowship and financial sponsorship from VI Plan Propio de Investigación y Transferencia at Universidad de Sevilla.

**Conflicts of Interest:** The authors declare no conflict of interest.

## References

1. García-González, D.L.; Aparicio-Ruiz, R.; Aparicio, R. Olive oil. In *Gourmet and Health-Promoting Specialty Oils*, 1st ed.; Kamal-Eldin, A., Moreau, R., Eds.; AOCS Press: Urbana, IL, USA, 2009; pp. 33–72.
2. Visioli, F.; Franco, M.; Toledo, E.; Luchsinger, J.; Willett, W.C.; Hu, F.B.; Martínez-González, M.A. Olive oil and prevention of chronic diseases: Summary of an International conference. *Nutr. Metab. Cardiovasc. Dis.* **2018**, *28*, 649–656. [CrossRef] [PubMed]
3. Frankel, E.N. Nutritional and biological properties of extra virgin olive oil. *J. Agric. Food Chem.* **2011**, *59*, 785–792. [CrossRef] [PubMed]
4. Servili, M.; Esposto, S.; Fabiani, R.; Urbani, S.; Taticchi, A.; Mariucci, F.; Selvaggini, R.; Montedoro, G.F. Phenolic compounds in olive oil: Antioxidant, health and organoleptic activities according to their chemical structure. *Inflammopharmacology* **2009**, *17*, 76–84. [CrossRef] [PubMed]
5. Perona, J.S.; Botham, K.M. Olive oil a Functional Food: Nutritional and Health Benefits. In *Handbook of Olive Oil: Analysis and Properties*, 2nd ed.; Aparicio, R., Harwood, J., Eds.; Springer: New York, NY, USA, 2013; pp. 677–714.
6. Garcia-Gonzalez, D.L.; Morales, M.T.; Aparicio, R. Chapter 43. Olive and Olive. In *Handbook of Fuit and Vegetable Flavors*, 1st ed.; Hui, Y.H., Ed.; John Wiley and Sons, Inc.: Hoboken, NJ, USA, 2010; pp. 821–847.
7. Aparicio, R.; Morales, M.T.; Alonso, M.V. Relationship between volatile compounds and sensory attributes by statistical sensory wheels—An application to virgin olive oil. *J. Am. Oil Chem. Soc.* **1996**, *73*, 1253–1264. [CrossRef]
8. Kotti, F.; Jaziri, K.; Arab, F.; Mater, Y.; Sifi, S.; Fares, N.; Hammami, M.; Gargouri, M. Lipoxygenase: Optimization of extraction and evaluation of its contribution to virgin olive oil aroma. *Food Biotechnol.* **2010**, *24*, 95–105. [CrossRef]
9. Morales, M.T.; Luna, G.; Aparicio, R. Comparative study of virgin olive oil sensory defects. *Food Chem.* **2005**, *91*, 293–301. [CrossRef]
10. Garcia-oliveira, P.; Jimenez-Lopez, C.; Lourenço-Lopes, C.; Chamorro, F.; Gonzalez Pereira, A.; Carrera-Casis, A.; Fraga-Corral, M.; Carpena, M.; Simal-Gandara, J.; Prieto, M.A. Evolution of Flavors in Extra Virgin Olive Oil Shelf-Life. *Antioxidants* **2021**, *10*, 368. [CrossRef]
11. Cecchi, L.; Migliorini, M.; Giambanelli, E.; Rossetti, A.; Cane, A.; Mulinacci, N. New Volatile Molecular Markers of Rancidity in Virgin Olive Oils under Nonaccelerated Oxidative Storage Conditions. *J. Agric. Food Chem.* **2019**, *67*, 13150–13163. [CrossRef]

12. Esposto, S.; Selvaggini, R.; Taticchi, A.; Veneziani, G.; Sordini, B.; Servili, M. Quality evolution of extra-virgin olive oils according to their chemical composition during 22 months of storage under dark conditions. *Food Chem.* **2020**, *311*, 126044. [CrossRef]
13. Morales, M.T.; Przybylski, R. Olive oil oxidation. In *Handbook of Olive Oil: Analysis and Properties*, 2nd ed.; Aparicio, R., Harwood, J., Eds.; Springer: New York, NY, USA, 2013; pp. 479–522.
14. Caponio, F.; Durante, V.; Varva, G.; Silleti, R.; Previtali, M.A.; Viggiani, I.; Squeo, G.; Summo, C.; Pasqualone, A.; Gomes, T.; et al. Effect of infusion of spices into the oil vs. combined malaxation of olive paste and spices on quality of naturally flavoured virgin olive oils. *Food Chem.* **2016**, *202*, 221–228. [CrossRef]
15. Caporaso, N.; Paduano, A.; Nicoletti, G.; Sacchi, R. Capsaicinoids, antioxidant activity, and volatile compounds in olive oil flavored with dried chili pepper (*Capsicum annuum*). *Eur. J. Lipid Sci. Technol.* **2013**, *115*, 1434–1442. [CrossRef]
16. Moldao-Martins, M.; Beirao-da-Costa, S.; Neves, C.; Cavaleiro, C.; Salgueiro, L.; Beirao-da-Costa, M.L. Olive oil flavoured by the essential oils of Mentha x piperita and *Thymus mastichina* L. *Food Qual. Prefer.* **2004**, *15*, 447–452. [CrossRef]
17. Sacchi, R.; Della Medaglia, D.; Paduano, A.; Caporaso, N.; Genovese, A. Characterisation of lemon-flavoured olive oils. *J. Food Sci. Technol.* **2017**, *79*, 326–332. [CrossRef]
18. Akçar, H.H.; Gümüskesen, A.S. Sensory evaluation of flavored extra virgin olive oil. *GIDA—J. Food.* **2011**, *36*, 249–253.
19. Romaniello, R.; Baiano, A. Discrimination of flavoured olive oil based on hyperspectral imaging. *J. Food Sci. Technol.* **2018**, *55*, 2429–2435. [CrossRef] [PubMed]
20. Cecchi, L.; Balli, D.; Urciuoli, S.; Urciuolo, A.; Bordiga, M.; Travaglia, F.; Zanoni, B.; Mulinacci, N. Co-millin of sound olives with fresh chili peppers improves the volatile compound, capsaicinoid and sensory profiles of flavoured olive oil with respect to the typical infusion. *Food Chem.* **2023**, *404*, 134696. [CrossRef]
21. Zellama, M.S.; Chahdoura, H.; Zairi, A.; Cherif Ziani, B.E.; Boujbiha, M.A.; Snoussi, M.; Ismail, S.; Flamini, G.; Mosbah, H.; Selmi, B.; et al. Chemical characterization and nutritional quality investigations of healthy extra virgin olive oil flavored with chili pepper. *Environ. Sci. Pollut. Res.* **2022**, *29*, 16392–16403. [CrossRef]
22. Lamas, S.; Rodrigues, N.; Peres, A.M.; Pereira, J.A. Flavoured and fortified olive oils—Pros and cons. *Trends Food Sci. Technol.* **2022**, *124*, 108–127. [CrossRef]
23. Pateiro, M.; Munekata, P.E.S.; Sant'Ana, A.S.; Dominguez, R.; Rodriguez-Lazaro, D.; Lorenzo, J.M. Application of essential oils as antimicrobial agents against spoilage and pathogenic microorganisms in meat products. *Int. J. Food Microbiol.* **2021**, *337*, 108966. [CrossRef]
24. Zhong, X.; Wang, X.; Zhou, N.; Li, J.; Liu, J.; Yue, J.; Hao, X.; Gan, M.; Lin, P.; Shang, X. Chemical characterization of the polar antibacterial fraction of the ethanol extract from Rosmarinus officinalis. *Food Chem.* **2021**, *344*, 128674. [CrossRef]
25. Gambacorta, G.; Faccia, M.; Pati, S.; Lamacchia, C.; Baiano, A.; La Notte, E. Changes in the chemical and sensorial profile of extra virgin olive oils flavored with herbs and spices during storage. *J. Food Lip.* **2007**, *14*, 202–215. [CrossRef]
26. Sousa, A.; Casal, S.; Malheiro, R.; Lamas, H.; Bento, A.; Pereira, J.A. Aromatized olive oils: Influence of flavouring in quality, composition, stability, antioxidants, and antiradical potential. *Food Sci. Technol.* **2015**, *60*, 22–28. [CrossRef]
27. Perestrelo, R.; Silva, C.; Silva, P.; Câmara, J.S. Global volatile profile of virgin olive oils flavoured by aromatic/medicinal plants. *Food Chem.* **2017**, *227*, 111–121. [CrossRef] [PubMed]
28. Arfaoui, M.; Boulares, M.; Bezzezi, A.; Ayachi, S.; Ghrab, M.; Jouini, N.E.; Hassouna, M.; Boudiche, S. Effect of the enrichment with natural antioxidants obtained by maceration or ultrasound-assisted extraction from olive leaves on organic extra virgin olive oil. *Riv. Ital. Sostanze Grasse* **2022**, *99*, 131 140.
29. Baiano, A.; Gomes, T.; Severini, C. Effects of herbs on hydrolytic and oxidative degradation of olive oil in canned tomatoes. *J. Am. Oil Chem. Soc.* **2005**, *82*, 759–765. [CrossRef]
30. Commission Delegated Regulation (EU) 2022/2104 of 29 July 2022 supplementing Regulation (EU) No 1308/2013 of the European Parliament and of the Council as regards marketing standards for olive oil, and repealing Commission Regulation (EEC) No 2568/91 and Commission Implementing Regulation (EU) No 29/2012. *Off. J. Eur. Union* **2022**, *284*, 1.
31. Kasimoglu, Z.; Tontul, I.; Soylu, A.; Gulen, K.; Topuz, A. The oxidative stability of flavoured virgin olive oil: The effect of the water activity of rosemary. *J. Food Meas. Charact.* **2018**, *12*, 2080–2086. [CrossRef]
32. *COI/T.15/Doc. No 3/Rev.19*; Trade Standard Applying to Olive Oils and Olive-Pomace Oils. International Olive Council: Madrid, Spain, 2022; pp. 1–16.
33. *COI/T.20/Doc. No 15/Rev.10*; Sensory Analysis of Olive Oil Method for the Organoleptic Assessment of Virgin Olive Oil. International Olive Council: Madrid, Spain, 2018; pp. 1–20.
34. Angerosa, F.; Campestre, C. Sensory Quality: Methodologies and Applications. In *Handbook of Olive Oil: Analysis and Properties*, 2nd ed.; Aparicio, R., Harwood, J., Eds.; Springer: New York, NY, USA, 2013; pp. 423–560.
35. Aparicio-Ruiz, R.; Garcia-Gonzalez, D.L.; Morales, M.T.; Lobo-Prieto, A.; Romero, I. Comparision of two analytical methods validated for the determination of volatile compounds in virgin olive oil: GC-FID vs GC-MS. *Talanta* **2018**, *187*, 133–141. [CrossRef] [PubMed]
36. Aparicio, R. Characterization: Mathematical procedures for chemical analysis. In *Handbook of Olive Oil: Analysis and Properties*, 1st ed.; Aparicio, R., Harwood, J., Eds.; Aspen: Gaitherburg, MD, USA, 2000; pp. 285–354.
37. Díaz-Montaña, E.J.; Barbero-López, M.; Aparicio-Ruiz, R.; Morales, M.T. Does A Flavoured Extra Virgin Olive Oil Have Higher Antioxidant Properties? *Antioxidants* **2022**, *11*, 550. [CrossRef]

38. Morales, M.T.; Aparicio-Ruiz, R.; Aparicio, R. Chromatographic methodologies: Compounds for olive oil odor Issues. In *Handbook of Olive Oil: Analysis and Properties*, 2nd ed.; Aparicio, R., Harwood, J., Eds.; Springer: New York, NY, USA, 2013; pp. 261–309.
39. Genovese, A.; Caporaso, N.; Leone, T.; Paduano, A.; Mnea, C.; Perez-Jimenez, M.A.; Sacchi, R. Use of odorant series for extra virgin olive oil aroma characterization. *J. Sci. Food Agric.* **2019**, *99*, 1215–1224. [CrossRef]
40. Caporaso, N.; Genovese, A.; Pérez-Jiménez, M.A.; Olivero-David, R.; Sacchi, R. Impacto f Olive Harvesting Date on Virgin Olive Oil Volatile Composition in Four Spanish Varieties. *Eur. J. Lipid Sci. Technol.* **2021**, *123*, 2000350. [CrossRef]
41. Monteleone, E.; Langstaff, S. *Olive Oil Sensory Science*, 1st ed.; John Wiley & Sons, Ltd.: West Sussex, UK, 2014; pp. 33–80.
42. Neugebauer, A.; Granvogl, M.; Schieberle, P. Characterization of the Key Odorants in High-Quality Extra Virgin Olive Oils and Certified Off-Flavor Oils to Elucidate Aroma Compounds Causing a Rancid Off-Flavor. *J. Agric. Food Chem.* **2020**, *68*, 5927–5937. [CrossRef] [PubMed]
43. Sezer Kiralan, S.; Goksu Karagoz, S.; Ozkan, G.; Kiralan, M.; Ketenoglu, O. Changes in Volatile Compounds of Virgin Olive Oil Flavored with Essential Oils During Thermal and Photo-Oxidation. *Food Anal. Methods* **2021**, *14*, 883–896. [CrossRef]
44. Benkhoud, H.; M'Rabet, Y.; Gara ali, M.; Mezni, M.; Hosni, K. Essential oils as flavoring and preservative agents: Impact on volatile profile, sensory attributes, and the oxidative stability of flavored extra virgin olive oil. *J. Food Process. Preserv.* **2021**, *46*, e15379. [CrossRef]
45. Chouliyoudi, E.; Ganiari, S.; Tsironi, T.; Ntzimani, A.; Tsimogiannis, D.; Taoukis, P.; Oreopoulou, V. Edible coating enriched with rosemary extracts to enhance oxidative and microbial stability of smoed eel fillets. *Food Packag. Shelf Life* **2017**, *12*, 107–113. [CrossRef]
46. Soltanbeigi, A. Qualitative variations of Lavandin Essential Oil under Various Storage Conditions. *J. Essent. Oil-Bear. Plants* **2020**, *23*, 1237–1252. [CrossRef]
47. Zhou, Z.; Xian, J.; Wei, W.; Xu, C.; Yang, J.; Zhan, R.; Ma, D. Volatile metabolic profiling and functional characterization of four terpene synthases reveal terpenoid diversity in different tissues of *Chrysanthemum indicum* L. *Phytochemistry* **2021**, *185*, 112687. [CrossRef]

*antioxidants*

MDPI

*Article*

# Effective Improvement of the Oxidative Stability of *Acer truncatum* Bunge Seed Oil, a New Woody Oil Food Resource, by Rosemary Extract

**Yue Qi [1,2,3], Yeqin Huang [1,2,3], Yanmei Dong [1,2], Wenying Zhang [1,2,3], Fei Xia [1,2], Hongtong Bai [1,2], Zora Dajic Stevanovic [4], Hui Li [1,2,*] and Lei Shi [1,2,*]**

1 Key Laboratory of Plant Resources, Institute of Botany, Chinese Academy of Sciences, Beijing 100093, China; qiyue@ibcas.ac.cn (Y.Q.); huangyeqin@ibcas.ac.cn (Y.H.); dongyanmei@ibcas.ac.cn (Y.D.); wyzhang@ibcas.ac.cn (W.Z.); xiafei@ibcas.ac.cn (F.X.); baiht@ibcas.ac.cn (H.B.)

2 China National Botanical Garden, Beijing 100093, China

3 University of Chinese Academy of Sciences, Beijing 100049, China

4 Department of Agrobotany, University of Belgrade Faculty of Agriculture, Nemanjina 6, 11080 Zemun, Serbia; zoradajic@agrif.bg.ac.rs

* Correspondence: lihui@ibcas.ac.cn (H.L.); shilei@ibcas.ac.cn (L.S.)

**Abstract:** *Acer truncatum* Bunge is a versatile, oil-producing, woody tree natively and widely distributed in northern China. In 2011, The People's Republic of China's Ministry of Health certified *Acer truncatum* seed oil (Aoil) as a new food resource. Unsaturated fatty acids account for up to 92% of the entire Aoil. When Aoil is processed or stored, it can easily oxidize. In this study, the effects of rosemary (*Rosmarinus officinalis* L.) extract on the oxidation stability of Aoil were analysed from multiple angles. The results of radical scavenging ability, malondialdehyde, and free fatty acid reveal that rosemary crude extract (RCE), rosmarinic acid (RA), and carnosic acid (CA) can significantly inhibit the oxidation of Aoil, and CA has the best oxidative stability for Aoil among the tested components of the crude rosemary. The delayed oxidation ability of CA for Aoil was slightly weaker than that of tert-butylhydroquinone (TBHQ), but stronger than that of butylated hydroxyanisole (BHA), butylated hydroxytoluene (BHT), and α-tocopherol (α-T), which was confirmed by microstructures, kinematic viscosity, Aoil weight change, and functional group. Additionally, CA-enriched Aoil had the smallest content of volatile lipid oxidation products. Moreover, lecithin-CA particles were added to enhance the oxidative stability of Aoil. These findings show that CA is a potent antioxidant, capable of successfully preventing Aoil oxidation.

**Keywords:** *Acer truncatum* seed oil; oxidation; rosemary extract; carnosic acid; lecithin

## 1. Introduction

Purpleblow maple (*Acer truncatum* Bunge, 2n = 2x = 26) is a diploid monoecious tree species of the family Aceraceae [1]. It is a versatile, oil-producing, woody tree that is a native species widely distributed in northern China, Korea, and Japan. Unsaturated fatty acids (FAs) account up to 92% of the entire *A. truncatum* seed oil (Aoil) [2]. In 2011, The People's Republic of China's Ministry of Health certified Aoil as a new food resource [3]. Aoil is the main plant source responsible for the large-scale production of nervonic acid (NA), found at a portion of 3–7% [4]. NA is a very-long-chain fatty acid (VLCFA) with great human health benefits, particularly relating central nervous system development. Substantial evidence has shown that a variety of neurological disorders can be linked with NA deficiency [5]. A simple method for the efficient bioaccessibility of NA is its direct intake [6]. Aoil is a novel type of plant drug with potential applications in treating the human brain and neurological problems, making it eligible for providing exogenous NA as well as other types of essential unsaturated fatty acids [7].

When Aoil is processed or stored, it can easily oxidize. The main oxidation by-products are hydroperoxides (HPs) [8]; the decomposition products of HPs can cause oxidative rancidity. However, due to their unpleasant flavour, HPs make edible oils less attractive to consumers. Additionally, consuming oils and goods containing HP can provoke food hazards and cause a number of ailments in humans, including cancer and atherosclerosis [9]. Antioxidants, either natural or synthetic, can be added to postpone lipid oxidation, in addition to preserving the oil quality and for extending the product shelf life by guarding against rancidity. Aiming at shielding fatty foods against rancidity and discoloration, commercially available synthetic antioxidants such as tert-butylhydroquinone (TBHQ), butylated hydroxyanisole (BHA), and butylated hydroxytoluene (BHT) are frequently utilized. Nevertheless, Szydowska-Czerniak reported that these antioxidant dietary additives are linked to liver damage and carcinogenesis [10]. Therefore, it is crucial to find new natural antioxidants or to enhance the antioxidant activity of the natural compounds which already exist.

Rosemary (*Rosmarinus officinalis* L.) is a unique spice that is commercially available for use in culinary, pharmacy, and cosmetics, acting as a strong antioxidant source [11]. Cyclic diterpene diphenols—rosmarinic acid (RA), carnosic acid (CA), and carnosol—have been reported as being among the most potent antioxidant components of rosemary extract [12]. CA is known as the most prevalent and is a potent natural antioxidant [13]. CA is a natural phenolic compound, containing the phenolic hydroxyl group. Numerous studies have documented how polyphenols prevent lipid oxidation [14,15]. In comparison with the more commonly used synthetic antioxidants BHT and BHA in sunflower oil, soybean oil, and bulk corn oil, CA was shown to have a better antioxidative activity [16]. It has been demonstrated that CA effectively reduces the development of HPs of particular fatty acids [17]. Additionally, CA is crucial for the pharmacological actions against inflammatory disorders, diabetes, cancer, and cardiovascular diseases [18]. However, due to the presence of several ring structures and a polar nature of phenolics, they are poorly soluble in fatty matrix, exhibiting a low processing stability [19]. As a result, phenolic compounds are less effective in preventing the lipid oxidation and related deterioration of oils and fats [20]. Vesicular carriers such as lecithin-based systems for polyphenols delivery (mainly the liposomes) are frequently used in different aqueous systems [19]. The lecithin, a co-enrichment substance that may improve the stability of polyunsaturated oils, can also increase the solubility of phenolics in oils [20].

To the best of our knowledge, the effect of rosemary extract on the oxidation stability of Aoil was not studied. Moreover, this is the first study on the combined influence of various CAs and lecithins on the oxidation processes of Aoil. In the present study, the effects of rosemary crude extract (RCE), RA, and CA on radical scavenging activity and Aoil oxidative stability were estimated. Our aim was to compare the antioxidant capacity of RCE, RA, and CA with those of TBHQ, BHA, BHT, and α-tocopherol (α-T) in Aoil. We also assessed the oxidative stability of Aoil enriched with lecithin-CA particles under accelerated heating conditions. This study provides insight into ways of maintaining Aoil quality and extend its shelf life.

## 2. Materials and Methods

### 2.1. Materials

Dried *Acer truncatum* seeds were dehulled and then pressed with a QYZ-460 automatic hydraulic oil press (Shandong Tai'an Liangjun Yiyou Machinery Co., Ltd., Taian, China) to obtain crude oil. The crude oil was refined using oil customized refining equipment (200 KG/D oil refining equipment, Jinan Yueheng Machinery Co., Ltd., Jinan, China). The refining process mainly included degumming, deacidification, dehydration, decolorization, deodorization, and impurity removal. After refining, the Aoil was obtained. Then Aoil was preserved in dark glass packaging under cold conditions for treatment and additional analysis.

Rosemary material (fresh shoots at the full flowering stage) was obtained from the aromatic plant germplasm nursery of the China National Botanical Garden, Beijing. Aerial parts of rosemary plants were dried using a vacuum freeze dryer (LCG-10, SONG YUNA FREEZE DRYER, Beijing, China), and powdered for treatment.

### 2.2. Fatty Acid Composition Analysis of Aoil

Aoil was first derivatized to fatty acid methyl esters [21] and was then analysed by gas chromatography-mass spectrometry (GC-MS, Agilent Technologies, Santa Clara, CA, USA) with HP-5MS column (30 m × 250 μm × 0.25 μm). The oven program was set up as follows: a 1 μL sample was injected in split mode at 50:1, the temperature was held at 50 °C for 2 min, then it was increased linearly to 175 °C at a rate of 20 °C/min, 200 °C at a rate of 3.5 °C/min, 210 °C at a rate of 1 °C/min, and finally 230 °C at a rate of 1.5 °C /min. Helium was employed as the carrier gas, with a flow rate of 1.2 mL/min across the column, and the transfer line temperature was 280 °C. Aoil fatty acid compounds were identified by comparing with 37 fatty acid standards according to the National Standards of the People's Republic of China (GB5009.168—2016) (Table S1). Based on the peak region, the content of Aoil fatty acid components was calculated [22].

### 2.3. Rosemary Crude Extract (RCE) Preparation

Samples of the rosemary powder (10 g) were mixed with 100 mL of 80% ethanol:20% water at 50 °C for 60 min using the 250 W ultrasonic bath and filtered. The solvents were evaporated using RV10 vacuum evaporation (Tusem, Shanghai, China) and a vacuum freeze dryer (LCG-10, SONG YUAN FREEZE DRYER, Beijing, China). High-Performance Liquid Chromatography (HPLC) (UltiMate 3000, Thermo Fisher Scientific, Boston, MA, USA) was used to determine the amount of RA and CA in RCE. The HPLC elution procedure is summarized in Table S2. The content of the total phenols and flavonoids in RCE was measured using the Plant Total Phenols (Tp) Test Kit and Plant Flavonoid Test Kit obtained from BAIGEN (Hubei, China). The RCE powder was stored at −20 °C until further use [23].

### 2.4. RA and CA Preparation

RA and CA were simultaneously separated using preparative liquid chromatography (Waters UPC2/I-Class UPLC/Xevo TQ MS, Boston, MA, USA) from obtained rosemary extract. The injection volume was determined according to the concentrations of RA and CA in the extract. The best separation conditions were achieved using a mixture of the 0.1% phosphoric acid in water (mobile phase A) and acetonitrile of the HPLC grade (mobile phase B). The C18 column was used for elution at a gradient flow rate. The detailed elution procedure is presented in Table S2. The solutions were evaporated by RV10 vacuum evaporation (Tusem, Shanghai, China) and a vacuum freeze dryer (LCG-10, SONG YUAN FREEZE DRYER, Beijing, China). A fraction collector was used to collect the CA and RA fractions. The purities of CA and RA in each dry sample were determined using analytical HPLC. The CA and RA powders were kept at −20 °C [23].

### 2.5. Determination of the Antioxidant Activity (AC)

The AC of 0.07% RCE, 0.07% RA, 0.07% CA, 0.07% α-T, 0.02% TBHQ, 0.02% BHA, and 0.02% BHT [24,25] was determined using DPPH, ABTS, FRAP, and PTIO, respectively. All solvents are methanol.

DPPH radical scavenging capacity was measured according to a Total Antioxidant Capacity Assay Kit with a Rapid DPPH method (BAIGEN, Hubei, China). ABTS free radical scavenging capacity was measured using the Total Antioxidant Capacity Assay Kit with a Rapid ABTS method (Beyotime, Shanghai, China). The antioxidant activity was elucidated by 2-Phenyl-4,4,5,5-tetramethylimidazoline-1-oxyl-3-oxide (PTIO) free radical scavenging capacity. A PTIO solution (0.015% *w*/*v*) was prepared using $H_2O_2$ [26]. The 20 μL of the sample was added to 180 μL of PTIO solution. The PTIO radical scavenging activity was calculated as %Radical scavenging = [1 − (absorbance sample/absorbance control)] × 100, where absorbance control is the absorbance of the PTIO solution and absorbance sample is the absorbance of the PTIO radical in the sample. FRAP was measured using the Total Antioxidant Capacity Assay Kit with FRAP method (Beyotime, Shanghai, China) [27].

Upon obtained antioxidant tests, the antioxidant potency composite index (APC) was determined. The following calculation was used: antioxidant index score = [(sample score/best score) × 100]; the average values of all four tests were taken for the APC [28].

### 2.6. Preparation of Aoil Enriched with Antioxidants

The RCE, RA, CA, TBHQ, BHA, BHT, and α-T were added to the 40 g Aoil. The samples were added to a glass bottle with a diameter of 6.5 cm and a volume of 350 mL. For comparative research, RCE, RA, and CA were directly added to the Aoil at a legal limit of 0.07%. TBHQ, BHA, and BHT were directly added to the Aoil at a legal limit of 0.02%. The 0.07% α-T was added immediately for comparison experiments [24,25]. To ensure the antioxidant compound incorporation, each sample was stirred in an ultrasonic clear bath (TUC-220, Ultrasonic Cleaner, JIEKANG UL TRASONIC, Dongguan, China) with an ultrasound input power of 500 W for 30 min.

### 2.7. Enrichment of Aoil with Lecithin-CA

To determine how lecithin-CA will affect the Aoil oxidative stability, 0.8 g lecithin (20 mg/g oil, the maximal amount of lecithin that could be solubilized in the oil) (from soybean, >98%; Macklin, Shanghai, China) and 0.028 g CA (0.07%) were added directly to 40 g Aoil [24,25,29]. The samples were added to a glass bottle with a diameter of 6.5 cm and a volume of 350 mL. Only CA was added to the Aoil as a control. Then these samples were sonicated (TUC-220, Ultrasonic Cleaner, JIEKANG UL TRASONIC, Dongguan, China) for 30 min and analysed by HPLC (procedure according to 2.3) before accelerating oxidation.

### 2.8. Accelerated Oxidation Assays

A glass bottle with a diameter of 6.5 cm and a volume of 350 mL containing 40 g of enriched and non-enriched samples was placed in an oven for accelerated oxidation. The surface area of Aoil in direct contact with air was 33.18 $cm^2$. The Aoil surface-to-volume ratio used was 0.77 $cm^{-1}$. Therefore, the height of the Aoil is approximately 1.3 cm. For that, the glass bottles containing the Aoil sample were heated at 60 °C in a laboratory drying electric thermostatic blast drying oven (DGG-9140B, Shanghai Senxin Experimental Instrument Co., Ltd., Shanghai, China) for 30 days (d). It is worth noting that the glass bottles containing the oil sample is open in the oven. Oil samples were taken out of the heater at various points over the course of the 30 d heating treatment (0, 5, 10, 15, 20, 25, and 30 d) and were stored at −80 °C for subsequent analyses. The amount of oil withdrawn in the samplings was 5 g. The experiment was repeated three times.

### 2.9. Determination of Malondialdehyde (MDA) Value and Free Fatty Acid (FFA) Value

The level of secondary oxidation in the oil was determined using the thiobarbituric acid reactive substance (TBARS) measurement. MDA content was measured using a Malondialdehyde Content Detection Kit (Solarbio, Beijing, China).

The copper reagent method was used to measure the copper ion content for determination of the free fatty acids level. The FFA levels were quantified (Solarbio, Beijing, China) using the free fatty acid (FFA) assay kit [30].

### 2.10. Cryo-Scanning Electron Microscopy (Cryo-SEM) Analysis of the Oil Samples Microstructure

Cryo-SEM was used to examine the impact of the added different antioxidants on the microstructure of the Aoil samples [31]. The cryogenic preparation and transmission system was used to release the oil samples from the liquid nitrogen environment to prepare the chamber for fracture sublimation (90 °C, 10 min) and gold-plating (10 mA current, 60 s). The micromorphology of the samples was observed using SEM (Regulus8100, Remote PEO, Tokyo, Japan).

### 2.11. Viscosity and Weight Change of the Oil Samples Analysis

The viscosity of the oil samples was measured using an automatic folding tube kinematic viscometer (SL-SF01D, HUNAN SHELI ELECTRON SCIENCE & TECHNOLOGY Co., Ltd., Hunan, China) [32]. A 5 mL measuring cylinder (Beijing Jianqiang Weiye Technology Co., Ltd., Beijing, China) was used to accurately measure 1 mL of oil. The weight of the 1 mL oil sample was measured using an electronic analytical balance (d = 0.1 mg) (AB204-N, METTLER TOLEDO, Shanghai, China) [33].

### 2.12. Analysis of Oil Volatile Lipid Oxidation Products by GC–MS

The volatile lipid oxidation products were extracted from the Aoil samples (1 g) using solid-phase microextraction (SPME). Gas chromatography-mass spectrometry (GC–MS) was then performed using the 7890A-7000B GC–MS (Agilent Technologies, Santa Clara, CA, USA) equipped with an HP-5MS column (30 m × 250 μm × 0.25 μm; Agilent Technologies). The internal standard was added, the 3-octanol (Sigma, Berlin, Germany). The following conditions of the SPME were included: samples were processed at 70 °C for 40 min with a DVB/CAR/PDMS fibre for 30 min (Bellefonte, PA, USA), and then desorbed for 240 s at 250 °C. The GC heater temperature was configured to rise from a starting temperature of 35 °C for 3 min to temperatures of 35–100 °C at a rate of 2 °C/min, 140 °C at a rate of 2 °C/min, and 280 °C at a rate of 10 °C/min. The helium gas at a flow rate of 3 mL/min as a vector was used. The MS parameters were as follows: ionization energy of 70 eV, electronic impaction source temperature of 230 °C, quadrupole temperature of 280 °C, and mass range of 30–500 u. No samples were split before injection. Retention time (RT) and retention index (RI) values were compared using reference standards and published data in order to identify the volatile lipid oxidation products, as well as by matching the spectral and RI values with NIST 14.0 [34].

### 2.13. Nuclear Magnetic Resonance (NMR) Analysis of Oxidized Oil for Functional Group

After 5 h of heating at 60 °C, the oil samples were analysed. The method used for calculation of the olefinic and bisallylic protons loss was previously described in [35]. A quantity of 50 μL of each oil sample was dissolved in 0.5 mL of $CDCl_3$ containing 20 mg of benzoic acid before being placed in a 5 mm NMR tube. At 25 °C, all spectra were captured. Tetramethylsilane (TMS) is used to mark the chemical shifts. Using the AVANCE NEO (AVANCE NEO 600 MHz, Bruker Biospin, Hersbruck, Germany) operating at a proton frequency of 600 MHz, 1D proton NMR spectra were collected. The following acquisition parameters were utilized to obtain the 600 MHz proton NMR spectra: size 262,144, line broadening 0.1 Hz, spectral width 19.8356 ppm, acquisition time 4 s, pulse width 9.65 s, and 16 scans for 17 min. The program "MestReNova.9.0.1" was used for data processing [36].

### 2.14. Statistical Analyses

Data are expressed as mean ± standard deviation (SD, $n$ = 3). Statistical analyses were performed with one-way analysis of variance (ANOVA) followed by Duncan's test (IBM SPSS Statistics software; IBM Corp., Armonk, NY, USA), and $p$-values < 0.05 were considered statistically significant.

## 3. Results

### 3.1. Fatty Acid Composition

*Acer truncatum* is an indigenous species of China. In China, this tree species is referred to as 'yuan bao feng' because of its gold ingot-shaped fruits (Figure 1A,B). It is a versatile oil-producing woody tree that is a native species widely distributed in northern China, Korea, and Japan. The findings of the GC-MS analysis of Aoil's fatty acid composition are displayed in Table S3. Considering the overall Aoil fatty acid composition, the following order was determined: Monounsaturated fatty acids (MUFA) > Polyunsaturated fatty acids (PUFA) > Saturated fatty acid (SFA). Unsaturated fatty acids (UFAs) were the most abundant in the Aoil, accounting for up to 90% of the total fatty acids (Figure 1C) [2].

Linoleic acid (LA, C18:2, cis (n6)) was the main fatty acid found in the Aoil (Figure 1D), with a content of 34.859 ± 0.2436 g/100 g. It is important to note that Aoil has a rather high content of nervonic acid, with 6.327 ± 0.1549 g/100 g.

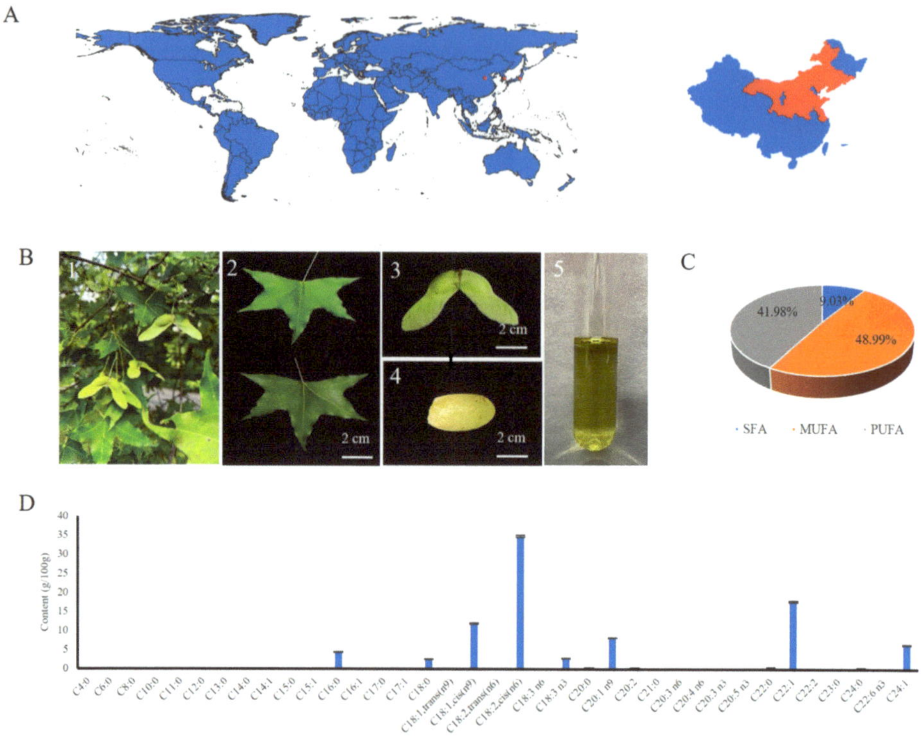


**Figure 1.** (**A**), Geographic distribution map of *Acer truncatum* in the world and China. (**B**), *Acer truncatum* (**B1**), leaves (**B2**), fruit (**B3**), seed (**B4**), and oil (**B5**) of *Acer truncatum*. (**C**), Percentage of different types of fatty acids in the Aoil. (**D**), Fatty acid composition of the Aoil.

### 3.2. Analysis of the Composition of RCE and the Purity of RA, CA

RCE was prepared by ultrasound method, whereas the RA and CA were obtained by preparative liquid chromatography (Figure 2A–C). In the present study, according to the standard curve (Figure S1), the RCE had a significant content of the RA and CA, which are the most effective antioxidant constituents of rosemary (Figure 2D). It also contains a large amount of total phenols and flavonoids (Figure 2D). The purity of RA and CA determined by HPLC was 73% and more than 99%, respectively (Figure 2E,F).

### 3.3. The Antioxidant Properties of RCE, RA, and CA In Vitro and in Aoil

RCE, RA, and CA showed an effective efficiency in free radical scavenging capacity, where the free radical scavenging capacity increased with raising concentrations of a tested compound (Figure S2). The antioxidant capacity of RCE, RA, and CA was compared with the synthetic antioxidants the TBHQ, BHA, BHT, and the natural antioxidant α-T using DPPH, ABTS, FRAP, and PTIO analyses. The results of DPPH and PTIO testing showed that RCE, RA, and CA had higher antioxidant activity than BHA and BHT, but a slightly lower effect compared to TBHQ. The results of DPPH and ABTS showed that RCE, RA, and CA had similar antioxidant activities to α-T (Figure 3A,B). FRAP analyses showed that RCE and RA had higher antioxidant activities than CA, TBHQ, BHA, BHT, and α-T (Figure 3C). The results of the antioxidant potency composite showed that RCE and CA had a higher antioxidant activity than BHA, BHT, and α-T and a slightly lower antioxidant activity than TBHQ. However, RA had higher antioxidant activity than TBHQ, BHA, BHT, and α-T (Figure 3E). Overall, DPPH, ABTS, FRAP, and PTIO analyses showed that RA has a higher antioxidant activity in vitro.

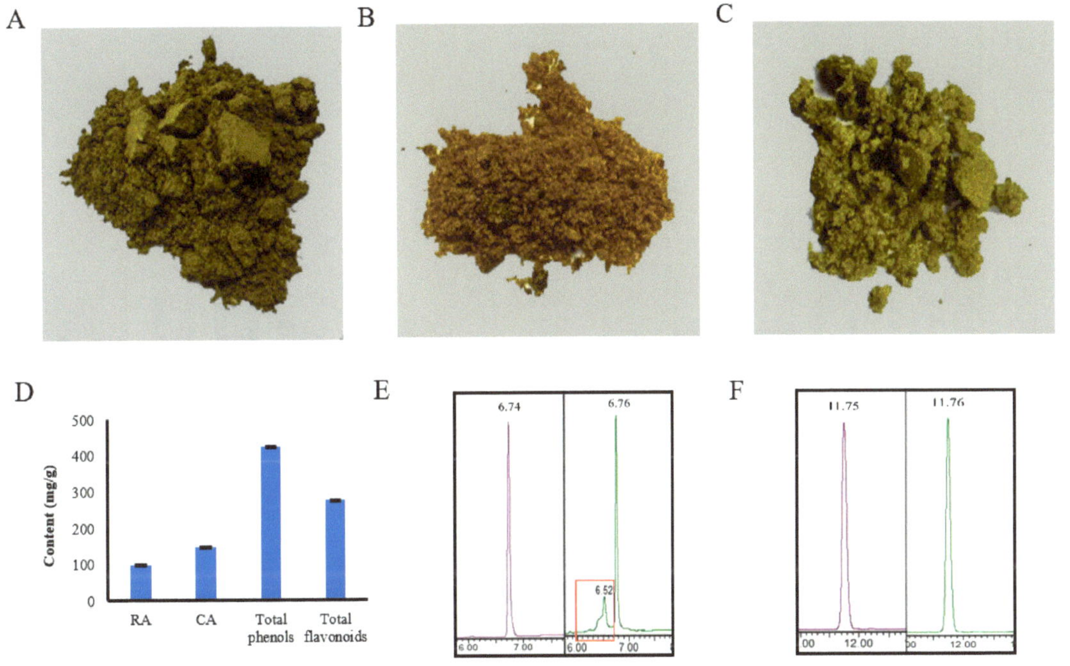


**Figure 2.** Analysis of the RCE (**A**), RA (**B**), and CA (**C**) powder. (**D**), RCE composition analysis. (**E**), Purple: HPLC data for RA standard; green: HPLC data for RA obtained by the preparative liquid chromatograph from the RCE; red square: HPLC data for impurities in RA. (**F**), Purple: HPLC data for CA standard; green: HPLC data for CA obtained by the preparative liquid chromatograph from the RCE.

The antioxidant property of RCE, RA, and CA in Aoil were preliminarily assessed by determination of the oxidation products. The greater the MDA content, the more serious the oxidation of the oil. The MDA values demonstrated that total MDA concentration in the Aoil consequently rises due to accelerated oxidation during the period from 0 to 30 d (Figure 3F). After an oxidation period of 30 d, the oils enriched with RCE and CA had lower MDA content than those enriched with BHA, BHT, and α-T, but had a slightly higher MDA content than those enriched with TBHQ oil. Based on the results on MDA inhibition rate, the CA had the strongest protective effect on Aoil among RCE, RA, and CA (Figure 3G). The FFA assay results were similar to those of the MDA assay. The concentration of the total FFA in Aoil increased with the accelerated oxidation time from 0 to 30 d (Figure 3H). After oxidation for 30 d, the oils enriched with RCE and CA had a lower FFA content than oils enriched with BHA, BHT, and α-T, as well as a slightly higher FFA content than oil enriched with TBHQ. The results of the inhibition rate of FFA showed that CA had the strongest protective effect on Aoil among RCE, RA, and CA (Figure 3I). Both MDA and FFA results indicated that RCE, RA, and CA can significantly inhibit the oxidation of Aoil. However, CA has better oxidative stability to Aoil among RCE, RA, and CA.

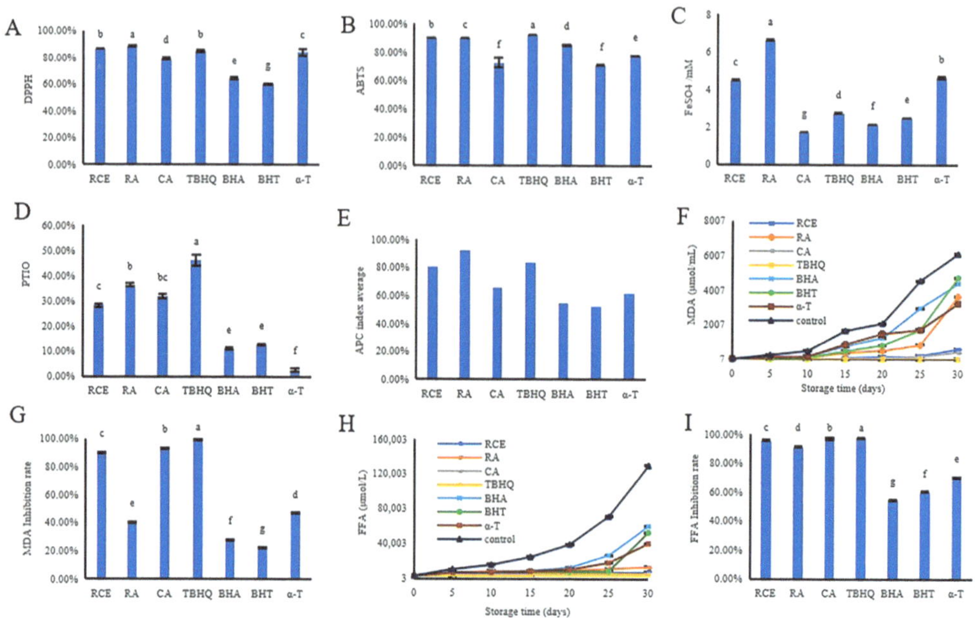

**Figure 3.** Comparison of the antioxidant capacity of RCE, RA, CA, TBHQ, BHA, BHT, and α-T. Determination of the antioxidant activity by DPPH (**A**), ABTS (**B**), FRAP (**C**), and PTIO (**D**). (**E**), Antioxidant potency composite index. (**F**), Determination of MDA. (**G**), Antioxidant inhibition rate of MDA production compared to control. (**H**), Determination of FFA. (**I**), Antioxidant inhibition rate of FFA production compared to control. a–g means with different superscripts differ significantly ($p < 0.05$).

### *3.4. The Physical Characteristic of the Aoil Enriched with Antioxidants*

The influence of antioxidants on the microstructure of Aoil after oxidation for 30 d was assessed using cryo-SEM. The microstructures of the oil samples after rapid freezing, fracturing, sublimation, and gold plating are shown in Figure 4A. The microstructures of the oils enriched with CA and TBHQ after exposure to 30 d oxidation were the same as those of the oil sample at the beginning of the oxidation process. The microstructures of the oils enriched with BHA, BHT, and α-T after 30 d of oxidation were similar to those of the non-enriched oil sample exposed to 30 d oxidation. The microstructures of the oils enriched with CA and TBHQ were less solid than those enriched with BHA, BHT, and α-T. Next, the kinematic viscosities of the differently treated oil samples were compared. The results of the kinematic viscosity assay showed that the lower the degree of oxidation of the oil, the lower the kinematic viscosity. The kinematic viscosity of oils enriched with CA and TBHQ after oxidation for 30 d was much lower than that of oils enriched with BHA, BHT, and α-T after oxidation for 30 d (Figure 4B). After accelerated oxidation, the weight of Aoil increased significantly. The weight of oil samples enriched with CA and TBHQ after oxidation for 30 d was smaller than oils enriched with BHA, BHT, and α-T after oxidation for 30 d (Figure 4B). The results showed that CA had a stronger protective effect on Aoil oxidation.

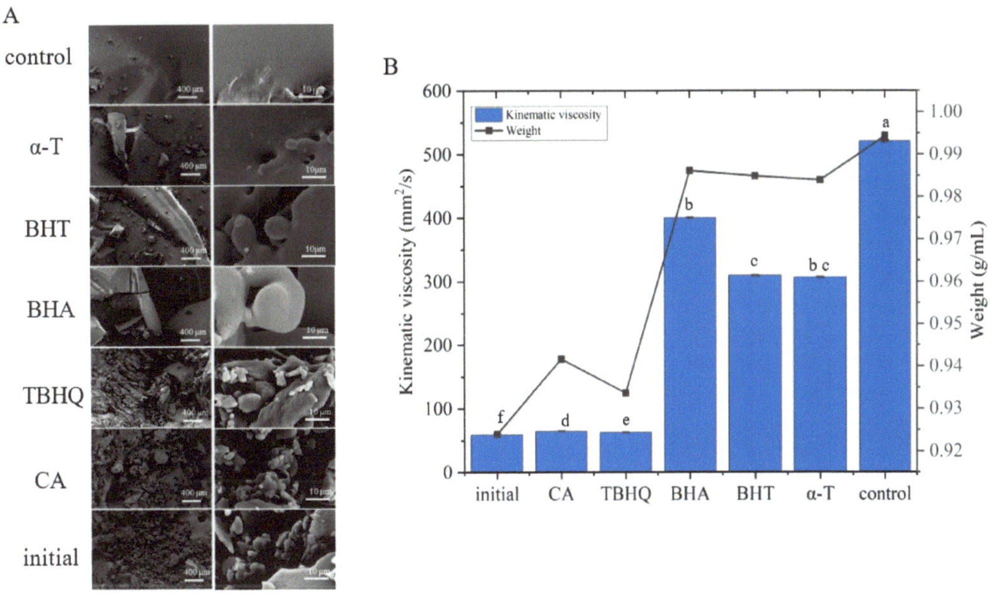


**Figure 4.** The physical characteristic of the Aoil enriched with different antioxidants. (**A**), Microstructure of Aoil at the beginning and storage at 60 °C for 30 d. (**B**), Weight and kinematic viscosity of Aoil at the beginning and storage at 60 °C for 30 d. a–f means with different superscripts differ significantly ($p < 0.05$).

### 3.5. Volatile Lipid Oxidation Products in Aoil Enriched with Different Antioxidants

The main volatile lipid oxidation products of Aoil were identified by GC-MS analysis as alkyls, aldehydes, and ketones (Figure S3B). The volatile lipid oxidation products in the headspace gas from Aoil and Aoil blended with antioxidants were measured using SPME–GC–MS during oxidation at 60 °C. Gas chromatograms of volatile lipid oxidation products from pure Aoil and Aoil blended with antioxidants after oxidation for 30 d are shown in Figure S3A. Pure Aoil contained more volatile lipid oxidation products than Aoil blended with antioxidants after oxidation for 30 d. As shown in Table 1, many varieties of volatile lipid oxidation products were detected in the oxidized Aoil (44 compounds). The main volatile lipid oxidation products in Aoil after 30 d of oxidation were identified as aldehydes (57.35%) (such as hexanal, 2-Octenal, (E)-, 2,4-Decadienal, (E,E)-, 2-Undecenal 2-Decenal, (E)-, etc.), alcohols (13.21%) (such as 2,5-Dimethylcyclohexanol, Cyclohexanol, 1-butyl-, 1-Octen-3-ol, 2-Ethylnon-1-en-3-ol, 1,7-Octadien-3-ol, 2,6-dimethyl-, etc.), alkyls (13.01%) (such as Bicyclo[2.2.2]octane, 1-methoxy-4-methyl-, Cyclotetrasiloxane, octamethyl-, Cyclotrisiloxane, hexamethyl-, etc.), ketones (7.78%) (such as 2-Sec-Butylcyclohexanone, 5-Undecen-4-one, Cyclononanone, etc.), anhydrides (3.88%) (such as Pentanoic acid, 2-methyl-, anhydride), acids (1.99%) (such as hexanoic acid, 3-Decanynoic acid, nonanoic acid, etc.), furans (1.28%) (such as Furan, 2-pentyl-, etc.), esters (0.98%) (such as Hexanoic acid, methyl ester, 5-Oxotetrahydrofuran-2-carboxylic acid, ethyl ester, n-Caproic acid vinyl ester, etc.), and phenols (0.52%) (such as 4a(2H)-Naphthalenol, octahydro-, trans-, etc.) (Figure S3C). Oil enriched with CA had fewer volatile lipid oxidation products than oils enriched with TBHQ, BHA, BHT, and α-T after oxidation for 30 d (Table 1). Several volatile lipid oxidation products were also detected in the oil enriched with CA and TBHQ after oxidation for 30 d, although their peak intensities were low. The volatile lipid oxidation products from the oil enriched with CA and TBHQ were quite different from those of the oxidised oil. The main volatile lipid oxidation products of the CA-enriched Aoil after oxidation for 30 d were identified by GC-MS analysis as aldehydes, alkyl, ketones, alcohols, and furan (Figure S3D). The main volatile lipid oxidation products in Aoil enriched with TBHQ after oxidation for 30 d were identified as quinones, alkyls, aldehydes, ketones, and alcohol compounds (Figure S3E). It is worth noting that oil enriched with TBHQ had more volatile lipid oxidation products than oil enriched with CA. The metabolite of TBHQ was the volatile substance 2-tert-butyl-1,4-benzo-quinone (TBQ) (Figure S3F). According to the SPME–GC–MS results, the CA had a stronger protective effect on Aoil than TBHQ, BHA, BHT, and α-T (Figure S3G).

### 3.6. Quantitative Evaluation of Aoil Enriched with Antioxidants by High Resolution Proton-NMR Spectroscopy ($^1$H-NMR)

Based on the MDA and FFA results, the protective effects of CA, TBHQ, BHA, BHT, and α-T on Aoil were assessed by $^1$H-NMR. The $^1$H-NMR spectrum of Aoil showed eight main peaks (Figure 5A). The functional groups represented by each peak are listed in Table 2. After oxidation for 5 h, the H atom species in Aoil did not change significantly before and after oxidation; only the number changed slightly (Figure 5A). Comparing the results of each peak in the $^1$H-NMR spectrum before and after oxidation, it can be seen that the number of unsaturated double bonds in the oxidized Aoil is reduced (Figure 5B). Furthermore, the number of bisallyl hydrogen protons and allyl hydrogen protons in oxidized Aoil was reduced (Figure 5C,D). This shows that Aoil (RH) was triggered to generate R· and H. The number of hydrogen atoms on the double bond and the terminal methyl group of Aoil decreased. This shows that after H· combines with the more active ROO· to form linolenic acid hydroperoxide (ROOH), part of the Aoil hydroperoxide is decomposed to form other small molecular substances, such as aldehydes, alcohols, and alkanes. Oils enriched with CA have a slightly lower amount of unsaturated double bonds, bisallyl hydrogen protons, and allyl hydrogen protons than oils enriched with TBHQ. However, oil enriched with CA had a higher amount of unsaturated double bonds, bisallyl hydrogen protons, and allyl hydrogen protons than oils enriched with BHA, BHT, and α-T (Figure 5B–D). $^1$H-NMR results showed that CA had a stronger protective effect on Aoil.

### 3.7. Improvement of the Oxidative Stability of Aoil by Enrichment with Lecithin-CA

To assess the effect of enriching Aoil with lecithin-CA on its oxidative stability, the lecithin (20 mg/g oil) [29] and CA (0.07%) were added directly to Aoil. The maximum amount of lecithin that could be solubilised in the oil was added to Aoil. The results of the solubility assay showed that the concentration of CA in the Aoil enriched with lecithin was higher than that in the Aoil non-enriched with lecithin (Figure 6A). The results of the influence of lecithin-CA on the oxidative stability of enriched Aoil showed that the oil enriched with lecithin-CA had a lower MDA and FFA levels than oils enriched with CA (Figure 6B,C). Compared with CA, the inhibition rate of lecithin-CA on MDA increased by 2.23%. Compared with CA, the inhibition rate of lecithin-CA on FFA was 0.19% higher (Figure S4). The microstructures of the oils enriched with lecithin-CA oxidized for 30 d were similar to those of the oil enriched with CA oxidised for 30 d (Figure 6D). The main volatile lipid oxidation products of oil enriched with the lecithin-CA after oxidation for 30 d were less presented than in oil enriched with CA after oxidation for 30 d (Figure 6E). Compared to CA, lecithin-CA has a reduction of 0.213 g in rancid volatile gases. Oil enriched with lecithin-CA had a higher amount of unsaturated double bonds, bisallyl hydrogen protons, and allyl hydrogen protons than oil enriched with CA (Figure 6F–H). In general, enrichment with lecithin-CA can improve the oxidative stability of Aoil.

**Table 1.** Volatile lipid oxidation products of 1 g Aoil at the beginning and storage at 60 °C for 30 d.

| Compound | Empirical Formula | RT | RI | Initial (µg) | CA (µg) | TBHQ (µg) | BHA (µg) | BHT (µg) | α-T (µg) | Control (µg) |
|---|---|---|---|---|---|---|---|---|---|---|
| 1-Pentanol | $C_5H_{12}O$ | 5.576 | 753.939 | - | - | - | 3.758 ± 1.293 $^{b}$ | 4.549 ± 1.293 $^{b}$ | 3.353 ± 0.377 $^{b}$ | 6.57 ± 0.44 $^{a}$ |
| Hexanal | $C_6H_{12}O$ | 6.750 | 800.665 | - | 0.615 ± 0.071 $^{b}$ | 0.094 ± 0.006 $^{b}$ | 169.806 ± 55.105 $^{abcd}$ | 121.044 ± 16.412 $^{bcd}$ | 111.657 ± 11.694 $^{bcd}$ | 259.331 ± 6.944 $^{a}$ |
| Cyclotrisiloxane, hexamethyl- | $C_6H_{18}O_3Si_3$ | 7.898 | 823.127 | 0.187 ± 0.001 $^{d}$ | 0.759 ± 0.139 $^{c}$ | 0.724 ± 0.062 $^{c}$ | 5.161 ± 1.852 $^{b}$ | 5.085 ± 1.163 $^{b}$ | 4.651 ± 0.062 $^{b}$ | 7.182 ± 0.873 $^{a}$ |
| 2-Heptanone | $C_7H_{14}O$ | 11.322 | 890.119 | - | - | - | 3.453 ± 0.678 $^{b}$ | 3.08 ± 0.568 $^{b}$ | 2.905 ± 0.254 $^{b}$ | 6.655 ± 0.494 $^{a}$ |
| n-Caproic acid vinyl ester | $C_8H_{14}O_2$ | 11.607 | 895.696 | - | - | - | 2.707 ± 0.88 $^{c}$ | 4.507 ± 0.288 $^{ab}$ | 3.77 ± 0.543 $^{bc}$ | 5.619 ± 0.082 $^{a}$ |
| Heptanal | $C_7H_{14}O$ | 11.927 | 901.519 | - | - | - | 3.324 ± 0.772 $^{b}$ | 2.528 ± 0.232 $^{b}$ | 2.538 ± 0.162 $^{b}$ | 6.753 ± 0.359 $^{a}$ |
| Oxirane, pentyl- | $C_7H_{14}O$ | 12.109 | 904.284 | - | - | - | - | 2.013 ± 0.288 $^{b}$ | 1.499 ± 0.098 $^{c}$ | 5.035 ± 0.008 $^{a}$ |
| Hexanal, 3-methyl- | $C_7H_{14}O$ | 12.364 | 908.157 | - | - | - | 9.511 ± 2.557 $^{b}$ | 12.709 ± 2.412 $^{b}$ | 9.153 ± 0.081 $^{b}$ | 19.773 ± 2.438 $^{a}$ |
| Hexanoic acid, methyl ester | $C_7H_{14}O_2$ | 13.481 | 925.125 | - | - | - | 4.348 ± 1.145 $^{b}$ | 2.586 ± 0.459 $^{b}$ | 3.037 ± 0.23 $^{b}$ | 10.52 ± 1.062 $^{a}$ |
| 2-Heptenal, (E)- | $C_7H_{12}O$ | 15.374 | 953.881 | 0.07 ± 0.004 $^{d}$ | 0.29 ± 0.075 $^{c}$ | 0.234 ± 0.055 $^{c}$ | 37.217 ± 9.266 $^{b}$ | 43.898 ± 6.141 $^{ab}$ | 34.834 ± 1.37 $^{b}$ | 50.771 ± 3.22 $^{a}$ |
| 1-Octen-3-ol | $C_8H_{16}O$ | 17.065 | 979.569 | - | - | - | 16.344 ± 4.223 | 22.283 ± 5.278 | 15.676 ± 3.47 | 16.419 ± 1.475 |
| Furan, 2-pentyl- | $C_9H_{14}O$ | 17.792 | 990.612 | - | 0.132 ± 0.033 $^{d}$ | - | 19.322 ± 3.573 $^{b}$ | 10.997 ± 0.929 $^{c}$ | 12.344 ± 0.695 $^{c}$ | 28.887 ± 3.205 $^{a}$ |
| Octanal | $C_8H_{16}O$ | 18.631 | 1003.111 | - | - | - | 9.447 ± 1.457 $^{b}$ | 6.041 ± 0.775 $^{bc}$ | 5.771 ± 0.275 $^{c}$ | 19.056 ± 2.499 $^{a}$ |
| Cyclotetrasiloxane, octamethyl- | $C_8H_{24}O_4Si_4$ | 18.973 | 1007.926 | 0.242 ± 0.082 $^{e}$ | 0.366 ± 0.2 $^{e}$ | 0.637 ± 0.273 $^{d}$ | 8.689 ± 1.975 $^{c}$ | 18.58 ± 0.618 $^{a}$ | 15.37 ± 0.816 $^{b}$ | 17.633 ± 6.116 $^{a}$ |
| Hexanoic acid | $C_6H_{12}O_2$ | 20.836 | 1034.155 | - | - | - | 5.178 ± 0.031 | 6.798 ± 0.799 | 10.624 ± 5.355 | 17.7 ± 4.71 |
| 3,5-Octadien-2-ol | $C_8H_{14}O$ | 21.155 | 1038.646 | - | - | - | 5.649 ± 1.214 $^{ab}$ | 3.847 ± 0.76 $^{b}$ | 3.844 ± 0.378 $^{b}$ | 6.341 ± 0.545 $^{a}$ |
| 5-Oxotetrahydrofuran-2-carboxylic acid, ethyl ester | $C_7H_{10}O_4$ | 22.059 | 1051.373 | - | - | - | 3.051 ± 0.664 $^{ab}$ | 2.457 ± 0.287 $^{b}$ | 2.028 ± 0.052 $^{b}$ | 5.989 ± 0.174 $^{a}$ |
| 2-Octenal, (E)- | $C_8H_{14}O$ | 22.508 | 1057.694 | - | 0.382 ± 0.087 | 0.227 ± 0.006 | 137.407 ± 29.366 | 178.448 ±16.335 | 129.053 ± 5.058 | 240.989 ± 28.915 |
| Pentanoic acid, 2-methyl-, anhydride | $C_{12}H_{22}O_3$ | 22.683 | 1060.158 | - | - | - | 49.564 ± 12.749 $^{b}$ | 79.554 ± 9.605 $^{a}$ | 59.212 ± 3.674 $^{b}$ | 87.662 ± 4.848 $^{a}$ |
| 1-Octanol | $C_8H_{18}O$ | 23.568 | 1072.617 | - | - | - | 3.573 ± 1.358 $^{b}$ | 2.851 ± 0.482 $^{b}$ | 2.644 ± 0.38 $^{b}$ | 6.076 ± 0.52 $^{a}$ |
| 2,5-Dimethylcyclohexanol | $C_8H_{16}O$ | 25.127 | 1094.566 | - | 0.15 ± 0.049 | 0.054 ± 0.001 | 72.127 ± 13.376 | 53.232 ± 3.127 | 53.992 ± 3.767 | 120.257 ± 5.245 |
| Nonanal | $C_9H_{18}O$ | 25.822 | 1104.376 | - | 0.606 ± 0.193 | 0.325 ± 0.023 | 37.056 ± 7.064 | 23.181 ± 2.039 | 23.211 ± 0.865 | 62.332 ± 1.466 |
| 9-Oxabicyclo[6.1.0]nonan-4-ol | $C_8H_{14}O_2$ | 26.016 | 1107.123 | - | - | - | 5.207 ± 1.198 $^{b}$ | 5.527 ± 0.74 $^{b}$ | 4.817 ± 0.247 $^{b}$ | 8.025 ± 0.088 $^{a}$ |
| 3-Nonen-2-one | $C_9H_{16}O$ | 28.300 | 1139.465 | - | - | - | 5.503 ± 1.361 $^{b}$ | 3.765 ± 0.614 $^{b}$ | 4.809 ± 0.61 $^{b}$ | 9.895 ± 0.681 $^{a}$ |
| 2-Nonenal, (E)- | $C_9H_{16}O$ | 29.646 | 1158.524 | - | - | - | 13.272 ± 2.551 | 10.291 ± 0.835 | 11.775 ± 0.293 | 20.375 ± 2.287 |
| 2-tert-Butyl-1,4-benzoquinone | $C_{10}H_{12}O_2$ | 32.328 | 1196.502 | - | - | 3.277 ± 0.079 | - | - | - | - |
| Decanal | $C_{10}H_{20}O$ | 32.949 | 1205.513 | - | - | - | 7.066 ± 1.135 $^{b}$ | 3.849 ± 0.33 $^{b}$ | 4.001 ± 0.444 $^{b}$ | 11.44 ± 2.362 $^{a}$ |
| 2,4-Nonadienal, (E, E)- | $C_9H_{14}O$ | 33.356 | 1211.512 | - | - | - | 21.42 ± 3.668 $^{b}$ | 21.546 ± 3.437 $^{b}$ | 23.512 ± 3.208 $^{b}$ | 33.922 ± 1.589 $^{a}$ |
| trans-2-undecenoic acid | $C_{11}H_{20}O_2$ | 34.336 | 1225.958 | - | - | - | 2.624 ± 0.518 | 1.421 ± 0.008 | 2.125 ± 0.409 | 5.535 ± 0.381 |
| Cyclononanone | $C_9H_{16}O$ | 35.051 | 1236.498 | - | - | - | 5.493 ± 1.021 $^{b}$ | 5.015 ± 0.37 $^{b}$ | 5.119 ± 0.286 $^{b}$ | 11.552 ± 0.837 $^{a}$ |
| 2-Sec-Butylcyclohexanone | $C_{10}H_{18}O$ | 35.754 | 1246.860 | 0.056 ± 0.006 | 0.211 ± 0.101 | 0.059 ± 0.001 | 80.737 ± 18.734 | 47.42 ± 5.39 | 47.313 ± 6.818 | 115.601 ± 2.401 |
| 2-Ethylnon-1-en-3-ol | $C_{11}H_{22}O$ | 36.267 | 1254.422 | - | - | - | 8.316 ± 3.154 $^{b}$ | 4.483 ± 1.276 $^{b}$ | 4.826 ± 1.609 $^{b}$ | 15.493 ± 2.473 $^{a}$ |
| 2-Decenal, (E)- | $C_{10}H_{18}O$ | 36.731 | 1261.262 | - | - | - | 75.964 ± 14.733 | 49.299 ± 4.179 | 52.391 ± 5.574 | 137.881 ± 3.164 |
| 4a(2H)-Naphthalenol, octahydro-, trans- | $C_{10}H_{18}O$ | 36.890 | 1263.606 | - | - | - | 5.811 ± 1.33 | 3.288 ± 0.15 | 3.335 ± 0.32 | 11.836 ± 0.673 |
| 1,7-Octadien-3-ol, 2,6-dimethyl- | $C_{10}H_{18}O$ | 37.308 | 1269.767 | - | - | - | 3.54 ± 0.91 | 3.555 ± 0.255 | 4.082 ± 0.586 | 10.067 ± 1.65 |
| 3-Decanynoic acid | $C_{10}H_{16}O_2$ | 37.784 | 1276.784 | - | - | - | 7.78 ± 2.156 $^{b}$ | 4.963 ± 0.502 $^{b}$ | 5.725 ± 0.752 $^{b}$ | 11.842 ± 1.304 $^{a}$ |
| Nonanoic acid | $C_9H_{18}O_2$ | 38.175 | 1282.547 | - | - | - | 7.812 ± 1.633 $^{a}$ | 4.033 ± 2.205 $^{b}$ | 1.805 ± 0.293 $^{b}$ | 9.862 ± 0.186 $^{a}$ |
| 5-Undecen-4-one | $C_{11}H_{20}O$ | 39.661 | 1304.699 | - | - | - | 6.272 ± 1.741 $^{b}$ | 6.119 ± 1.049 $^{b}$ | 7.123 ± 1.246 $^{b}$ | 18.581 ± 1.654 $^{a}$ |
| 2,4-Decadienal, (E, E)- | $C_{10}H_{16}O$ | 40.311 | 1314.813 | - | - | - | 144.685 ± 33.1 | 121.984 ± 9.562 | 154.824 ± 17.412 | 190.167 ± 6.62 |
| Bicyclo[2.2.2]octane, 1-methoxy-4-methyl- | $C_{10}H_{18}O$ | 42.082 | 1342.368 | - | - | - | 146.796 ± 33.312 | 72.405 ± 6.601 | 83.374 ± 12.952 | 264.051 ± 9.302 |
| 1,13-Tetradecadien-3-one | $C_{14}H_{24}O$ | 42.485 | 1348.639 | - | - | - | 5.17 ± 1.512 $^{b}$ | 2.806 ± 0.269 $^{c}$ | 3.244 ± 0.343 $^{bc}$ | 7.447 ± 0.833 $^{a}$ |

**Table 1.** *Cont.*

| Compound | Empirical Formula | RT | RI | Initial (μg) | CA (μg) | TBHQ (μg) | BHA (μg) | BHT (μg) | α-T (μg) | Control (μg) |
|---|---|---|---|---|---|---|---|---|---|---|
| Cyclohexanol, 1-butyl- | $C_{10}H_{20}O$ | 42.899 | 1355.080 | - | - | - | 70.784 ± 18.432 [b] | 34.188 ± 7.701 [c] | 38.015 ± 9.553 [c] | 109.367 ± 8.796 [a] |
| 2-Undecenal | $C_{11}H_{20}O$ | 43.439 | 1363.482 | - | - | - | 94.604 ± 25.497 | 44.549 ± 4.237 | 51.429 ± 5.972 | 166.774 ± 6.339 |
| trans-4,5-Epoxy-(E)-2-decenal | $C_{10}H_{16}O_2$ | 44.397 | 1378.388 | - | - | - | 53.754 ± 21.541 | 18.665 ± 9.526 | 27.748 ± 4.543 | 76.397 ± 4.575 |
| 4-Hydroxy-3-pentyl-cyclohexanone | $C_{11}H_{20}O_2$ | 44.754 | 1383.943 | - | - | - | 4.254 ± 1.494 | 6.357 ± 6.604 | 1.961 ± 0.447 | 6.011 ± 0.517 |
| total | | | | 0.555 ± 0.093 [f] | 3.511 ± 0.948 [e] | 5.631 ± 0.506 [d] | 1383.556 ± 341.879 [b] | 1085.796 ± 136.13 [c] | 1054.519 ± 113.573 [c] | 2259.67 ± 134.35 [a] |

Notes: RT, retention time; RI, Retention indices. Values are expressed as mean ± standard deviation of three replications; '-' means not detected. Compound identification was based on the NIST 14 mass spectral database and RI values, and several compounds were identified using authentic standard compounds. [a–f] means within a row with different superscripts differ significantly ($p < 0.05$).

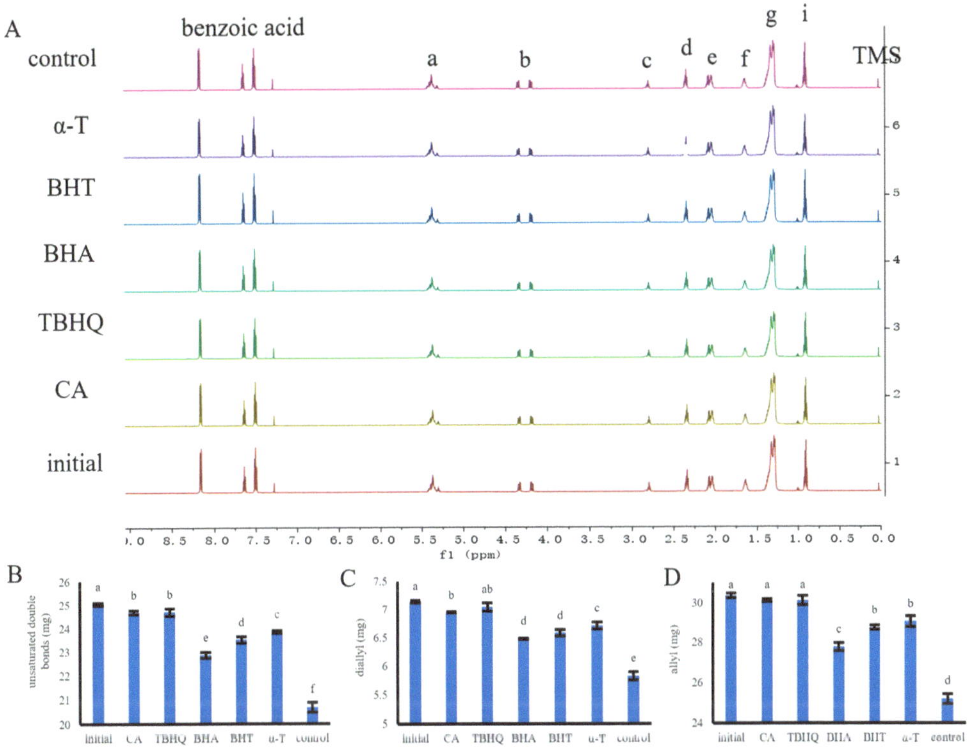


**Figure 5.** [1]H-NMR analyses of Aoil at the beginning and storage at 60 °C for 5 h. (**A**), [1]H-NMR spectrum of Aoil enriched with different antioxidants. a: unsaturated double bond; b: glyceryl; c: diallyl; d: alpha carbonyl peri-methyl; e: allyl; f: β-carbonyl vicinal methyl; g: saturated methylene; i: terminal methyl group (saturated acid, oleic acid, linoleic acid acyl). Quantification of unsaturated double bonds (**B**), diallyl (**C**), and allyl (**D**). a–f means with different superscripts differ significantly ($p < 0.05$).

**Table 2.** Chemical shift assignments of [1]H-NMR signals of oxidation products of Aoil.

| Peak Number | Chemical Shift (ppm) | Functional Group Structure | Functional Group Name |
|---|---|---|---|
| a | 5.55–5.20 | **—CH=CH—** | unsaturated double bond |
| b | 4.50–4.00 | $HOCH_2CHOHCHOH$— | glyceryl |
| c | 3.00–2.60 | **$CH_2$=CH—$CH_2$—CH=$CH_2$** | diallyl |
| d | 2.50–2.20 | —C=O—$CH_2$ | alpha carbonyl peri-methyl |
| e | 2.20–1.90 | **$H_2C$=CH—$CH_2$—** | allyl |
| f | 1.80–1.55 | —C=O—$CH_2$—$CH_3$ | β-carbonyl vicinal methyl |
| g | 1.50–1.15 | —$CH_2$— | saturated methylene |
| i | 1.10–0.75 | —$CH_3$ | terminal methyl group (saturated acid, oleic acid, linoleic acid acyl) |

Notes: The bold font indicates that the functional group contains unsaturated double bonds.

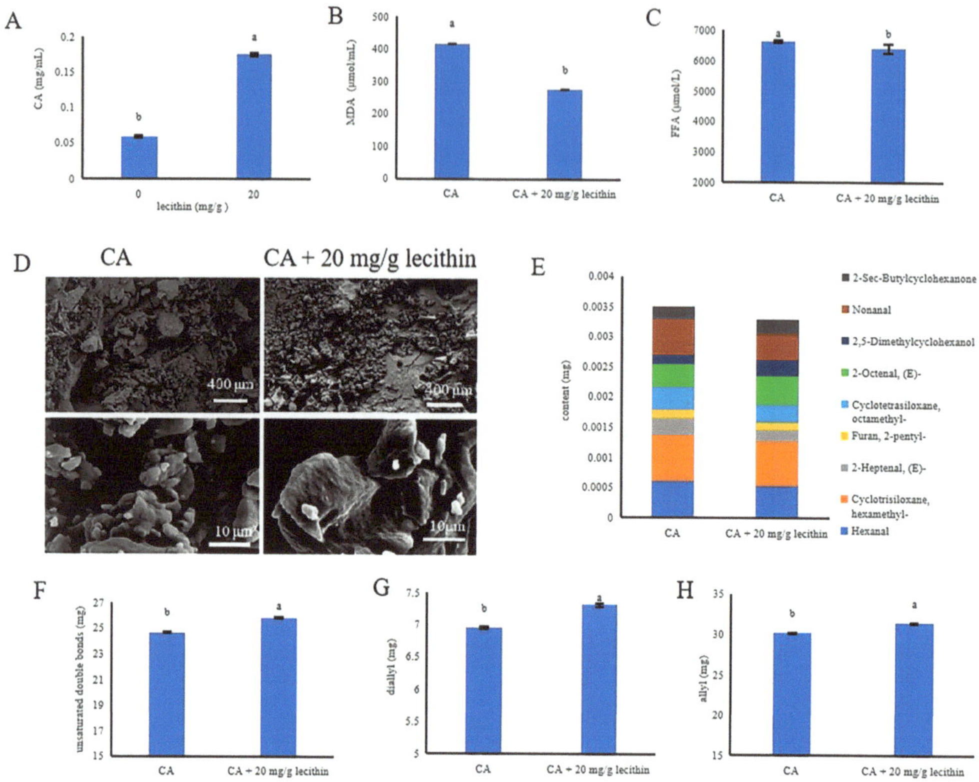

**Figure 6.** Improvement of the oxidative stability of Aoil by enrichment with lecithin-CA. Determination of the solubility of CA in Aoil (**A**). Determination of MDA (**B**) and FFA (**C**). Microstructure of Aoil enriched with CA and lecithin-CA (**D**). Volatile lipid oxidation product analyses of Aoil enriched with CA and lecithin-CA (**E**). Quantification of unsaturated double bonds (**F**), diallyl (**G**), and allyl (**H**). a,b means with different superscripts differ significantly ($p < 0.05$).

## 4. Discussion

Herbal extracts have a strong ability to give hydrogen, electrons, and free electrons. They also have antioxidant action [37]. Several studies have shown that rosemary is capable of antioxidant activity, which can effectively scavenge free radicals and reduce metal ions [20]. The RCE, RA, and CA antioxidant capacities were assessed utilizing the DPPH, ABTS, FRAP, and PTIO assays. The assays for DPPH, ABTS, FRAP, and PTIO revealed discrepancies among obtained results. The principle and detection conditions for each antioxidant detection method were different. Therefore, the results for various antioxidant activities were inconsistent, which has been reported in many similar studies [38–40]. In this study, to compare the antioxidant activity of RCE, RA, and CA, the detection results of four different methods were combined for a systematic evaluation. The results are shown in Figure 3. The order of the APC index for the comprehensive evaluation of antioxidant activity was RA > RCE > CA. However, the order of antioxidant activity in Aoil was CA > RCE > RA. The results of measuring the concentrations of CA and RA in oil showed that the solubility of CA in Aoil is greater than that of RA (Figure S5A). This may lead to a better oxidation stability of Aoil enriched with CA than oils enriched with RA. Therefore, CA had the strongest protective effect of the Aoil. CA was more effective in delayed Aoil oxidation. Lecithin was used to make RA more soluble in Aoil (Figure S5B).

In the future, whether lecithin can enhance the protective effect of RA on lipids remains to be further verified.

Fatty acids in the oil react with oxygen during the oxidation process to create primary oxidation products [41]. The secondary oxidation products giving the oil its foul odour is created when the main oxidation products are broken down to generate smaller molecular components such as aldehydes, ketones, and acids. MDA represent the decomposition products, and the greater the content of MDA, the more severe the oxidation of oil. FFA value is commonly used to monitor an oil quality. FFA is a product of oil hydrolysis [42]. During the heating of the oil, the di- and monoacylglycerols, glycerols, and FFA are produced because of the degradation of the ester linkage of triacylglycerols. The greater content of FFA means more hydrolysis of the oil. During the oxidation process, the viscosity of oil gradually increases. This is because the fatty acids in the oil undergo oxidation reactions, forming various oxidation products including carbonyl compounds with double bonds, alcohols, ketones, carboxylic acids, and others [43]. These products interact with the original molecules in the oil, such as fatty acids and glycerol, forming longer molecular chains, which increases the mutual attractive force between oil molecules and thus leads to an increase in oil viscosity [44]. Furthermore, the oxidation products in the oil can also form polymers and colloids, further increasing the viscosity of the oil. Additionally, these oxidation products may also react with metal ions in the oil, forming metal chelates, which can also affect the oil viscosity [33]. Maskan pointed out that when heated, oil undergoes a polymerization process as well as oxidation, hydrolysis, and isomerization [45]. Kalogianni pointed out that the changes in the viscosity of an oil during heating can be expected to depend on products generated during frying and differing in molecular size from triacylglycerols (i.e., oxidation, polymerization, hydrolysis, and fission products) [46]. Shin showed that when the oil is heated at high temperatures and is oxidized to form a polymer with a high molecular weight, viscosity increases as the amount of the polymer increases. Therefore, the higher the viscosity of frying oil, the more oxidation can be generally expected [47]. Overall, the principle of the increase in viscosity during oil oxidation is due to the interaction between oxidation products and original molecules in the oil, forming longer molecular chains, polymers, and colloids, which increase the mutual attractive force between oil molecules. Unsaturated esters are less viscous, and they quickly oxidize, whereas saturated esters have a high cloud point and high viscosity [48]. The preponderance of saturated fatty acids restricts the flow of oil. Therefore, the level of oxidation rises and the viscosity of the oil increases. In addition, it is well known that oils deteriorate due to the effects of heating and oxidation [49]. This deterioration is accompanied by odours that worsen the cooking environment. Edible oil degradation frequently produces disagreeable flavours [50]. By monitoring the change of volatile lipid oxidation products and proton peaks at the reactive sites of lipid molecules during the oxidation process, GC-MS and $^{1}$H-NMR can assess the oxidation state of lipids, because the oxidation of unsaturated fatty acids is accompanied by volatile lipid oxidation products and a series of changes in hydrogen atoms in different chemical environments [51].

Foods frequently contain lecithin, which has many beneficial functions such as emulsification, crystallization inhibition, non-stick releasing agents, wetting agents, and anti-spattering agents [52]. Lecithin may enhance the transport of polyphenols, which leads to increased antioxidant activity [19]. Suárez showed that the addition of lecithin improved the dispersion and stability of phenolic extracts in olive oil, thus enhancing the oxidation stability of concentrated oil [53]. Ramadan found that soybean lecithin increased the solubility of quercetin and could improve the antioxidant effect of sunflower seed oil [54]. Nikoo also pointed out that the solubility and antioxidant activity of epigallocatechin gallate in lipids can be enhanced through the carrier system or esterification with aliphatic acyl derivatives [20]. In the study, lecithin was added to the Aoil, which lead to the increase of the solubility of CA in the oil, and thus improves the protective effect of CA on the Aoil, and Aoil was supplemented with the highest amount of lecithin (20 mg/g oil) that could be dissolved in the oil [29]. In the future, the stability of Aoil will be tested using various concentrations of lecithin, applied in varying proportions. Lecithin is sometimes used as an antioxidant in foods. Several mechanisms of how lecithin could influence lipid oxidation have been proposed. In general, lecithin could bind prooxidative metals, produce antioxidative compounds through Maillard reactions during lipid oxidation, alter the location of other antioxidants [55], and regenerate primary antioxidants such as tocopherols [56]. However, lecithin could also serve as oxidation substrates themselves. Owing to their high degree of unsaturation, negative charges that attract prooxidant metals, and large surface area when they exist as dispersions, they can be an important substrate for oxidation in foods containing considerable amounts of biological membranes, such as meats [57]. In addition, there were also times when lecithin showed no antioxidant activity or even acted as prooxidants [58]. One possible prooxidant mechanism of phospholipids in bulk oil could be their formation of association colloids, such as reverse micelles, which can increase metal–lipid interactions [59]. Therefore, the oxidation effect of lecithin on oil needs to be further verified.

## 5. Conclusions

Aoil is rich in UFAs, and the sum of unsaturated fatty acids accounted for more than 90% of the total fatty acids. RCE, RA, and CA can significantly inhibit the oxidation of Aoil. CA has better oxidative stability for Aoil among RCE, RA, and CA. In addition, the solubility of CA in Aoil is successfully enhanced by lecithin enrichment via a solvent-free route. Enrichment with lecithin-CA can improve the oxidative stability of Aoil. CA is an excellent antioxidant that can significantly inhibit the thermal oxidation of Aoil and has the potential to be a substitute for TBHQ, BHA, and BHT in foods. Rosemary extract and lecithin-CA were first added to Aoil. This study provides a new natural antioxidant for the preservation of Aoil. It provides a theoretical basis for the application of rosemary extract in Aoil.

**Supplementary Materials:** The following supporting information can be downloaded at: https://www.mdpi.com/article/10.3390/antiox12040889/s1, Figure S1: Standards curve for RA(A) and CA (B). Figure S2: Antioxidant activities of different concentrations of RCE, RA and CA. Figure S3: SPME–GC–MS spectrum of 1 g Aoil at the beginning and storage at 60 °C for 30 d. Figure S4: Improvement of the oxidative stability of Aoil by enrichment with lecithin-CA. Figure S5: Determination of the solubility of CA and RA in Aoil. Table S1: 37 fatty acid standards. Table S2: Elution procedure. Table S3: Fatty acid composition of Aoil.

**Author Contributions:** Y.Q., conceptualization, methodology, investigation, formal analysis, writing—original draft; Y.H., methodology, writing—review and editing; Y.D., methodology, supervision; W.Z., methodology, supervision; F.X., supervision, resources; H.B., supervision, resources; Z.D.S., supervision; H.L., conceptualization, writing—review and editing, supervision; L.S., conceptualization, writing—review and editing, supervision. All authors have read and agreed to the published version of the manuscript.

**Funding:** This research was funded and supported by the International Partnership Program of the Chinese Academy of Science (Grant No. 063GJHZ2022038GC) and the Strategic Priority Research Program of the Chinese Academy of Sciences (Grant No. XDA23080603).

**Institutional Review Board Statement:** Not applicable.

**Informed Consent Statement:** Not applicable.

**Data Availability Statement:** Data is contained within the article and supplementary material.

**Acknowledgments:** We thank Yan Zhu, from the Key Laboratory of Plant Resources, Institute of Botany, the Chinese Academy of Sciences, for her excellent technical assistance of mass spectrum analysis. We thank Hang Su, from the Key Laboratory of Plant Resources, Institute of Botany, the Chinese Academy of Sciences, for her excellent technical assistance of HPLC analysis. We thank Qian Li for his excellent English grammar instruction.

**Conflicts of Interest:** The authors declare no conflict of interest.

## References


1. Ma, Q.; Sun, T.; Li, S.; Wen, J.; Zhu, L.; Yin, T.; Yan, K.; Xu, X.; Li, S.; Mao, J.; et al. The *Acer truncatum* genome provides insights into nervonic acid biosynthesis. *Plant J.* **2020**, *104*, 662–678. [CrossRef]
2. Chen, D.J.; Yan, L.H.; Li, Q.; Zhang, C.J.; Si, C.L.; Li, Z.Y.; Song, Y.J.; Zhou, H.; Zhang, T.C.; Luo, X.G. Bioconversion of conjugated linoleic acid by *Lactobacillus plantarum* CGMCC8198 supplemented with *Acer truncatum* bunge seeds oil. *Food Sci. Biotechnol.* **2017**, *26*, 1595–1611. [CrossRef]
3. Ma, Q.; Wang, Y.; Zhu, L.; Bi, C.; Li, S.; Li, S.; Wen, J.; Yan, K.; Li, Q. Characterization of the complete chloroplast genome of *Acer truncatum* Bunge (Sapindales: Aceraceae): A new woody oil tree species producing nervonic acid. *BioMed Res. Int.* **2019**, *2019*, 7417239. [CrossRef] [PubMed]
4. Xing, W.; Jin, F.; Shu, W.; Sun, R.C. A new resource of nervonic acid from purpleblow maple (*Acer truncatum*) seed oil. *Forest Prod. J.* **2006**, *56*, 147–150. [CrossRef]
5. Chivandi, E.; Davidson, B.C.; Erlwanger, K.H. A comparison of the lipid and fatty acid profiles from the kernels of the fruit (nuts) of *Ximenia caffra* and *Ricinodendron rautanenii* from zimbabwe. *Ind. Crop. Prod.* **2008**, *27*, 29–32. [CrossRef]
6. Li, Q.; Chen, J.; Yu, X.; Gao, J.M. A mini review of nervonic acid: Source, production, and biological functions. *Food Chem.* **2019**, *301*, 125286. [CrossRef] [PubMed]
7. Yang, R.; Zhang, L.; Li, P.; Yu, L.; Mao, J.; Wang, X.; Zhang, Q. A review of chemical composition and nutritional properties of minor vegetable oils in China. *Trends Food Sci. Technol.* **2018**, *74*, 26–32. [CrossRef]
8. Karami, H.; Rasekh, M.; Mirzaee-Ghaleh, E. Qualitative analysis of edible oil oxidation using an olfactory machine. *J. Food Meas. Charact.* **2020**, *14*, 2600–2610. [CrossRef]
9. Corsinovi, L.; Biasi, F.; Poli, G.; Leonarduzzi, G.; Isaia, G. Dietary lipids and their oxidized products in Alzheimer's disease. *Mol. Nutri. Food Res.* **2011**, *55*, S161–S172. [CrossRef]
10. Szydłowska-Czerniak, A.; Rabiej, D.; Kyselka, J.; Dragoun, M.; Filip, V. Antioxidative effect of phenolic acids octyl esters on rapeseed oil stability. *LWT-Food Sci. Technol.* **2018**, *96*, 193–198. [CrossRef]
11. Nieto, G.; Ros, G.; Castillo, J. Antioxidant and antimicrobial properties of rosemary (*Rosmarinus officinalis*, L.): A review. *Medicines* **2018**, *5*, 98. [CrossRef] [PubMed]
12. Tai, J.; Cheung, S.; Wu, M.; Hasman, D. Antiproliferation effect of rosemary (*Rosmarinus officinalis*) on human ovarian cancer cells in vitro. *Phytomedicine* **2012**, *19*, 436–443. [CrossRef] [PubMed]
13. Park, J.; Rho, S.J.; Kim, Y.R. Enhancing antioxidant and antimicrobial activity of carnosic acid in rosemary (*Rosmarinus officinalis* L.) extract by complexation with cyclic glucans. *Food Chem.* **2019**, *299*, 125119. [CrossRef] [PubMed]
14. Huber, G.M.; Rupasinghe, H.V.; Shahidi, F. Inhibition of oxidation of omega-3 polyunsaturated fatty acids and fish oil by quercetin glycosides. *Food Chem.* **2009**, *117*, 290–295. [CrossRef]
15. Zhou, L.; Elias, R.J. Antioxidant and pro-oxidant activity of (-)-epigallocatechin-3-gallate in food emulsions: Influence of pH and phenolic concentration. *Food Chem.* **2013**, *138*, 1503–1509. [CrossRef]
16. Luis, J.; Johnson, C. Seasonal variations of rosmarinic and carnosic acids in rosemary extracts. Analysis of their in vitro antiradical activity. *Span. J. Agric. Res.* **2005**, *3*, 106–112. [CrossRef]
17. Erkan, N.; Ayranci, G.; Ayranci, E. A kinetic study of oxidation development in sunflower oil under microwave heating: Effect of natural antioxidants. *Food Res. Int.* **2009**, *42*, 1171–1177. [CrossRef]
18. Ercan, P.; El, S.N. Bioaccessibility and inhibitory effects on digestive enzymes of carnosic acid in sage and rosemary. *Int. J. Biol. Macromol.* **2018**, *115*, 933–939. [CrossRef]
19. Ghanbarzadeh, B.; Babazadeh, A.; Hamishehkar, H. Nano-phytosome as a potential food-grade delivery system. *Food Biosci.* **2016**, *15*, 126–135. [CrossRef]
20. Nikoo, M.; Regenstein, J.M.; Ahmadi Gavlighi, H. Antioxidant and antimicrobial activities of (-)-epigallocatechin-3-gallate (EGCG) and its potential to preserve the quality and safety of foods. *Compr. Rev. Food Sci. Food Saf.* **2018**, *17*, 732–753. [CrossRef]
21. Devi, A.; Khatkar, B. Effects of fatty acids composition and microstructure properties of fats and oils on textural properties of dough and cookie quality. *J. Food Sci. Technol.* **2018**, *55*, 321–330. [CrossRef] [PubMed]
22. Shi, R.; Guo, Y.; Vriesekoop, F.; Yuan, Q.; Zhao, S.; Liang, H. Improving oxidative stability of peanut oil under microwave treatment and deep fat frying by stearic acid–surfacant–tea polyphenols complex. *Eur. J. Lipid Sci. Technol.* **2015**, *117*, 1008–1015. [CrossRef]
23. Maleki, M.; Ariaii, P.; Fallah, H. Effects of celery extracts on the oxidative stability of canola oil under thermal condition. *J. Food Process. Pres.* **2016**, *40*, 531–540. [CrossRef]
24. Li, Y.; Pan, L.; Zhang, Y. Effects of combination of natural antioxidants on oxidative stability and sensory characteristics of flaxseed oil. *China Oils Fats* **2018**, *43*, 118–123. [CrossRef]
25. Chen, P.; Sun, D.; Zheng, X.M. Preparation, structure and antioxidant activity of EGCG palmitate. *J. Zhejiang Univ. Sci. Ed.* **2003**, *30*, 422–425. [CrossRef]
26. Li, X. 2-Phenyl-4, 4, 5, 5-tetramethylimidazoline-1-oxyl 3-oxide (PTIO•) radical scavenging: A new and simple antioxidant assay in vitro. *J. Agr. Food Chem.* **2017**, *65*, 6288–6297. [CrossRef]
27. Thaipong, K.; Boonprakob, U.; Crosby, K.; Cisneros-Zevallos, L.; Byrne, D.H. Comparison of ABTS, DPPH, FRAP, and ORAC assays for estimating antioxidant activity from guava fruit extracts. *J. Food Compos. Anal.* **2006**, *19*, 669–675. [CrossRef]

28. López-Froilán, R.; Hernández-Ledesma, B.; Cámara, M.; Pérez-Rodríguez, M.L. Evaluation of the antioxidant potential of mixed fruit-based beverages: A new insight on the folin-ciocalteu method. *Food Anal. Methods* **2018**, *11*, 2897–2906. [CrossRef]
29. del Pilar Garcia-Mendoza, M.; Espinosa-Pardo, F.A.; Savoire, R.; Harscoat-Schiavo, C.; Cansell, M.; Subra-Paternault, P. Improvement of the oxidative stability of camelina oil by enrichment with phospholipid-quercetin formulations. *Food Chem.* **2021**, *341*, 128234. [CrossRef]
30. Wang, R.; Yuan, J.; Zhang, C.; Wang, L.; Liu, Y.; Song, L.; Zhong, W.; Chen, X.; Dong, J. Neuropeptide Y-positive neurons in the dorsomedial hypothalamus are involved in the anorexic effect of Angptl8. *Front. Mol. Neurosci.* **2018**, *11*, 451. [CrossRef]
31. Nishino, Y.; Miyazaki, K.; Kaise, M.; Miyazawa, A. Fine cryo-SEM observation of the microstructure of emulsions frozen via high-pressure freezing. *Microscopy-JPN* **2022**, *71*, 60–65. [CrossRef]
32. Yin, H.; Sathivel, S. Physical properties and oxidation rates of unrefined menhaden oil (*Brevoortia patronus*). *J. Food Sci.* **2010**, *75*, E163–E168. [CrossRef]
33. Van Aardt, M.; Duncan, S.E.; Long, T.E.; O'Keefe, S.F.; Marcy, J.E.; Sims, S.R. Effect of antioxidants on oxidative stability of edible fats and oils: Thermogravimetric analysis. *J. Agric. Food Chem.* **2004**, *52*, 587–591. [CrossRef]
34. Łyczko, J.; Jałoszyński, K.; Surma, M.; Masztalerz, K.; Szumny, A. HS-SPME analysis of true lavender (*Lavandula angustifolia* Mill.) leaves treated by various drying methods. *Molecules* **2019**, *24*, 764. [CrossRef] [PubMed]
35. Hwang, H.S.; Winkler-Moser, J.K.; Liu, S.X. Reliability of $^1$H NMR analysis for assessment of lipid oxidation at frying temperatures. *J. Am. Oil Chem. Soc.* **2017**, *94*, 257–270. [CrossRef]
36. Tyl, C.E.; Brecker, L.; Wagner, K.H. $^1$H NMR spectroscopy as tool to follow changes in the fatty acids of fish oils. *Eur. J. Lipid Sci. Technol.* **2008**, *110*, 141–148. [CrossRef]
37. Delfanian, M.; Esmaeilzadeh Kenari, R.; Sahari, M.A. Influence of extraction techniques on antioxidant properties and bioactive compounds of loquat fruit (*Eriobotrya japonica* Lindl.) skin and pulp extracts. *Food Sci. Nutr.* **2015**, *3*, 179–187. [CrossRef]
38. Du, L.; Sun, G.; Zhang, X.; Liu, Y.; Prinyawiwatkul, W.; Xu, Z.; Shen, Y. Comparisons and correlations of phenolic profiles and anti-oxidant activities of seventeen varieties of pineapple. *Food Sci. Biotechnol.* **2016**, *25*, 445–451. [CrossRef]
39. de Oliveira, I.R.; Teófilo, R.F.; de Oliveira, E.B.; Ramos, A.M.; de Barros, F.A.; Maia, M.D.P.; Stringheta, P.C. Evaluation of potential interfering agents on in vitro methods for the determination of the antioxidant capacity in anthocyanin extracts. *Int. J. Food Sci. Technol.* **2017**, *52*, 511–518. [CrossRef]
40. Al-Laith, A.A.; Alkhuzai, J.; Freije, A. Assessment of antioxidant activities of three wild medicinal plants from Bahrain. *Arab. J. Chem.* **2019**, *12*, 2365–2371. [CrossRef]
41. Zhang, N.; Li, Y.; Wen, S.; Sun, Y.; Chen, J.; Gao, Y.; Sagymbek, A.; Yu, X. Analytical methods for determining the peroxide value of edible oils: A mini-review. *Food Chem.* **2021**, *358*, 129834. [CrossRef] [PubMed]
42. Crosa, M.J.; Skerl, V.; Cadenazzi, M.; Olazábal, L.; Silva, R.; Suburú, G.; Torres, M. Changes produced in oils during vacuum and traditional frying of potato chips. *Food Chem.* **2014**, *146*, 603–607. [CrossRef]
43. Akaza, I.; Aota, N. Colorimetric determination of lipid hydroperoxides in oils and fats with microperoxidase. *Talanta* **1990**, *37*, 925–929. [CrossRef] [PubMed]
44. Kanner, J. Dietary advanced lipid oxidation endproducts are risk factors to human health. *Mol. Nutr. Food Res.* **2007**, *51*, 1094–1101. [CrossRef] [PubMed]
45. Maskan, M. Change in colour and rheological behaviour of sunflower seed oil during frying and after adsorbent treatment of used oil. *Eur. Food Res. Technol.* **2003**, *218*, 20–25. [CrossRef]
46. Kalogianni, E.P.; Karapantsios, T.D.; Miller, R. Effect of repeated frying on the viscosity, density and dynamic interfacial tension of palm and olive oil. *J. Food Eng.* **2011**, *105*, 169–179. [CrossRef]
47. Shin, J.A.; Kim, Y.M.; Duan, B.; Lee, K.T. Evaluation of oxidation parameters of caffeic acid phenethyl ester comparing with caffeic acid, ascorbyl palmitate, and a-Tocopherol on the thermal oxidation model. *J. Oleo Sci.* **2020**, *69*, 93–103. [CrossRef]
48. Liu, J.; Yang, F.; Xia, J.; Wu, F.; Pu, C. Mechanism of ultrasonic physical–chemical viscosity reduction for different heavy oils. *ACS Omega* **2021**, *6*, 2276–2283. [CrossRef]
49. Baig, A.; Zubair, M.; Sumrra, S.H.; Nazar, M.F.; Zafar, M.N.; Jabeen, K.; Hassan, M.B.; Rashid, U. Heating effect on quality characteristics of mixed canola cooking oils. *BMC Chem.* **2022**, *16*, 1–11. [CrossRef]
50. Fullana, A.; Carbonell-Barrachina, A.A.; Sidhu, S. Comparison of volatile aldehydes present in the cooking fumes of extra virgin olive, olive, and canola oils. *J. Agric. Food Chem.* **2004**, *52*, 5207–5214. [CrossRef]
51. Castejón, D.; Herrera, A.; Heras, Á.; Cambero, I.; Mateos-Aparicio, I. Oil quality control of culinary oils subjected to deep-fat frying based on NMR and EPR spectroscopy. *Food Anal. Methods* **2017**, *10*, 2467–2480. [CrossRef]
52. Zhao, F.; Li, R.; Liu, Y.; Chen, H. Perspectives on lecithin from egg yolk: Extraction, physicochemical properties, modification, and applications. *Front. Nutr.* **2022**, *9*, 1082671. [CrossRef] [PubMed]
53. Suárez, M.; Romero, M.P.; Ramo, T.; Motilva, M.J. Stability of a phenol-enriched olive oil during storage. *Eur. J. Lipid Sci. Technol.* **2011**, *113*, 894–903. [CrossRef]
54. Ramadan, M.F. Antioxidant characteristics of phenolipids (quercetin-enriched lecithin) in lipid matrices. *Ind. Crop. Prod.* **2012**, *36*, 363–369. [CrossRef]
55. Hidalgo, F.J.; León, M.M.; Zamora, R. Antioxidative activity of amino phospholipids and phospholipid/amino acid mixtures in edible oils as determined by the Rancimat method. *J. Agric. Food Chem.* **2006**, *54*, 5461–5467. [CrossRef]

56. Cui, L.; McClements, D.J.; Decker, E.A. Impact of phosphatidylethanolamine on the antioxidant activity of α-tocopherol and trolox in bulk oil. *J. Agric. Food Chem.* **2015**, *63*, 3288–3294. [CrossRef]
57. Min, B.; Ahn, D. Mechanism of lipid peroxidation in meat and meat products-a review. *Food Sci. Biotechnol.* **2005**, *14*, 152–163.
58. Cardenia, V.; Waraho, T.; Rodriguez-Estrada, M.T.; Julian McClements, D.; Decker, E.A. Antioxidant and prooxidant activity behavior of phospholipids in stripped soybean oil-in-water emulsions. *J. Am. Oil Chem. Soc.* **2011**, *88*, 1409–1416. [CrossRef]
59. Kittipongpittaya, K.; Panya, A.; Cui, L.; McClements, D.J.; Decker, E.A. Association colloids formed by multiple surface active minor components and their effect on lipid oxidation in bulk oil. *J. Am. Oil Chem. Soc.* **2014**, *91*, 1955–1965. [CrossRef]

*antioxidants*

MDPI

*Review*

# Partitioning of Antioxidants in Edible Oil–Water Binary Systems and in Oil-in-Water Emulsions

**Sonia Losada-Barreiro [1,2], Fátima Paiva-Martins [2] and Carlos Bravo-Díaz [1,*]**

1 Departamento Química-Física, Facultad de Química, Universidade de Vigo, 36310 Vigo, Spain
2 REQUIMTE-LAQV, Departamento de Química e Bioquímica, Faculdade de Ciências, Universidade do Porto, 4169-007 Porto, Portugal
* Correspondence: cbravo@uvigo.gal

**Abstract:** In recent years, partitioning of antioxidants in oil–water two-phase systems has received great interest because of their potential in the downstream processing of biomolecules, their benefits in health, and because partition constant values between water and model organic solvents are closely related to important biological and pharmaceutical properties such as bioavailability, passive transport, membrane permeability, and metabolism. Partitioning is also of general interest in the oil industry. Edible oils such as olive oil contain a variety of bioactive components that, depending on their partition constants, end up in an aqueous phase when extracted from olive fruits. Frequently, waste waters are subsequently discarded, but their recovery would allow for obtaining extracts with antioxidant and/or biological activities, adding commercial value to the wastes and, at the same time, would allow for minimizing environmental risks. Thus, given the importance of partitioning antioxidants, in this manuscript, we review the background theory necessary to derive the relevant equations necessary to describe, quantitatively, the partitioning of antioxidants (and, in general, other drugs) and the common methods for determining their partition constants in both binary ($P_W^{OIL}$) and multiphasic systems composed with edible oils. We also include some discussion on the usefulness (or not) of extrapolating the widely employed octanol–water partition constant ($P_W^{OCT}$) values to predict $P_W^{OIL}$ values as well as on the effects of acidity and temperature on their distributions. Finally, there is a brief section discussing the importance of partitioning in lipidic oil-in-water emulsions, where two partition constants, that between the oil-interfacial, $P_O^I$, and that between aqueous-interfacial, $P_w^I$, regions, which are needed to describe the partitioning of antioxidants, and whose values cannot be predicted from the $P_W^{OIL}$ or the $P_W^{OCT}$ ones.

**Keywords:** partition constant; antioxidant; edible oils; emulsion; intermolecular interactions

## 1. Introduction

The increasing consumer demand for natural food additives and ingredients has stimulated the search for different sources of antioxidants (AOs) [1–5]. An enormous number of plants and biomass wastes, including those from natural food sources, have been screened as potential antioxidants [6]. Most of these studies have focused on the identification of bioactive components in order to reach a (more) sustainable world by searching for new bio-renewable sources for valuable components [7] and adding value to waste [8,9]. These bio-renewable sources are of interest for food, cosmetic, pharmaceutical, and medicinal industries, and have the additional advantage of being largely available, widely distributed, and inexpensive, particularly when their uncontrolled disposal would create serious environmental problems, especially when they are geographically concentrated [1,10–12].

In recent years, there has been increased concern regarding the reduction of food loss and waste, and consumers have been demanding more sustainable food systems. Alternative pathways for food waste management include the valorization of by-products as a source of phenolic compounds for functional food formulations [13]. This is the case for phenolic antioxidants, an important group of compounds with enormous potential as a result of their beneficial properties in health and in the prevention of the oxidative deterioration of edible oils [14]. Phenolic extracts from natural sources also exhibit a wide range of physiological properties, including anti-thrombotic, anti-atherogenic, anti-allergenic, anti-inflammatory, anti-microbial, vasodilatory, and cardioprotective effects [4,15]. They are present in a number of agro-industrial by-products (husks, peels, seeds, etc.), fruit peels, olive mill waste, olive leaves, and residue from grapes and winemaking are all considered important natural sources of antioxidants [4]. In particular, olive mill wastewaters have been identified as a promising source of antioxidants to retard lipid oxidation in fish oil-enriched food products, and the use of active packaging films containing antioxidants is being explored [16–19].

Partitioning is, thus, an important phenomenon not only in medicinal chemistry and drug design, but also in the recovery of antioxidants from wastewaters, biomass, etc. Common procedures for the extraction of antioxidants from wastewaters require, among others, the use of aqueous and organic solvents [20], where antioxidants are partitioned according to their hydrophobicity [21,22]. Determining the partition constant values of antioxidants (and, in general, of other compounds) in binary oil–water mixtures and in multiphasic systems such as emulsions is, thus, of great interest to the industry for, among others, the following reasons:

(1) The partition constant value of a molecule (e.g., antioxidant (AO)) is frequently employed as a "measure" of its relative affinity to be incorporated in a given region (Figure 1), and its "tendency" to be dissolved in such a region depends on the value and sign of the Gibbs free energy of transfer from one region to another one ($\Delta G_{transfer} < 0$) [23–27]. In binary oil–water mixtures, the distribution of an antioxidant can be described by one partition constant $P_W{}^O$, while in emulsified systems, two partition constants are needed, $P_W{}^I$ (that between the aqueous and interfacial regions) and $P_O{}^I$ (that between the oil and interfacial regions), Figure 1.

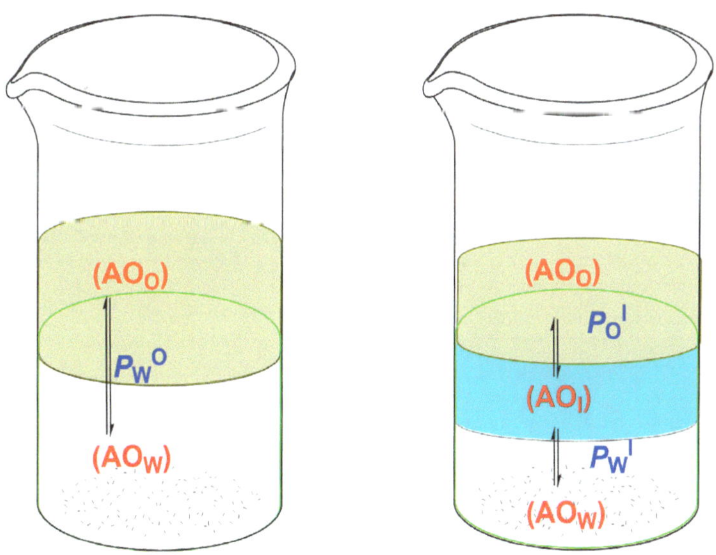


**Figure 1.** Basic representation of the distribution of an antioxidant (AO) in a binary oil–water system (**left**) and in an emulsified system (**right**) between the oil (O), interfacial (I), and aqueous (W) regions. The solvation spheres of the antioxidant in the different regions are not displayed for the sake of clarity.

Once extracted from the raw material, antioxidants (or their extracts) are added to different delivery systems that range from simple aqueous solutions to complex emulsified systems, and where they partition between the different phases/regions, thus affecting their biological or chemical activity [5,28–32].

(2) Research work carried out during recent years has shown that olive oil–water partition constants are better correlated with drug efficiency and their partitioning into human tissues [33–36]. For instance, Poulin et al. [37] showed that the partition constants determined in olive oil–water systems ($P_W{}^{OIL}$) are clearly superior to those determined in octanol/water systems ($P_W{}^{OCT}$) for predicting adipose tissue plasma partition coefficients, because olive oil better represents the lipid composition of the adipose tissue content, which is primarily composed of triglycerides. In addition, antioxidants need to pass through lipid membranes to reach their targets, and these lipid membranes can also cause them to partition [38].

(3) Partition constant values depend on the solvation properties of a particular region, usually computed in terms of hydration/solvation free energies [39,40]. Thus, one can obtain insights on the molecular environments of antioxidants by determining their partition constants in edible oil–water systems and, by extrapolation, to determine their bioconcentrations in living systems such as fish [34,35].

(4) In multiphasic systems such as emulsions, antioxidants are added to minimize the oxidation of lipids. The partition constant values between the various regions define their distribution within the system, and such distribution is crucial for assessing their efficiency in inhibiting lipid peroxidation as mass transfers affect chemical reactivity because the "local" or "effective" concentrations of antioxidants, which depend on the partitioning of the reactant, may be orders of magnitude different from one region to the other and much higher or lower than the stoichiometric concentrations [41–45].

(5) Partitioning of antioxidants is also crucial for developing new encapsulation systems employed to protect them and to deliver them in a controlled manner to target environments. These technologies have attracted great interest in preparing functional foods, nutraceuticals, and pharmaceutical formulations because of the important health benefits that antioxidants have in the prevention of human diseases [4,30,46,47].

(6) A literature search carried out by employing Scifinder (CAS, accessed on 3 March 2023) revealed that in the 2012–2023 period, the number of published papers containing the searching criteria "partitioning " were higher than 5000; however, it decreased significantly to only 236 when the criteria "partitioning + antioxidants" were used. The number of published papers in the period was even lower (98) when searching for publications containing emulsion (partitioning + antioxidant + emulsion) or the keywords "partitioning + antioxidant + oil-in-water" (60 papers). These numbers reveal that much of the published works on partitioning employ molecules other than antioxidants and thus reinforce the need for a review paper updating and enhancing the published information on antioxidant partitioning.

We noticed, when searching in the literature, that the terms "partition constant" and "partition coefficient" are often used, sometimes leading to confusion regarding the proper use of these terms. IUPAC recommendations [48] suggest that $P_W{}^O$ should be better defined as "partition ratio"; however, in many scientific discussions and in specialized literature, the term partition constant or partition coefficient is commonly employed [21,23,49]. Thus, for convenience and for the sake of clarity, hereafter we will employ the term "partition constant" to describe the equilibrium partitioning of neutral antioxidants between two (essentially) immiscible phases, namely oil (O) and water (W), so that the ratio between the effective concentrations in each of the two phases, i.e., the $P_W{}^O$ value, is a concentration-independent constant at given experimental conditions. Examples of these systems include edible oil–water and the widely employed *n*-octanol–water binary systems, where partitioning molecules may be present both as neutral and charged species. Some authors employ the term $D_W{}^O$ instead of $P_W{}^O$ to denote the apparent partition constants in the pH range where ionic species coming from ionizable drugs are present [25,50]. For the sake of clarity, consistency, and simplicity, in these cases, we will use the term $P_W{}^O$ (app). Other terms that can also be found in the specialized literature such as "distribution constants", "distribution coefficients", or "distribution ratio" are less common and will not be employed here.

## 2. Physicochemical Basis of Partitioning: Balance of Intermolecular Interactions

Partitioning is intimately related to the concept of hydrophobicity, which is commonly understood to be the tendency of non-polar molecules to reduce its contact with polar molecules, such as water, but to increase that with other non-polar molecules [51]. Daily manifestations include the simple observable macroscopic phenomena of immiscibility between oil and water and constitute the foundations of some modern analytical techniques such as chromatographic separation. A central point in the study of partitioning phenomena are the intermolecular forces: the same set of forces and interactions that partition a molecule between polar and hydrophobic solvent phases, i.e., determines its hydrophobicity, can also be employed to explain the physical states of matter and even biological interactions, including small molecule binding and protein folding.

To fully understand the partitioning phenomena, its significance, and, most importantly, to be able to predict partition constant values in different binary oil–water systems, we first must be aware of the different kinds of interactions that may exist between solutes, solute–solvents, and their relative strengths in both phases. It will be the balance of all these attractive and repulsive forces that, in the end, will dictate the relative affinity of the antioxidant (or other molecules of interest) for the two competing phases. We therefore need to identify the nature, number, and position of functional groups (e.g., -OH, -$CH_2$-, and -COOH) present in the antioxidants that partition and in the solvents to identify the kinds of intermolecular forces (or interactions) that are "formed" or "broken" during the partitioning process. Attractive forces between molecules stem from uneven electron distributions, deserving some discussion considering the regions of the solute and solvent molecules where electron-rich or electron-deficient molecules may be located [52,53].

The process of transfer of an antioxidant (AO; or, in general, any other molecule) from within the oil phase to within the aqueous phase (or vice versa) implies that intermolecular attractions are "formed" and "broken" during the exchange. These interactions are much weaker (1–12 kJ/mol) than covalent bond strengths (50–300 kJ/mol), but they are able to control critical characteristics of matter such as the boiling and melting points, vapor pressure, and viscosity. For the sake of simplicity, hereafter, we will consider that there is not physical restriction to the transfer of antioxidants between phases, and that this partitioning is governed exclusively by the thermal motion of molecules so that partitioning takes place essentially at the time of bulk mixing solutes and solvents.

The resulting interactions between uncharged solutes or solvent molecules arise from the attractive or repulsive electrostatic forces between electron deficient regions in a molecule that are attracted by electron-rich or repelled by electron-deficient counterparts in neighboring molecules. The overall affinity of molecules for each other results from the balance of all these attractions/repulsions. A summary of the main intermolecular forces is given in Table 1 and further details on their origin, nature, and energetics can be found elsewhere [54,55].

**Table 1.** Summary of the major intermolecular forces. For the sake of energetic comparisons, the interionic ion–ion force is also included.

| Type of Force | Energy (kJ/mol) | Interacting Species |
|---|---|---|
| Ionic bond | 300–600 | Ions/Ions |
| Hydrogen bonding | 20–40 | Polar molecules containing N-H, O-H, or F-H, the link is a shared H atom |
| Ion–Dipole | 10–20 | Ion/polar molecule |

**Table 1.** *Cont.*

| Type of Force | Energy (kJ/mol) | Interacting Species |
|---|---|---|
| Dipole–Dipole (Keeson) | 1–5 | polar molecule/polar molecule |
| Dipole–Induced dipole (Debye) | 2–10 | Stationary polar molecules/all types of molecules |
| Induced dipole–Induced dipole (London dispersion forces) | <2 | All types of molecules/all types of molecules |

For descriptive purposes, the intermolecular forces can be divided in two main groups:

(1) Nonspecific interactions, which are those that exist between any kind of molecules, no matter their chemical structures. They are globally known as the van der Walls forces (vdW) and include dipole–dipole interactions (Keeson), dipole–induced dipole interactions (Debye), and induced dipole–induced dipole interactions resulting from uneven electron distributions over time (London dispersive forces) [55].

The dipole–dipole (Keeson) interactions arise from the electrostatic interaction between molecules with permanent dipoles that makes them orient so that the dipoles face each other so as to minimize repulsive forces, that is, electrostatic attractive forces between the positive end of one polar molecule and the negative end of another polar molecule. The strengths of these attractions are proportional to the product of the permanent dipole moments of the two interacting molecules and depend on the orientation of the interacting partners, ranging from 1 to 5 kJ per mole. They are, therefore, much weaker than ionic or covalent bonds and have a significant effect only when the molecules involved are close to each other.

Hydrogen bonding is a special case of dipole–dipole interaction that takes place between the hydrogen atom in polar N-H, O-H, or F-H bonds and electronegative atoms such as O, N, or F. Hydrogen bonding (H-bonding) is a highly directional, noncovalent interaction, present in many organic molecules that is, notably, responsible for supramolecular ordering in biological systems, because proteins and nucleic acids are composed of numerous -NH and -OH groups that can donate hydrogen bonds, and C=O and other groups that can accept them. These forces are responsible, for example, for the microscopic structure of water [55] or for the strength of cellulose fibers.

Dipole–induced dipole interactions (Debye) occur when a polar molecule with a permanent dipole (e.g., molecules containing bonded atoms with different electronegativities (e.g., O-H)) disturbs the arrangement of electrons in nearby non-polar molecules, inducing a dipole.

The London dispersion forces (induced dipole–induced dipole attractive forces) are related to the repellency of electrons of non-polar molecules when they approach each other, resulting in the formation of two induced dipoles leading to intermolecular attractions. At any instant, the electron distribution may be unsymmetrical and hence produce an instantaneous dipole. This can cause an induced transient dipole in the neighboring molecule and can cause the molecules to be attracted. The London dispersion force is, by far, the weakest intermolecular force. In spite of this, it is very important because it is universal and causes, for instance, non-polar substances to be liquids or solids if the temperature is lowered sufficiently [55].

When electron donor–acceptor interactions are negligible, London dispersive forces are the main contributors to the overall attraction of many molecules to their neighbors. If molecules show stronger interactions than those expected from these universal interactions, then it means that other intermolecular forces make an important contribution to the overall balance. It is worth noting that the London interactions are universal because their origin arises from momentary displacements of electrons within the structure of the molecule (in the order of the femtosecond time scale), so that there is a continuous presence of short-lived dipoles in the structure, and these fleeting dipoles are felt by neighboring molecules whose electrons react in a complementary fashion, thus inducing instant dipoles [55].

A particular and important type of dispersion force from van der Waals forces is $\pi$–$\pi$ stacking interactions. As the electron distributions in aromatic systems are relatively easily distorted, they can engage in atypically strong induced dipole–induced dipole interactions called $\pi$–$\pi$ stacking interactions. These interactions are named this because they occur when the planes of aromatic rings are stacked parallel to one another. This parallel stacking can occur in either a sandwich or a displaced stacking arrangement. Noncovalent interactions involving aromatic rings are pivotal to protein–ligand recognition and concomitantly to drug design. Indeed, the vast majority of X-ray crystal structures of protein complexes with small molecules reveal bonding interactions involving aromatic amino acid side chains of the receptor and/or aromatic and heteroaromatic rings of the ligand.

(2) Specific (or polar) interactions that result from molecular structures that enable attractions between permanent electron-poor parts of a molecule (e.g., H—attached to oxygen) and electron-rich sites in another molecule (e.g., non-bonded electrons of atoms such as O and N). These polar interactions are, thus, only possible when molecules bear complementary structural moieties: one moiety acts as an electron donor (H-acceptor) and the other one as an electron acceptor (H-donor). Both electron donor–acceptor (EDA) and hydrogen donor–acceptor (HAD) are terms that are widely used in the literature. Hydrogen acceptors are electronegative atoms (N, O, or F) of a neighboring molecule or ion that contain lone electron pairs. Typical H-donor groups are molecular moieties containing electronegative atoms such as N, O, and F that are covalently bonded to a hydrogen atom because they pull the covalently bonded electron pair closer to its nucleus, but away from the H atom. The H atom, consequently, has a partial positive charge, creating a dipole–dipole attraction between the hydrogen atom bonded to the donor and the lone electron pair of the acceptor [55].

## 3. Energetics of Partitioning: Thermodynamic Equations

The distribution of a solute between two phases is an equilibrium condition that can be described in terms of the energetics (thermodynamics) of the process [49,56]. Several situations can be found depending on the nature of the two phases involved: liquid–liquid, liquid–solid, gas–liquid, and gas–solid. Here, we focus on liquid–liquid biphasic systems because they are the basis for solvent extraction, constituting one of the most important extraction methods commonly employed to separate compounds based on their relative solubilities in the immiscible liquids [22].

To simplify the mathematics as much as possible, several reasonable assumptions were made. We assumed that (i) the antioxidant of interest is chemically stable, partitions between two immiscible bulk phases that are in contact with each other at a given temperature and pressure, and does not react with any component of the solvent mixture. (ii) The oil and water phases are at equilibrium with each other with respect to the amounts of all chemical species present in each phase and that there is no physical barrier that prevents the free movement of molecules between them, as illustrated in Figure 1. If the phases (e.g., oil and water) are not strictly immiscible, it is assumed that each phase is saturated with the molecules of the other one after bulk mixing the phases. (iii) We presumed, as a first approach, that when a small amount of an antioxidant C is added to the binary mixture, (a) the bulk properties of both phases are not significantly disturbed by the introduction of the antioxidant molecules and (b) the solubility limit of the antioxidant in any of the phases is not reached.

Under these conditions, the chemical potentials of the antioxidant in water ($\mu_C{}^W$) and oil ($\mu_C{}^O$) are given by Equations (1) and (2), respectively, where R is the universal gas constant [21,23,49,57]. Equations (1) and (2) show that the chemical potentials depend on the activity of the solute ($a_C = \gamma_C C_C$, where $\gamma_C$ and $C_c$ stand for the activity coefficient and the concentration of the antioxidant in units mol/L, respectively), and when the concentrations of the solute are low enough to make solute intermolecular interactions negligible (i.e., dilute solutions), it is safe to use the concentrations instead of activities (i.e., $\gamma_I \approx 1$) [21,49].

$$\mu_C^W = \mu_C^{0,W} + RT \ln a_C^W \tag{1}$$

$$\mu_C^O = \mu_C^{0,O} + RT \ln a_C^O \tag{2}$$

The Gibbs free energy change accompanying the transfer of the antioxidant from one phase (e.g., water) to the other phase (e.g., oil) in equilibrium with each other, at a constant temperature $T$ and pressure, is given by Equation (3), where $\Delta G$, $\Delta H$, and $\Delta S$ are the Gibbs free energy, enthalpy, and entropy changes, respectively, and $T$ is the absolute temperature.

$$\Delta G^{0,W\to O} = \Delta H^{0,W\to O} - T\Delta S^{0,W\to O} = \mu_C^{0,O} - \mu_C^{0,W} \tag{3}$$

Thus, when the change in Gibbs free energy is negative, the antioxidant is transferred spontaneously from the water to the oil, otherwise it is not, and needs some energy input to be transferred after some time. Assuming that the transfer process is spontaneous, after a short time, some antioxidant molecules are transferred from one phase to the other until the equilibrium is attained, so that the chemical potentials of the antioxidant in the oil and water phases are equal each other. At equilibrium, $\Delta G = 0$ and Equation (4) can be derived, where $(C_o)$ and $(C_W)$ are the effective concentrations of the antioxidant C in the oil and water phases (moles of the antioxidant C per liter of phase volume), respectively. Further details on the assumptions and equations involved can be found elsewhere [21,27,58].

The partition coefficient $P_W{}^O$ is, thus, defined as the ratio of the effective concentrations of the solute in the oil and water phases (moles of antioxidant per unit volume of each phase). In some textbooks, the partition constants are expressed as mole fractions, and one can easily convert mole fractions to molar concentrations, bearing in mind the molar volume of the mixture or solution, $C = x_i$ (mol i/total mol)/$V_M$ (L (total mol)$^{-1}$) assuming that the volumes are additive to a first approximation [27,49].

$$\Delta G^{0,W\to O} = -RT \ln \frac{(C_O)}{(C_W)} = -RT \ln P_w^O \tag{4}$$

To avoid future (potential) misunderstandings, we would like to recall here that the effective concentrations of the antioxidant in each of the phases, e.g., $(C_O)$ and $(C_W)$, are, in general, different from the stoichiometric concentration, $[C_T]$, which is expressed in moles per liter of the total volume solution (oil + water). To avoid misconceptions, and for the sake of clarity, hereafter, stoichiometric concentrations (in moles per liter of total volume of the oil–water mixture) will be indicated by the commonly employed brackets [ ] meanwhile parentheses ( ) will indicate molar concentrations in moles per liter of a particular phase.

Several factors contribute to $P_W{}^O$ values, including the molar volume of the solute, solvation parameters (H-bond donor acidity/acceptor basicity), and polarizability (orientation and induction forces). All of these contributions affect the energy costs required for exposing non polar solutes to water molecules and create hydrophobic–water contacts. It has also been argued that the formation of ion pairs between the ionized antioxidant and the strong electrolytes present in the aqueous phase also need to be considered. Equations for limiting cases, where the formation of ion pairs is negligible, where there is a high tendency to form ion pairs and where no ion pair and no dissociation takes place in the oily region have been derived. Certainly, comparisons of the three models may help to better understand the fate of ionizable antioxidants and the interested reader can obtain further information on the various possibilities elsewhere [25,50].

## 4. Methods to Measure Partition Constants

Numerous methods are available to estimate the partition constants of molecules of interest in binary oil–water systems, all of them having strengths and weakness, and limitations regarding the nature of the compounds that can be used. Here, we describe some of the most common methods, and for convenience, we have divided them into those requiring laboratory set-ups (experimental methods, Figure 2), and those that employ in silico calculations (computer software). It is necessary to recall that computer methods are always estimative and that the estimated values need to always be corroborated with experimental, accurately determined, $P_W{}^O$ values.

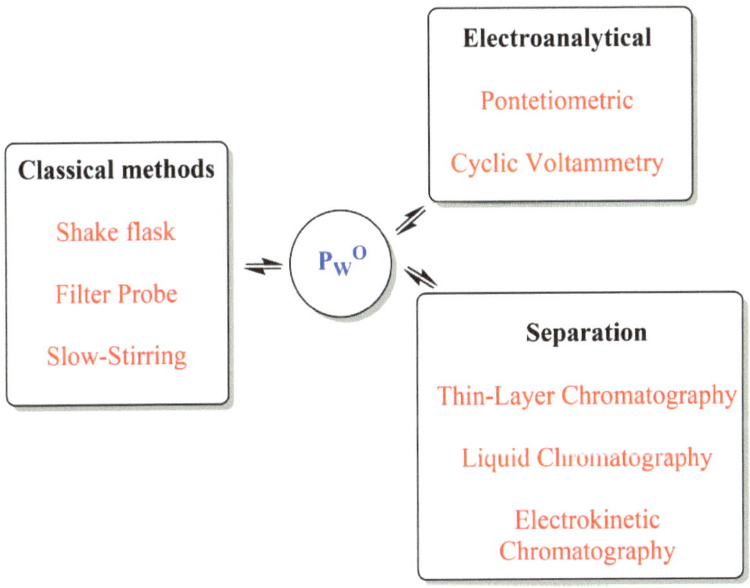


**Figure 2.** Common experimental methods employed to determine the partition constant values.

Table 2 summarizes some of the most common experimental and computer packages employed. We will only describe some of them briefly and the interested reader is referred to specialized reports [23,27,59–61].

**Table 2.** Some current experimental and computational methods employed to estimate partition constant values in oil–water and in octanol–water binary systems.

| Experimental | | | |
|---|---|---|---|
| **Method** | **Advantage** | **Weakness** | **Applicability** |
| Shake-flask | Most realistic, reliable, low experimental demand | Time consuming, large amounts of mutually saturated solvents required, emulsification may be a problem | Molecules of moderate hydrophobicity. Usually not recommended for very hydrophobic or hydrophilic compounds and when the tested substance dissociates |
| Slow-stirring methods | Avoid formation of emulsions | Time consuming, requires large amounts of solvent and product | Similar to the shake-flask method |

**Table 2.** *Cont.*

| Experimental | | | |
|---|---|---|---|
| **Method** | **Advantage** | **Weakness** | **Applicability** |
| Reverse-phase chromatography | Rapid, does not require large amounts of product nor solvent | Poor reproducibility because of different retention mechanisms, requires HPLC instrumentation | |
| Micellar electrokinetic chromatography | Good agreement with shake-flask method | | Applicable to ionic substances |
| Filter probe methods | Rapid | Expensive lab set-up, time consuming | |
| Software packages | | | |
| Name | Company | Freeware | Comments |
| ACD/logD | Advanced Chemistry Development (www.acdlabs.com) | No | Fragment-based |
| ADMET predictor | Simulation Plis Inc. (www.simulationsplus.com) | No | Neural network |
| AlogP | Virtual Computation Chemistry Laboratory (www.vcclab.org) | Yes | Neural network |
| Hyperchem | Hypercube Inc. (www.hypercube.com) | No | Atom-additive method |
| MolInspiration | Molinspiration Cheminformatics, (https://www.mlinspiration.com) | Yes | Fragment based |
| SPARC | Univeristy of Georgia (http://www.ibmlc2.chem.uga.edu/sparc/) | Yes | Allows calculations under different ionic strength conditions |

### 4.1. Experimental Methods

Most likely, the simplest (and traditional) method to determine $P_W{}^O$ is the shake-flask method [22,23], where an oil–water mixture is spiked with the antioxidant and shaken at constant temperature for some time until equilibrium is achieved [49]. Aliquots of the organic and aqueous phases are then extracted or, otherwise, phases are separated (by centrifugation, for example) and the concentration of the chemical in each phase is determined by employing any suitable analytical technique from previously prepared calibration plots. The method presents some drawbacks, mainly concerned with the necessary time to carry out the experiment and the lack of reliability when employing very hydrophobic or hydrophilic chemicals. Deviations may also occur when the tested chemical dissociates and the partition constant becomes pH and/or concentration dependent. The main advantage is its experimental simplicity and reliability [22,27].

Other (indirect) methods are based on the partition of antioxidants between a non-polar liquid organic phase and a polar aqueous phase in an HPLC column [22,62]. This is probably the most rapid and accurate method, but requires that the laboratory possesses a reverse-phase HPLC, which is a relatively expensive instrument—indeed much more than the simple separator funnels employed in the shake-flask method. In addition, solvents for the mobile phase are required and the results may be affected by the operating retention mechanisms. Frequently, multivariate techniques and molecular descriptors might need to be used for antioxidants with quite different chemical structures from that of the reference substances employed. Variations of the RP-HPLC method, including micellar electrokinetic chromatography and counter-current chromatography, have also been employed to determine partition constants [63–65].

*4.2. Computational Methods: Extrathermodynamic Approaches Based on Linear Free Energy Relationships (LFERs) to Predict and/or to Evaluate Partition Coefficients*

The importance of predicting log $P_W{}^O$ values arises because they are usually taken as a measure of the lipophilicity of the antioxidants, and food, pharmaceutical, and biological events depend on the lipophilic characteristics of the chemical species involved. Despite its importance, it is rather common that researchers find situations in which the partitioning behavior of a compound between organic matter (in its broad sense) and water is needed, but either cannot be easily determined experimentally or, for any other reason, some of the data required are not available. In these cases it is thus important to develop predictive methods to allow researchers to obtain estimates of the partition coefficients [23,24,58,66].

The basic idea behind the most common approaches used for predicting partition constants is to express the (unknown) free energy of transfer of the molecule of interest $\Delta G_{transfer}$ in the two-phase system by one or several other known free energy terms, chosen in a way that can be linearly related to $\Delta G_{transfer}$. The relationships between the unknown $\Delta G_{transfer}$ and the known free energy terms are usually called linear free energy relationships (LFERs; Table 3) [23,24,57,66]. Such relationships need to take into consideration the molecular interactions that exit between solute molecules and solute–solvent molecules.

**Table 3.** Examples of simple one-parameter linear free energy relationships (LFERs) proposed for connecting partition constants in various two-phase systems.

| Partition Constant/Coefficients Correlated | LFER |
|---|---|
| Octanol–water (oct–w)/aqueous solubility (sat) | $\log K_{i,w}^{oct} = -a \log K_{i,w}^{sat} + b$ |
| Organic carbon–water (oc–w)/octanol–water (oct–w) | $\log K_{i,w}^{oct} = -a \log K_{i,w}^{oct} + b$ |
| Lipid–water (lip–w)/octanol–water (oct–w) | $\log K_{i,w}^{lip} = -a \log K_{i,w}^{oct} + b$ |

Pretty good correlations can be obtained when considering similar systems, particularly when choosing groups of compounds that undergo the same type of interactions in a given phase. However, as we will see later (Section 5), it should not be surprising that poor relationships are found when trying to relate partition constants of a series of compounds of different polarities between two systems that contain phases with different solvent properties, for example, when attempting to predict partition constant values in edible oil–water binary mixtures from those obtained in octanol–water systems [67,68].

A second, conceptually different approach, assumes that $\Delta G$ of transfer for the whole molecule can be expressed by the sum (linear combination) of the terms describing the free energy of transfer of the various parts of the molecules [69]. To describe intramolecular interactions between different parts of the molecule that cannot be accounted for when considering the transfer of the isolated parts, it is necessary to include special interaction terms. The major advantage of this approach is that it allows one to estimate a partition constant based solely on the structure of the considered compound, and good results can be anticipated particularly in those cases where predictions of the partition constant of a structurally similar compound is known, so that only the contributions of the parts that are different between the two compounds have to be added and/or subtracted.

The most advanced and most widely used method that is based on this concept is the structural group contribution method for estimating partition constants. The method was first proposed by Hansch and Leo in a seminal paper published in 1971 [56], where they explain the fundamentals of the partitioning phenomena, providing detailed descriptions of the theory, the various uses of the partition constants, and a very comprehensive tabulation of a large number of $P_W{}^O$ values for a variety of substances. However, most interestingly, they also provided a discussion on additive–constitutive properties describing the use of various known linear-free relationships (e.g., Hammet equation) to calculate partitioning free energy and the various stereoelectronic effects on the partition constant values. The importance of the quantification of hydrophobic interactions energies has been, and still is, the key to drug design projects, as well as the routines included in many software currently employed in predictive studies and in computational chemistry [49,56,58].

Fragment-Based, Atom-Based, and Molecular Methods for Estimating Partition Constants

Theoretical prediction of the $P_W{}^O$ values is a convenient procedure because it is time-saving with respect to experimental, time-consuming methods and because in many instances, the number of existing experimental data are negligible compared with the number of chemicals for which it is necessary. Most of the models that employ current software packages are based on the fragmentation of the compound of interest in substructures and/or in the calculation properties of each fragment. Several statistical parameters are considered to assess the predictive values of $P_W{}^O$, and most of them are based on the use of quantitative structure–reactivity relationship (QSARs) models, but some structural or physicochemical properties such as surface-activity, ionization, and poor solubility make it difficult to determine $P_W{}^O$ values accurately, or perhaps the compound analyzed is chemically unstable and undergoes rapid degradation [70,71]. We will describe some of them briefly.

Fragment-based computational methods to estimate $P_W{}^O$ values were first proposed by Mannhold [72]. According to this approach, the molecule is essentially broken down into fragments so that the actual $P_W{}^O$ value can be calculated as the sum of the contributions of each fragment. However, correction factors are usually needed to compensate for intramolecular interactions because the various molecular environments may affect the effective contribution of each. The C-LOGP and ACD/LogP software packages, which are commercially available, employ this fragment approach.

An extension of the fragment contribution method is the atom contribution approach, which assumes that, instead of a contribution by fragments, the value of $P_W{}^O$ can be determined by the contribution of each individual atom in the molecule. The reduced dependence on corrections is likely a major advantage of these methodologies implemented in computer software, including the well-known Ghose-Crippen, XLOGP, and VEGA-MlogP methods.

Quantum mechanical calculations have been increasingly employed to estimate the interactions between solvent and solute molecules. Predictive methods that employ charge densities and molecular electrostatic potentials with an H-bonding capability have been employed for estimating molecular hydrophobicity. Neural networks have also been employed to predict $P_W{}^O$ values from a training set of electrotopological descriptors of a large data base of drugs.

## 5. Partition Constants of Homologous Series of Antioxidants in Different Oils: Can the $P_W{}^O$ Values Determined in Octanol–Water Systems Be Employed to Predict Those in Edible Oil–Water Systems?

Frequently, researchers need to predict partition coefficients for series of homologous antioxidants where the parent compound is grafted with inert residues to modify their solubility or hydrophobicity properties. This is the case, for instance, for ascorbic acid and ascorbyl palmitate (E-300 and E-304, respectively), and for food-approved propyl, octyl, and lauryl gallates (E310, E-311 and E312, respectively). In these cases, the focus is placed on the contribution that the methylene groups makes to $P_W{}^O$ values, under the assumption that, for a given binary oil–water mixture, is constant and independent of the nature of the parent chemical, which makes a constant contribution to $P_W{}^O$.

Freiría-Gándara et al. [67] and Costa et al. [68] determined the values of the partition coefficients for the distributions of several series of homologous in oil–water binary mixtures and, for the sake of comparisons, in octanol-water. The chemical structures of the AOs employed and the determined partition constant values are displayed in Table 4.

**Table 4.** Experimental partition constants for the distributions of AOs of different hydrophobicities in oil–water systems and theoretical octanol–water values, $T$ = 25 °C. Data extracted from Freiría-Gándara et al. [67].

| ANTIOXIDANT | | | Log $P_W^O$ | | | |
|---|---|---|---|---|---|---|
| **Structure** | **-R** | **$n(CH_2)$** | **Olive** | **Soybean** | **Corn** | **Octanol** |
| 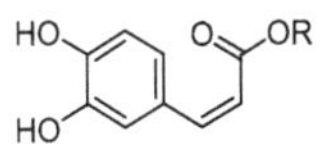 | $-CH_3$ (C1) | 0 | 0.40 | 0.45 | 0.52 | 1.56 |
| | $-CH_2CH_3$ (C2) | 1 | 0.89 | 0.92 | 1.04 | 1.93 |
| | $-(CH_2)_2CH_3$ (C3) | 2 | 1.45 | 1.48 | 1.58 | 2.44 |
| Caffeates | $-(CH_2)_7CH_3$ (C8) | 7 | 2.23 | 2.25 | 2.81 | 5.02 |
| 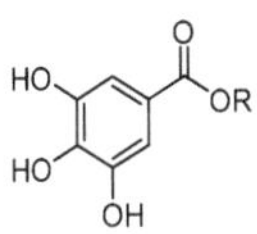 | $-CH_3$ (C1) | 0 | −1.30 | −1.40 | −0.94 | 0.85 |
| | $-CH_2CH_3$ (C2) | 1 | −0.70 | — | −0.60 | 1.23 |
| | $-(CH_2)_2CH_3$ (C3) | 2 | −0.07 | −0.07 | 0.10 | 1.73 |
| | $-(CH_2)_3CH_3$ (C4) | 3 | 0.48 | 0.51 | 0.60 | 2.29 |
| Gallates | $-(CH_2)_7CH_3$ (C8) | 7 | 1.88 | 2.29 | | 4.31 |
| 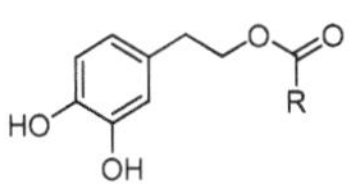 | $CH_3$ (C2) | 0 | −0.20 | — | | 1.22 |
| | $CH_2CH_3$ (C3) | 1 | 0.34 | — | | 1.58 |
| | $(CH_2)_4CH_3$ (C6) | 4 | 1.50 | — | | 3.15 |
| Hydroxytyrosol Derivatives | $(CH_2)_6CH_3$ (C8) | 6 | 1.61 | — | | 4.16 |
| 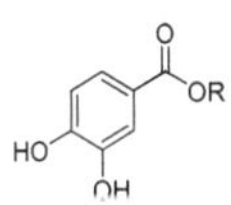 | $-CH_3$ (C1) | 0 | −0.11 | −0.11 | | 1.14 |
| | $-CH_2CH_3$ (C2) | 1 | 0.27 | 0.36 | | 1.52 |
| | $-(CH_2)_2CH_3$ (C3) | 2 | 0.91 | 0.91 | | 2.02 |
| | $-(CH_2)_3CH_3$ (C4) | 3 | — | 1.60 | | 2.58 |
| Protocatechuates | $-(CH_2)_5CH_3$ (C6) | 5 | 1.72 | 1.71 | | 3.59 |

According to the fragment approach, the contributions to $P_W^O$ come from three fragments. One is that of the polyphenolic moiety, the second one is that of terminal methyl group, and the third one is that from the methylene groups. Both the aromatic moiety and the terminal methyl group are the same for all individuals of each series, and thus make a constant contribution to $P_W^O$. However, the contribution coming from methylene groups is not constant because the different species contain a variable number of carbon atoms. Thus, the $\log(P_W^O)$ values of each series of homologous antioxidants (Table 4) can be computed as the sum of two main contributions, in Equation (5), where parameter $a_O$ stands for the (constant) contribution of the non-alkyl part of the molecule to $P_W^O$, $b_O$ stands for the contribution of each methylene group, and $n_{CH_2}$ is the number of methylene groups in the alkyl chain of the antioxidant. This linear relationship between $\log(P_W^O)$ and the contribution of methylene groups in the series of homologous compounds is usually known as the Colander relationship [73], predicting that an increase in the number of C atoms in the alkyl chain will increase the hydrophobicity of the AOs and thus its solubility in oil [67].

$$\log\left(P_W^O\right) = a_O + b_O n_{CH_2} \quad (5)$$

Figure 3A–D shows plots of the variations of $\log(P_W^O)$ (olive and octanol) vs. the number of $-CH_2$ groups for the esters derived from gallic (GA; Figure 3A), caffeic (CA; Figure 3B), and protochateuic (PT; Figure 3C) acids and hydroxytyrosol (HT, Figure 3D), which are linear except for the most hydrophobic AOs ($n > 7$), where deviations from the linearity are evident. These curved or even two-phasic relationships are obtained when including most hydrophobic or hydrophilic compounds, because their solubility in one of the phases is rather low, yielding $P_W^O$ values with large errors. Similar deviations are commonly observed in the studies of homologous series of compounds and their values are not usually considered when determining the slopes and intercepts (Table 5). The goodness of fit for any particular homologous series can, however, be quantified in terms of the standard deviation of the least squares fit of these data if necessary.

The contribution of the methylene group was, as expected, very similar, no matter the oil or the AO series (differences less than 10%) considered, with an average value of $b_O = 0.53 \pm 0.02$. This result should not be surprising, because the composition of the oils was quite similar (though the percentage and nature of the fatty acids may be different). Remarkably, this average value for the slope was quite similar to the average value obtained in octanol, $b_{oct} = 0.50 \pm 0.01$, suggesting that the contribution of the methylene group to the hydrophobicity was very similar in relative apolar solvents.

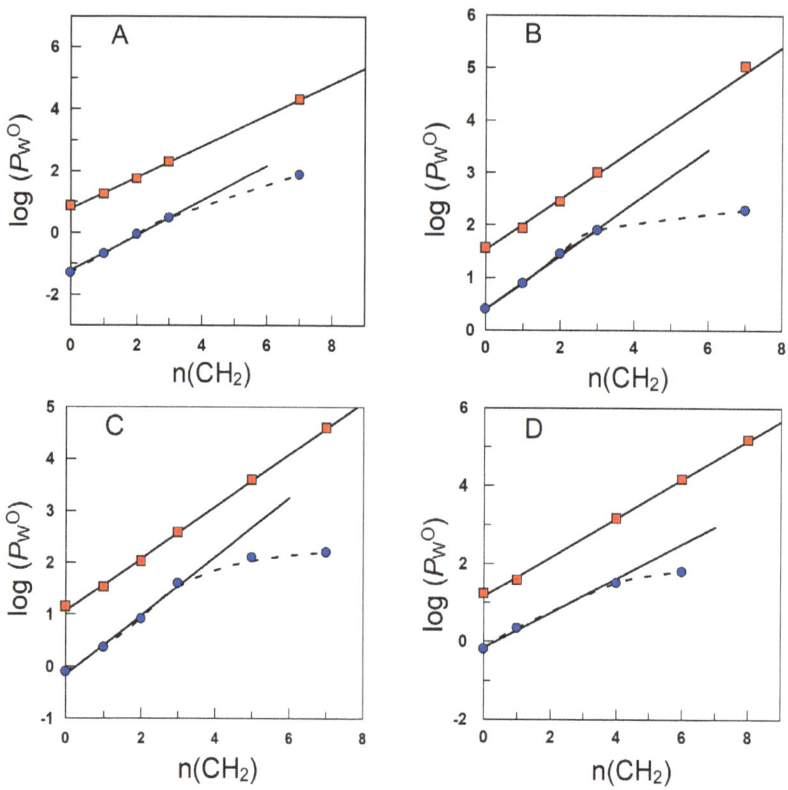


**Figure 3.** Changes in log($P_W^O$) with the number of methylene groups in the alkyl chain (•) of ester derivatives of gallic (**A**), caffeic (**B**), and protocatechuic (**C**) acids and hydroxytyrosol (**D**) in olive oil–water binary systems. $P_W^O$ values are determined experimentally by employing the shake–flask method. For the sake of comparison, the partition coefficients of the same antioxidants in octanol–water systems ($P_W^{OCT}$, ■) are also displayed. $P_W^{OCT}$ values are estimated by employing the Molinspiration (fragment-based) computer software [74] (see Table 2 for details). Figure adapted from Freiría et al. [67] with permission, copyright American Chemical Society.

The different structures for the aromatic moieties of the AOs employed made the intercepts different for each homologous series, following the order $a_{GA} < a_{PT} \approx a_{HT} < a_{CA}$. The non-alkyl moieties of the antioxidants contain different groups and the $a_O$ values refer to different oils with different compositions, hence they cannot be compared directly to each other. However, for a given oil (e.g., olive), the comparison of the $a_O$ values for GA and PT highlights the big effect of additional -OH groups on the overall hydrophobicity of the molecules. The addition of a single (HT) or double bond (CA) to the alkyl chain also has a big effect on the lipophilicity of the molecules. Further details and discussion can be found elsewhere [67,68].

**Table 5.** Slopes and intercepts for the variations of $\log(P_W^O)$ with the number of methylene groups according to Equation (5) (Figure 3) for ester derivatives of different antioxidants such as gallic acid (GA), caffeic acid (CA), protocatechuic acid (PT), and hydroxytyrosol (HT). Values from Freiría-Gándara et al. [67].

| OIL | Antioxidants | Gallic | Caffeic | Protocatechuic | Hydroxytyrosol |
|---|---|---|---|---|---|
| Octanol (OCT) | $a_{OCT}$ | $0.77 \pm 0.03$ | $1.53 \pm 0.05$ | $1.06 \pm 0.03$ | $1.15 \pm 0.04$ |
| | $b_{OCT}$ | $0.51 \pm 0.01$ | $0.49 \pm 0.01$ | $0.50 \pm 0.01$ | $0.50 \pm 0.01$ |
| Olive (OL) | $a_{OL}$ | $-1.15 \pm 0.04$ | $0.40 \pm 0.03$ | $-0.18 \pm 0.07$ | $-0.16 \pm 0.02$ |
| | $b_{OL}$ | $0.50 \pm 0.01$ | $0.51 \pm 0.01$ | $0.55 \pm 0.04$ | $0.44 \pm 0.02$ |
| Soybean (SO) | $a_{SO}$ | $-1.18 \pm 0.06$ | $0.44 \pm 0.03$ | — | — |
| | $b_{SO}$ | $0.58 \pm 0.03$ | $0.52 \pm 0.03$ | — | — |
| Corn (CO) | $a_{CO}$ | $-1.03 \pm 0.10$ | $0.52 \pm 0.01$ | — | — |
| | $b_{CO}$ | $0.54 \pm 0.06$ | $0.53 \pm 0.01$ | — | — |

Extrathermodynamic relationships between $P_W^O$ values in different solvents can also be established for predictive purposes [56,75] because the partition constants are, at the end, equilibrium constants, and the linear relationships between $P_W^O$ and $P_W^{OCT}$ with the length of the alkyl chain allows for deriving a relationship for the type $\ln P_W^O = A + B\ln P_W^{OCT}$. The high relevance of $P_W^{OCT}$ for drug discovery, design, and development, means that a large number of databases containing values for this key physicochemical parameter are available in the literature. These databases include several thousands of compounds and, if for whatever reason $P_W^{OCT}$ values are not available for a particular species, it can be easily calculated with the aid of various computer programs that employ different descriptors (see Section 4.2 and Tables 2 and 3). Thus, for predictive and practical purposes, it is was deemed interesting to investigate whether there was some kind of relationship between $P_W^{OCT}$ and $P_W^O$ that allowed for the prediction of $P_W^O$ values in edible oils from the $P_W^{OCT}$ ones.

Figure 4 shows the experimental $P_W^O$ values determined for a variety of antioxidants in different oils and, for comparisons the computer estimated $P_W^{OCT}$ values for the same antioxidants. As can be observed, there were significant differences between the $P_W^{OCT}$ and $P_W^O$ values for the different antioxidants and oils. Such a difference can be interpreted in terms of the expected higher hydrogen bond donating capacity of octanol compared with that of common edible oils, because the main component of octanol has one hydroxyl group, meanwhile edible oils are composed of glycerol esters present in different percentages. Thus, the intermolecular forces present in, for example, octanol, should be very different in size from those in edible oils. Hence, the results in Figure 4 clearly show that $P_W^{OCT}$ values cannot be used to predict the hydrophobicity of antioxidants in edible oils and that $P_W^O$ values should be determined for each antioxidant and edible oil.

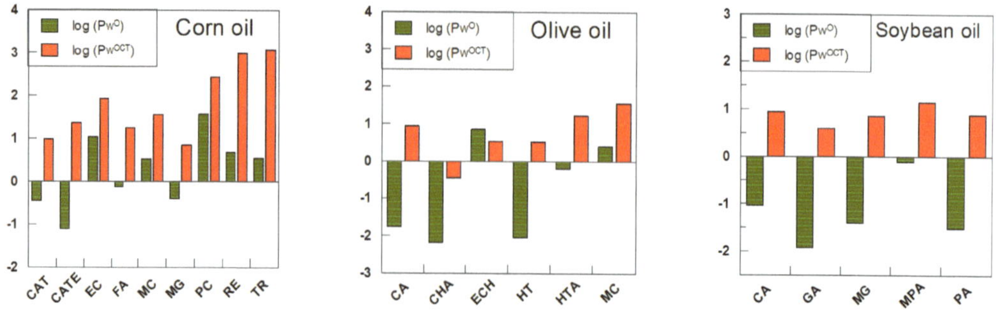


**Figure 4.** Comparison between the experimentally determined $P_W{}^O$ values in oil–water systems [76] and the theoretical $P_W{}^{OCT}$ values determined by employing the Molinspiration software (accessed on 23 February 2023) [74] for series of antioxidants with different molecular structures. CA: caffeic acid; CAT: catechol; CATE: catechin; CHA: chlorogenic acid; EC: ethyl caffeate; ECH: ethyl chlorogenate; FA: ferulic acid; GA: gallic acid; HT: hydroxytyrosol; HTA hydroxol acetate; MC: methyl caffeate; MG: methyl gallate; PA: protocatechuic acid; MPA: methyl protocatechuate; PC: propyl caffeate; RE: resveratrol; TR: trolox.

Similar results can be found when considering the partitioning of antioxidants in binary systems other that oil–water. For instance, the association constant $K_D$ of antioxidants with, for example, sodium dodecyl sulfate micelles, SDS micellar aggregates, Figure 5, can be, formally, considered similar to the oil–water partition constant $P_w{}^O$, Figure 1, because both are two-state systems where the antioxidant is located in the aqueous or in the micellar phases.

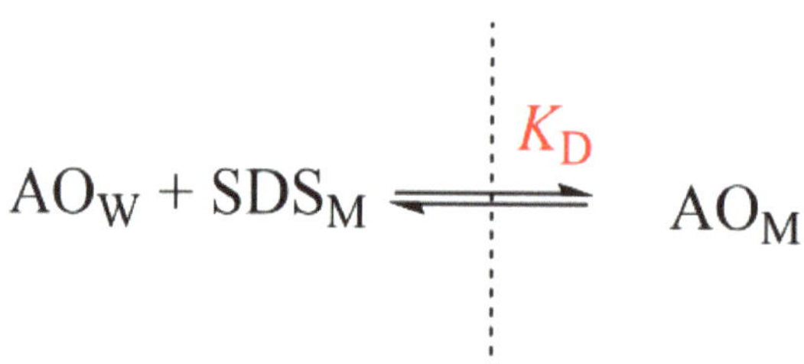


**Figure 5.** Representation of the association process of antioxidants to micellar aggregates. $AO_W$ stands for the antioxidant present in the water phase, $AO_M$ stands for that in the micellar phase, and $SDS_M$ stands for the micellized surfactant.

Vañova et al. [77] determined, by employing micellar electrokinetic chromatography, the association constants of a series of phenolic acids between water and SDS micelles ($K_D$), and, when comparing them with the theoretically determined octanol-water ($P_W{}^{OCT}$) partition constants (Molinspiration, accessed on 23 February 2023; Figure 6,) no relationship could be established between them, confirming that the values of the association constant $K_D$ could not be predicted from the $P_W{}^{OCT}$ values.

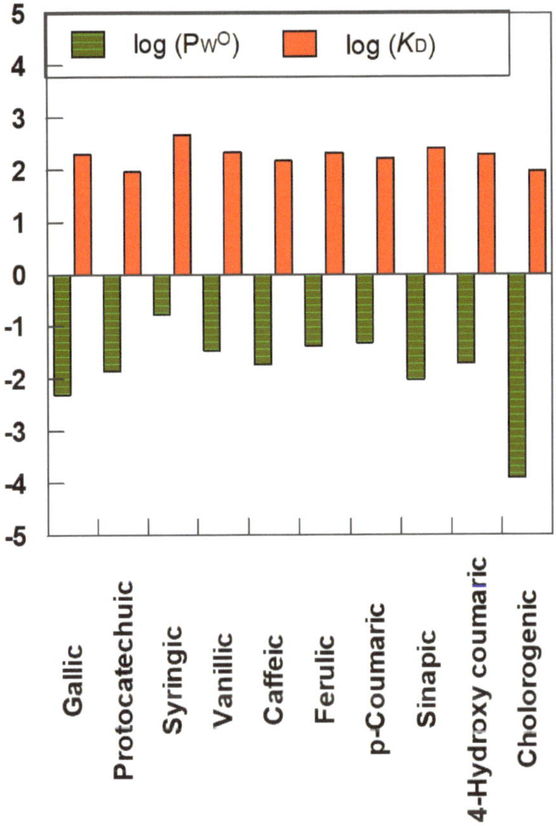


**Figure 6.** Values of experimentally determined association constants $K_D$ between antioxidants and SDS micelles [77] (pH = 7.4, buffered solution) and the theoretically determined partition constants $P_W^{OCT}$ at the same pH for a series of phenolic acids.

In summary, the large differences between the experimentally determined $P_W^O$ values in oil–water systems and the theoretical $P_W^{OCT}$ values proves that, in general, $P_W^O$ values cannot be predicted from $P_W^{OCT}$ values and thus should be determined for each antioxidant and oil.

## 6. Effects of Acidity: Apparent Partition Coefficients of Ionizable Antioxidants

When considering phenolic acids (or any other ionizable antioxidant), one must bear in mind that ionic and neutral species exhibit different polarities and thus their intermolecular interactions with the solvent should be different [50,59,78]. The partition constant $P_W^O$, defined by Equation (4), only considers the concentration of neutral molecules, but if the antioxidant is partially or fully ionized under some experimental conditions (as it may occur to most phenolic antioxidants in the acidity range of most foods, pH 4–7), then $P_W^O$ values become pH-dependent and need to be considered as "apparent" values. These partition constants are commonly designed by $P_W^O$ (app), which is defined in terms of all ionized and neutral forms present at a particular pH. In principle, the antioxidants (and any other ionizable compound) can be ionized in both the aqueous and oil phases, as shown in Figure 7.

**Figure 7.** Distribution of an ionizable antioxidant AO in a biphasic oil–water system. (**A**) Assuming that ionization can take place in both phases. (**B**) Assuming that ionization only takes place in the aqueous phase. $K_a{}^W$ and $K_a{}^O$ are the ionization constant of the antioxidant in the aqueous and oil phases, respectively.

However, in general, ionization of the antioxidants in the oil phase can be neglected because the ionization constants of weak acids in oils ($K_a{}^O$) are much smaller, by 5–6 orders of magnitude, than those in the aqueous phase $K_a{}^W$ [79], so that the ionization of the solute in the oil can be neglected ($pK_a(O) >> pK_a(W)$). Thus, the distribution of the ionizable antioxidants can be simplified as indicated in Figure 7B, and the apparent partition coefficient is given by Equation (6) [67,68]. Some literature values for the ionization constants of phenolic acids are displayed in Table 6.

$$P_W^O(\text{app}) = \frac{(AO_O)}{(AO_W) + (AO_W^-)} = \frac{P_W^O}{1 + \frac{K_a^W}{[H^+]}} \quad (6)$$

**Table 6.** Acidity constants for some phenolic acids for the carboxylic acid [pKa(1)] and that for the -OH group in the *p*-position [(pKa(2)].

| Phenolic Acid | Molecular Structure | p$K_a$(1) | p$K_a$(2) | Ref |
|---|---|---|---|---|
| p-Coumaric | 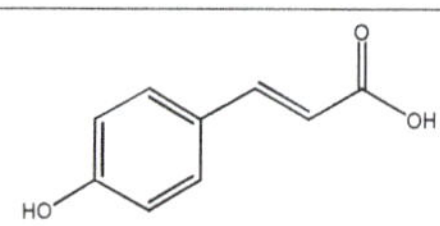 | 4.37 | 9.20 | [80] |
| Caffeic | 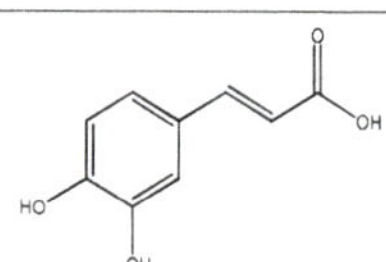 | 3.94<br>4.30 | 8.47<br>8.51 | [80]<br>[81] |
| Ferulic | 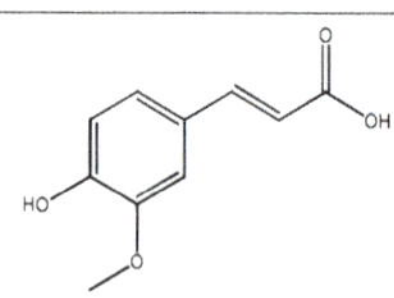 | 4.50<br>4.30 | 9.21<br>8.81 | [80]<br>[81] |
| 3,4 Dihydroxy phenylacetic | 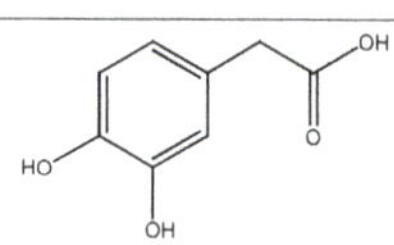 | 3.21 | 9.33 | [80] |

**Table 6.** *Cont.*

| Phenolic Acid | Molecular Structure | p$K_a$(1) | p$K_a$(2) | Ref |
|---|---|---|---|---|
| Gallic acid | 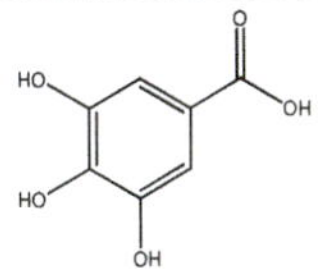 | 4.10<br>4.11 | 8.38<br>8.47 | [80]<br>[81] |
| Vanillic | 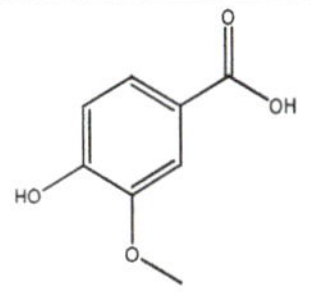 | 4.58<br>4.17 | 9.39<br>8.81 | [80]<br>[81] |
| p-Hydroxybenzoic | 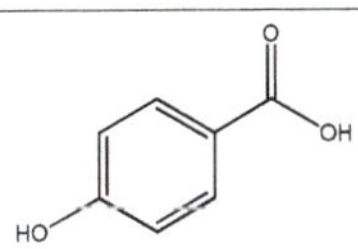 | 4.40<br>4.26 | 9.54<br>8.84 | [80]<br>[81] |
| Trolox | 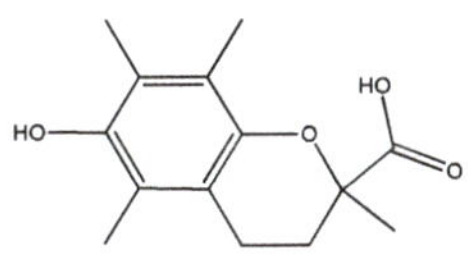 | 3.89 | 11.92 | [82] |

Equation (6) predicts that the variation of $P_W{}^O$ (app) with the acidity is sigmoidal, as shown in Figure 8A,C. Note that with the limit of low pH (pH at least two units below the p$K_a$ of the ionizable antioxidant), the apparent partition coefficient approaches that of the neutral form, $P_W{}^O$, because [H$^+$] >>> $K_a$. Thus, the upper limit value of $P_W{}^O$ (app) is that corresponding to the neutral molecule ($P_W{}^O$).

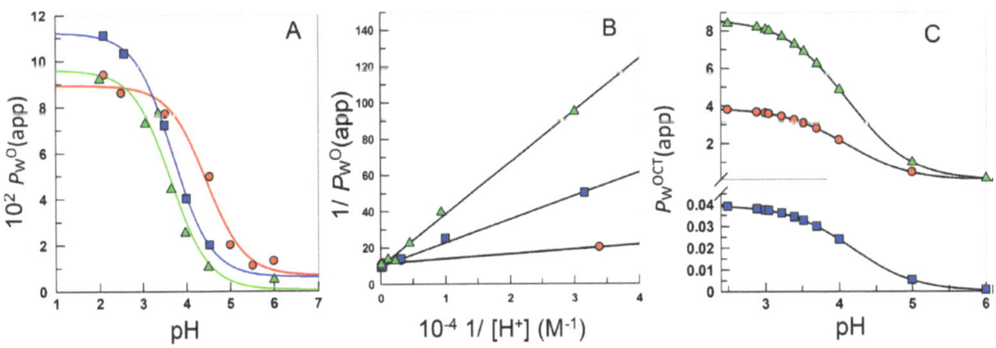


**Figure 8.** Experimental and theoretical variations of the apparent partition constants of ionizable antioxidants (gallic acid (●), ascorbic acid (■), and caffeic acid (▲)) with the acidity (pH) of the aqueous phase in olive oil–water (**A**) and linear plots of $1/P_W{}^O$ (app) with pH according to Equation (6) (**B**). For the sake of comparisons, in octanol-water (**C**) binary systems. The solid line is the theoretical curve obtained by fitting the experimental data to a sigmoidal Henderson–Hasselbach type (Equation (6)). Simulations are done assuming that the ionization of the antioxidant in the oil phase is negligible. Figure adapted from Freiría et al. [67], with permission, copyright the American Chemical Society.

In the case that, for any reason, the value of $P_W{}^O$ is unknown and/or cannot be determined experimentally, Equation (6) can be linearized so that the variation of $1/P_W{}^O$ (app) with $1/[H^+]$ should be linear from where the value of $P_W{}^O$ can be determined, as illustrated in Figure 8B.

When necessary, one can easily determine the effective local concentrations of the antioxidant in the oil and aqueous phases by bearing in mind the concentration of the antioxidant, in terms of the total volume $V_T$ of the system, using Equation (7). Note that brackets mean concentrations in moles per liter of the total volume, meanwhile parentheses indicate the effective concentrations in moles per liter of the corresponding phase.

$$[AO_T]V_T = (AO_{oil})V_{oil} + (AO_W)V_W + (AO_W^-)V_W \tag{7}$$

The concentration of the neutral weak antioxidant in the aqueous phase of the binary oil–water system ($AO_W$) relative to the total (stoichiometric) concentration is given by the Henderson–Hasselbach type Equation (8), where $\Phi_O = V_{oil}/(V_{oil} + V_{water})$ and $\Phi_W = V_{water}/(V_{oil} + V_{water})$.

$$\frac{(AO_W)}{[AO_T]} = \frac{1}{\Phi_W + P_W^O\Phi_O + \frac{K_a}{[H^+]}} \tag{8}$$

Equation (8) simulates the ratio of the effective concentration of the neutral antioxidant relative to the total (stoichiometric) concentration in terms of the pH of the aqueous phase, the volume fractions of each phase ($\Phi_O$ and $\Phi_W$), the partition constant of the neutral molecule, and the ionization constant ($K_a$) of the antioxidant, as illustrated in Figure 9. In 1:1 oil–water mixtures, the concentration in the water phase is almost twice that of the stoichiometric concentration in the high acidity limit (e.g., at low pH), meanwhile in 1:9 oil–water mixtures, the concentration of antioxidant in the water phase is very similar to the stoichiometric concentration.

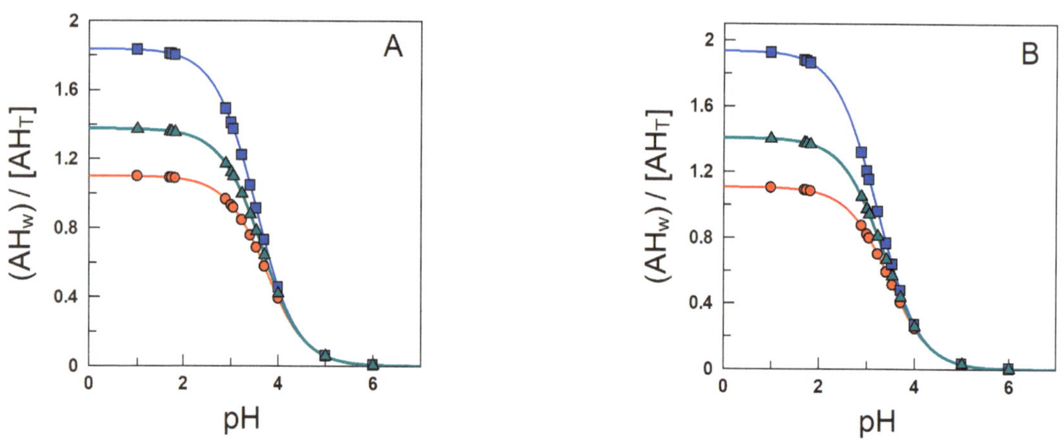


**Figure 9.** Changes in the effective concentration of the neutral forms of the antioxidants caffeic acid (**A**) and gallic acid (**B**) in the aqueous region relative to the stoichiometric one as a function of pH in 1:1 (blue), 3:7 (teal), and 1:9 (red) v:v oil–water binary mixtures. Values calculated by employing Equation (8).

In cases in which the monoprotic phenol undergoes partial ionization in the oil and aqueous phases (Figure 7A), the corresponding ionization constants are defined by Equations (9) and (10), respectively, and thus the apparent partition constant is given by Equation (11).

$$K_a^W = \frac{(AO_W^-)(H_W^+)}{(AO_W)} \tag{9}$$

$$K_a^O = \frac{(AO_O^-)(H_O^+)}{(AO_O)} \tag{10}$$

$$P_W^O(app) = \frac{(AO_O) + (AO_O^-)}{(AO_W) + (AO_W^-)} \tag{11}$$

**7. Effects of Temperature**

So far, we have considered the distribution of an antioxidant between two phases at a given temperature and pressure (Figure 1). In most cases of liquid–liquid partitioning, we can neglect the effects of pressure on equilibrium partitioning, and thus here we will limit our discussion to analyze the temperature effects on the partition constant. Predicting how partition constants change with temperature is useful because the temperatures of extraction procedures and/or storage conditions may change and so we may need to be able to extrapolate values at a given temperature to other conditions. Considering relatively small temperature changes, we can safely assume, to a first approximation, that both the volumes and densities of the oil and water will remain constant, or their variations, if any, will be so small that could be considered negligible.

From an energetic point of view, the variation in the partition constant $P_W{}^O$ of a given species between water and oil with temperature is related to thermodynamic changes, specifically the Gibbs energy, molar enthalpy, and molar entropy of the transfer of molecules from one liquid to the other. The variation in the partition constant with the temperature is given by the van't Hoff Equation (12), where $R$ stands for the universal gas constant, which predictsΔ that plots of ln $P_W{}^O$ vs. $(1/T)$ should be linear, with a slope equal to $\Delta H^{0,W\to O}/R$. Figure 10 shows typical van't Hoff plots obtained for propyl gallate and α-tocopherol.

At a given temperature, the variation in the Gibbs free energy of transfer from an aqueous phase to an organic phase is given by Equation (13). According to the Gibbs equation, the Gibbs free energy $\Delta G^{0,W\to O}$ is given by the difference between the enthalpy changes ($\Delta H^{0,W\to O}$) and the entropy changes ($\Delta S^{0,W\to O}$) in the partitioning process (Equation (14)), from where the entropy contribution can be determined.

$$\Delta H_T^{0,W\to O} = R\left[\frac{\partial(\ln P_W^O)}{\partial\left(\frac{1}{T}\right)}\right]_P \tag{12}$$

$$\Delta G^{0,W\to O} = -RT \ln P_W^O \tag{13}$$

$$\Delta G^{0,W\to O} = \Delta H^{0,W\to O} - T\Delta S^{0,W\to O} \quad (14)$$

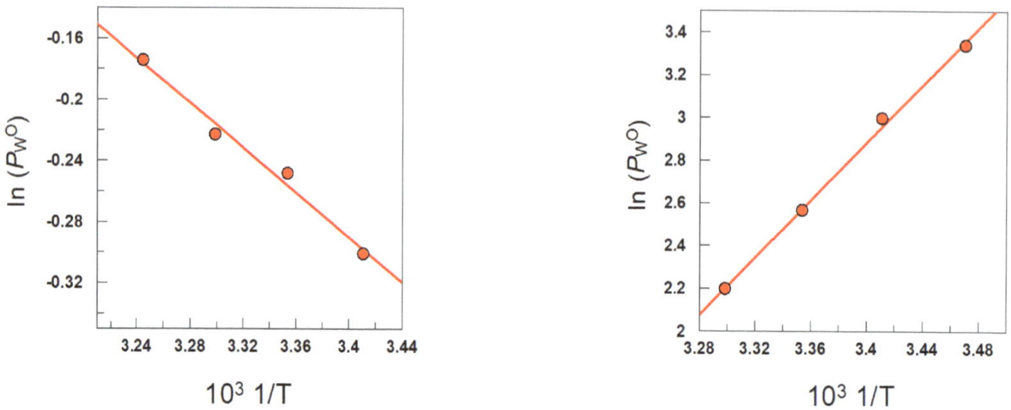


**Figure 10.** Variation of the partition constant $P_W{}^O$ of propyl gallate (PG, **left**) and α-tocopherol (TOC, **right**) with $1/T$ ($K^{-1}$) in binary corn oil–water mixtures according to the van't Hoff equation. Data from [83] (PG) and [84] (TOC).

## 8. Partitioning in Emulsions

When binary oil–water mixtures are shaken in the presence of surfactants or proteins, spherical emulsion droplets of 0.1–100 µm are formed. The surfactant is adsorbed in the oil-water interface, generating a three-dimensional, loose, boundary region of 2–20 nm thick, separating the water and oil regions that kinetically stabilize the emulsion [85,86]. The volume of this interfacial region is usually much smaller (3–5%) than the total emulsion volume, but plays a crucial role in many biological, health, and food systems. For example, oil-water emulsions are widely employed in many industries, among others, in the food, pharmaceutical, agrochemical, cosmetic, painting, printing, as well as petroleum industries. Their use as vehicles for drug delivery is an interesting approach that is currently being explored [4,30,68].

Emulsions containing edible oils are prone to oxidation, and to minimize this undesirable reaction, antioxidants are usually added to the system. As in the binary systems, antioxidants partition thermodynamically between the aqueous, interfacial, and oil regions according to their solubility (Figure 11), and their distribution is defined by the partition constants between the water-interfacial, $P_W{}^I$, and oil-interfacial, $P_O{}^I$, regions, as shown in Equations (15) and (16), respectively [42,45,68,87].

$$P_W^I = \frac{(AO_I)}{(AO_W)} \quad (15)$$

$$P_O^I = \frac{(AO_I)}{(AO_O)} \quad (16)$$

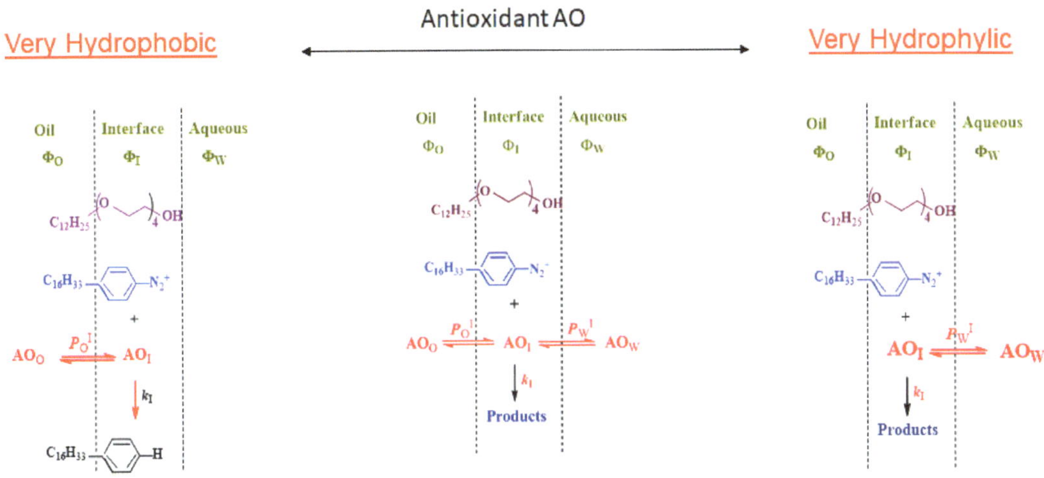


**Figure 11.** Partitioning of antioxidants in emulsions according to their polarity. In general, two partition constants are needed to define their distributions, but two limiting situations where only one partition constant is needed can be found. The first limiting situation when considering very hydrophilic antioxidants (e.g., phenolic acids) is that they are essentially oil-insoluble and only $P_W{}^I$ is needed to define their distribution. The second limiting situation is found when considering water-insoluble antioxidants (i.e., very hydrophobic antioxidants such as tocopherol). In these cases, the antioxidants only partition between the interfacial (I) and oil (O) regions and their distributions are described by the partition constant $P_O{}^I$ [88–92]. Figure reproduced from S. Losada-Barreiro, PhD Thesis, Universidad de Vigo, 2014, with permission.

Note that the ratio $P_W{}^I/P_O{}^I = (AO_O)/(AO_W) = P_W{}^O$. Thus, $P_W{}^I/P_O{}^I$ is numerically equal to the partition constant in a binary oil–water system because it can be considered as a limiting case of oil-in-water emulsions where no surfactant is added. This means that the partition constants in multiphasic systems cannot be extrapolated from that in the simpler binary system, and need to be determined in the intact emulsions to avoid disruptions of the existing equilibria. We also want to note that the partition constants, defined by Equations (15) and (16), do not consider the size of the droplets in the emulsion, i.e., it is assumed that they are independent of the size of the droplets. The point has been experimentally proven by determining the distributions of various antioxidants in emulsions containing droplets of different size (macro and nano-emulsions): no changes in the partition constants were detected [41,42,44,45,93–95]. Table 7 lists some partition constants for common antioxidants determined in intact emulsions prepared with edible oils under different experimental conditions. Details on the methodology employed and experimental conditions can be found elsewhere and in the references therein [41,42,44,45].

**Table 7.** Values of the partition constants of antioxidants (AO) between the aqueous-interfacial ($P_W{}^I$) and oil-interfacial ($P_O{}^I$) regions of emulsions. Partition constant values are determined in the intact emulsions by employing a kinetic methodology [42,45] to avoid disruption of the existing equilibria to avoid biasing of values. TW20 = Tween 20, TW80 = Tween 80, SP80 = Span 80. BG: butyl gallate; CA: caffeic acid; CAT: catechol; CGA: chlorogenic acid; CATE: catechin; DCG: dodecyl chlorogenate; EC: ethyl caffeate; ECG: ethyl chlorogenate; EG: ethyl gallate; EHT: hydroxytyrosol ethanoate; GA: gallic acid; HET: hydroxytyrosol hexanoate; HHT: hydroxytyrosol hexadecanoate; HT: hydroxytyrosol; HTA hydroxytyrosol acetate; LCG: lauryl chlorogenate; LG: lauryl gallate; LHT: hydroxytyrosol laurate; MC: methyl caffeate; MG: methyl gallate; OC: octyl caffeate; OCG: octyl chlorogenate; HCG: hexadecyl chlorogenate; OG: octyl gallate; OHT: hydroxytyrosol octanoate; PC: propyl caffeate; PCG: propyl chlorogenate; PG: propyl gallate; RES: resveratrol; TOC: α-tocopherol; TR: trolox.

| Antioxidant | Surfactant | Oil/Water Ratio (v:v) | pH | $P_W{}^I$ | $P_O{}^I$ | Reference |
|---|---|---|---|---|---|---|
| | | | Fish oil | | | |
| GA | TW80 | 4:6 | 3.7 | 118 | - - - | [95] |
| GA | TW80 | 1:9 | 3.0 | 85 | - - - | [96] |
| EG | TW80 | 4:6 | 3.7 | 233 | 706 | [95] |
| PG | TW80 | 1:9 | 3.0 | 154 | 101 | [96] |
| BG | TW80 | 4:6 | 3.7 | 559 | 253 | [95] |
| OG | TW80 | 4:6 | 3.7 | - - - | 183 | [95] |
| LG | TW80 | 4:6 | 3.7 | - - - | 142 | [95] |
| HT | TW80 | 1:9 | 3.7 | 34 | - - - | [93] |
| EHT | TW80 | 1:9 | 3.7 | 207 | 115 | [93] |
| HET | TW80 | 1:9 | 3.7 | - - - | 89 | [93] |
| OHT | TW80 | 1:9 | 3.7 | - - - | 119 | [93] |
| LHT | TW80 | 1:9 | 3.7 | - - - | 97 | [93] |
| HHT | TW80 | 1:9 | 3.7 | - - - | 75 | [93] |
| | | | Corn oil | | | |
| CATE | TW20 | 4:6 | 2.1 | 368 | — | [97] |
| CAT | TW20 | 4:6 | 3.7 | 57 | 170 | [98] |
| CA | TW20 | 4:6 | 3.7 | 268 | | [99] |
| PG | TW20 | 1:9 | 3.7 | 204 | 242 | [100] |
| OG | TW20 | 1:9 | 3.7 | - - - | 29.8 | [100] |
| LG | TW20 | 3:7 | 3.7 | - - - | 16.5 | [100] |
| RES | TW20 | 4:6 | 2.1 | 4076 | 860 | [101] |
| TOC | TW20 | 1:9 | 3.7 | - - - | 11.3 | [90] |
| | | | Olive oil | | | |
| HT | TW20/SP80 | 1:9 | 3.7 | 120 | - - - | [102] |
| HTA | TW20/SP80 | 1:9 | 3.6 | 204 | 331 | [102] |
| TR | TW20 | 4:6 | 2.2 | 5371 | 1773 | [103] |
| CGA | TW20 | 4:6 | 3.7 | 40 | - - - | [104] |
| ECG | TW20 | 4:6 | 3.7 | 78 | - - - | [104] |

**Table 7.** *Cont.*

| Antioxidant | Surfactant | Oil/Water Ratio (v:v) | pH | $P_W{}^I$ | $P_O{}^I$ | Reference |
|---|---|---|---|---|---|---|
| PCG | TW20 | 4:6 | 3.7 | 141 | --- | [104] |
| OCG | TW20 | 4:6 | 3.7 | --- | 111 | [104] |
| DCG | TW20 | 4:6 | 3.7 | --- | 124 | [104] |
| LCG | TW20 | 4:6 | 3.7 | --- | 159 | [104] |
| HCG | TW20 | 4:6 | 3.7 | --- | 89 | [104] |
| | | | Soybean oil | | | |
| GA | TW20 | 1:9 | 3.0 | 298 | | [105] |
| MG | TW20 | 1:9 | 3.0 | 329 | --- | [105] |
| PG | TW20 | 1:9 | 3.0 | 401 | 474 | [105] |
| BG | TW20 | 1:9 | 3.0 | 789 | 243 | [105] |
| OG | TW20 | 1:9 | 3.0 | 33 | --- | [105] |
| LG | TW20 | 1:9 | 3.0 | 23 | --- | [105] |
| CA | TW20 | 4:6 | 3.7 | 104 | --- | [106] |
| MC | TW20 | 4:6 | 3.7 | 445 | 150 | [106] |
| EC | TW20 | 4:6 | 3.7 | 1355 | 159 | [106] |
| PC | TW20 | 4:6 | 3.7 | 4727 | 164 | [106] |
| OC | TW20 | 4:6 | 3.7 | --- | 216 | [106] |

As in the case of binary systems, the distribution of ionizable antioxidants (e.g., phenolic acids) also depends on the acidity of the aqueous phase, and their ionization equilibrium needs to be taken into consideration (Figure 12). Both neutral and ionized species may be distributed between the interfacial and aqueous regions, but, as in binary systems, the solubility of ionized species in the oil region is negligible. In these cases, two apparent partition constants need to be considered, $P_O{}^I$ (app) and $P_W{}^I$ (app), defined by Equations (17) and (18), respectively [41,42,68].

$$P_O^I(\text{app}) = \frac{(\text{AO}_I) + (\text{AO}_I^-)}{(\text{AO}_O)} \tag{17}$$

$$P_W^I(\text{app}) = \frac{(\text{AO}_I) + (\text{AO}_I^-)}{(\text{AO}_W) + (\text{AO}_W^-)} \tag{18}$$

Losada-Barreiro et al. [90] investigated the driving force for the hydrophobic effect that makes antioxidants partition between the oil, aqueous, and interfacial regions of oil-in-water emulsions. They determined the Gibbs free energy, enthalpy, and entropy values for the transfer of hydrophilic and hydrophobic antioxidants from water to the interfacial and from oil to the interfacial regions of emulsions by employing Equations (19)–(22).

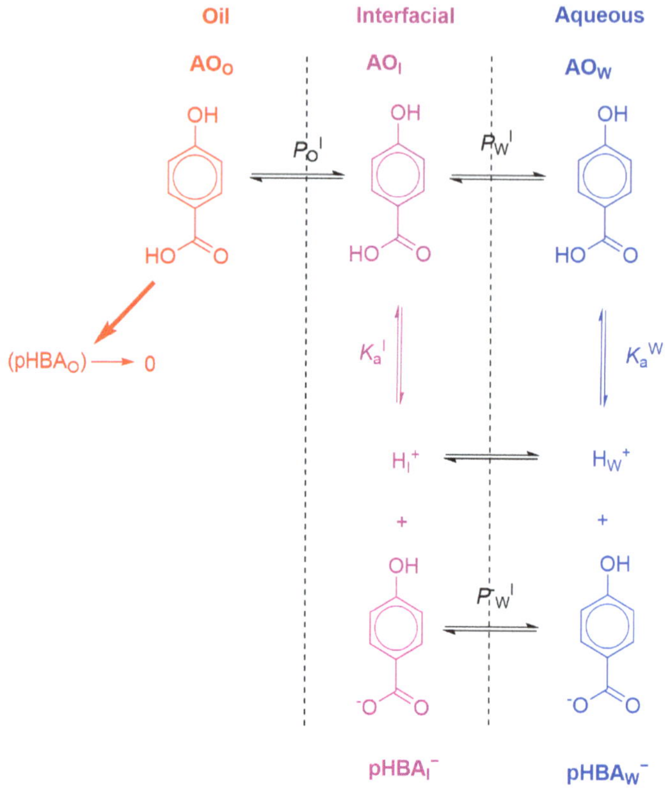


**Figure 12.** Equilibria for the distribution of a phenolic acid, AO, in emulsions. Phenolic acids only ionizes significantly at the typical acidities (pH = 2–6) of the emulsified foods. Adapted from Losada-Barreiro et al. [78] with permission, copyright J. Wiley and Sons.

$$\Delta G_T^{0,\,\mathrm{W}\rightarrow\mathrm{I}} = RT\ln\frac{\mathrm{V_m^W}}{\mathrm{P_W^I V_m^I}} \tag{19}$$

$$\Delta G_T^{0,\,\mathrm{O}\rightarrow\mathrm{I}} = RT\ln\frac{\mathrm{V_m^O}}{\mathrm{P_O^I V_m^I}} \tag{20}$$

$$\Delta H_T^{0,\mathrm{W}\rightarrow I} = R\left[\frac{\partial(\ln P_\mathrm{W}^I)}{\partial\left(\frac{1}{T}\right)}\right]_P \tag{21}$$

$$\Delta H_T^{0,\mathrm{O}\rightarrow I} = R\left[\frac{\partial(\ln P_\mathrm{O}^I)}{\partial\left(\frac{1}{T}\right)}\right]_P \tag{22}$$

The enthalpy and entropy changes for the transfer of CA from the aqueous to the interfacial region, $\Delta H_T^{0,\mathrm{W}\rightarrow I}$ and $\Delta S_T^{0,\mathrm{W}\rightarrow I}$, can be obtained from the variations of $P_\mathrm{W}{}^\mathrm{I}$ at a series of temperatures by using the van´t Hoff Equation (12) and the Gibbs Equation (14), respectively. Similar equations are employed to determine $\Delta H_T^{0,\mathrm{O}\rightarrow I}$ and $\Delta S_T^{0,\mathrm{O}\rightarrow I}$ for the transfer of TOC. Figure 4 illustrates the variation of $\ln(P_\mathrm{W}{}^\mathrm{I})$ and $\ln(P_\mathrm{O}{}^\mathrm{I})$ with $1/T$ according to the van´t Hoff equation, from where the enthalpy values of the transfer are obtained. An illustrative example of the determination of the thermodynamic parameters in emulsions is given in Figure 13 for caffeic acid and tocopherol.

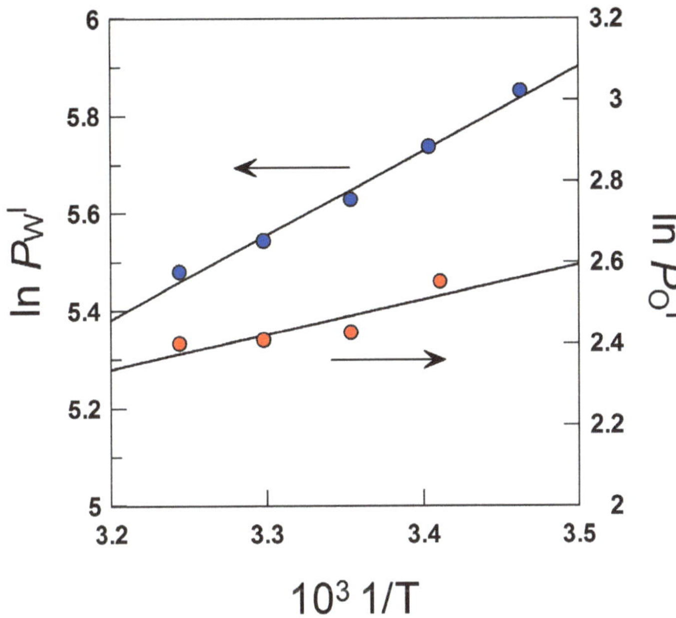


**Figure 13.** Variations of $\ln P_O{}^I$ and $\ln P_W{}^I$ with $1/T$ (Kelvin) according to the van't Hoff equation for caffeic acid (blue) and α-tocopherol (red) in corn oil emulsions. Figure reproduced from Losada-Barreiro et al. [90] with permission, copyright Royal Society of Chemistry.

## 9. Final Remarks and Conclusions

The goal of this manuscript was to have a first glimpse at the partitioning behavior of antioxidants between two immiscible liquids and of the molecular interactions that govern their partitioning, showing that simple thermodynamic concepts, in particular, chemical potential, can be used to quantify equilibrium partitioning. The interest in partitioning comes, in part, because the massive production of agri-food wastes and by-products generated in the agri-food industrial sector not only present safe disposal issues, but also contribute negative environmental impacts. This environmental stress is enormous in our globalized world, and a sustainable utilization of agri-food waste and/or by-products to produce value-added products for potential applications in cosmetic, pharmaceutical, or food industrial uses can provide considerable opportunities for earning an additional income for the dependent industry. In addition, adding economic value to environmentally undesirable agriculture and/or food wastes contributes to ensuring sustainable food production and security. To date, most of these wastes have been employed as organic fertilizers or as source of fuel, but modern technologies in combination with "green chemistry" principles lead to the effective and rational use of wastes and by-products producing value-added materials.

Bioactive components such as phenolic antioxidants, isolated from fruits, vegetables, and their wastes include polyphenols, tannins, flavonoids, and a variety of vitamins. In the marine sector, polyphenols with an antioxidant activity can be obtained from the fish and shellfish industries and from natural or harvested seaweeds. Interest in polyphenolic antioxidants has also increased notably because of their protective roles in food and pharmaceutical products against oxidative deterioration and in the body and against oxidative stress-mediated pathological processes. Indeed, the search for natural antioxidants in plants and plant-derived compounds, as well as their use in foods, cosmetics, pharmacological formulations, parenteral emulsions, etc., requires appropriate methods to determine their distributions in order to reveal their mechanisms of activity and to understand the kinetics of their reactions with free radicals.

Nevertheless, the valorization of wastes and by-products requires extraction procedures where the partition constants between two solvents play a major role. Determining partition constants has received much attention in the assessment of relative lipophilicity and hydrophilicity of a compound because it dictates formulation strategies for various biological, medical, and pharmaceutical processes. It is also important for synthesis and purification, and for understanding the release and distribution of drugs in human body. $P_W{}^O$ values are commonly used for the description of the hydrophobicity or lipophilicity of compounds and some of them, for example, the $P_W{}^{OCT}$ values are, in general, good descriptors for the characterization of the relationship between the structure and biological, pharmacological, and ecological effects of various drugs. It is not surprising, therefore, the enormous interest in determining the distributions of antioxidants in binary oil–water systems as they are one of the secondary metabolites present in nature. Their distribution (and in general, that of any other compound) depend on the balance of all intermolecular forces, mainly hydrogen bonding and van der Waals interactions between the solute and oil and water phases. Because the composition of the various alcohols commonly employed to mimic the partitioning of drugs in the human body (octanol, hexanol, etc.) is quite different from that of edible oils, the partition constant values determined in oil cannot be extrapolated to others, and hence need to be determined individually.

**Author Contributions:** Conceptualization, C.B.-D. and S.L.-B.; methodology, C.B.-D. and S.L.-B.; formal analysis, C. B-D., S.L.-B. and F.P.-M.; investigation, C.B.-D., S.L.-B. and F.P.-M.; resources, C.B.-D., S.L.-B. and F.P.-M.; data curation, S.L.-B.; writing—original draft preparation, C.B.-D.; writing—review and editing, S.L.-B., F.P.-M. and C.B.-D.; funding acquisition, S.L.-B., C.B.-D. and F.P.-M. All authors have read and agreed to the published version of the manuscript.

**Funding:** This work was funded by the University of Vigo, grants C11 and ABH1-V231, Xunta de Galicia (Cluster of Agricultural Research and Development—CITACA Strategic Partnership, grant ED431E 2018/07); the "Network for the Sustainable Use of Natural and Agrifood Resources" (REDUSO, grant ED431D 2017/18); and PT national funds (FCT/MCTES, Fundação para a Ciência e Tecnologia and Ministério da Ciência, Tecnologia e Ensino Superior) through projects UIDB/50006/2020 and UIDP/50006/2020.

**Institutional Review Board Statement:** Not applicable.

**Informed Consent Statement:** Not applicable.

**Data Availability Statement:** Not applicable.

**Acknowledgments:** We thank the University of Vigo and, particularly, the Physical-Chemistry Department for laboratory supplies.

**Conflicts of Interest:** The authors declare no conflict of interest.

## References

1. Conde, E.; Díaz Reinoso, B.; González-Muñoz, M.J.; Moure, A.; Domínguez, H.; Parajó, J.C. Recovery and Concentration of Antioxidants from Industrial Effluents and from Processing Streams of Underutilized Vegetal Biomass. *Food Public Health* **2013**, *3*, 69–91. [CrossRef]
2. Ben-Othman, S.; Jõudu, I.; Bhat, R. Bioactives from Agri-Food Wastes: Present Insights and Future Challenges. *Molecules* **2020**, *25*, 510. [CrossRef]
3. Lourenço, S.C.; Moldão-Martins, M.; Alves, V.D. Antioxidants of Natural Plant Origins: From Sources to Food Industry Applications. *Molecules* **2019**, *24*, 4132. [CrossRef] [PubMed]
4. Costa, M.; Sezgin-Bayindir, Z.; Losada-Barreiro, S.; Paiva-Martins, F.; Saso, L.; Bravo-Díaz, C. Polyphenols as Antioxidants for Extending Food Shelf-Life and in the Prevention of Health Diseases: Encapsulation and Interfacial Phenomena. *Biomedicines* **2021**, *9*, 1909. [CrossRef] [PubMed]
5. Costa, M.; Sezgin-Bayindir, Z.; Losada-Barreiro, S.; Paiva-Martins, F.; Saso, L.; Bravo-Díaz, C. Plant Antioxidants in Food Emulsions. In *Some New Aspects of Colloidal Systems in Foods*; Milani, J.M., Ed.; IntechOpen: London, UK, 2018.

6. Jideani, A.I.O.; Silungwe, H.; Takalani, T.; Omolola, A.O.; Udeh, H.O.; Anyasi, T.A. Antioxidant-rich natural fruit and vegetable products and human health. *Int. J. Food Prop.* **2021**, *24*, 41–67. [CrossRef]
7. Chandra, P.; Sharma, R.K.; Arora, D.S. Antioxidant compounds from microbial sources: A review. *Food Res. Int.* **2020**, *129*, 108849. [CrossRef]
8. Díaz-Reinoso, B.; González-López, N.; Moure, A.; Domínguez, H.; Parajó, J.C. Recovery of antioxidants from industrial waste liquors using membranes and polymeric resins. *J. Food Eng.* **2010**, *96*, 127–133. [CrossRef]
9. Fierascu, R.C.; Fierascu, I.; Avramescu, S.M.; Sieniawska, E. Recovery of Natural Antioxidants from Agro-Industrial Side Streams through Advanced Extraction Techniques. *Molecules* **2019**, *24*, 4212. [CrossRef]
10. Zhang, Z.F.; Zhang, X.; Zhang, X.; Sverko, E.; Smyth, S.A.; Li, Y.F. Diphenylamine Antioxidants in wastewater influent, effluent, biosolids and landfill leachate: Contribution to environmental releases. *Water Res.* **2021**, *189*, 116602. [CrossRef]
11. Justino, C.I.; Pereira, R.; Freitas, A.C.; Rocha-Santos, T.A.; Panteleitchouk, T.S.; Duarte, A.C. Olive oil mill wastewaters before and after treatment: A critical review from the ecotoxicological point of view. *Ecotoxicology* **2012**, *21*, 615–629. [CrossRef]
12. Moure, A.; Domínguez, H.; Parajó, J.C. Antioxidant properties of ultrafiltration-recovered soy protein fractions from industrial effluents and their hydrolysates. *Process Biochem.* **2006**, *41*, 447–456. [CrossRef]
13. Melini, V.; Melini, F.; Luziatelli, F.; Ruzzi, M. Functional Ingredients from Agri-Food Waste: Effect of Inclusion Thereof on Phenolic Compound Content and Bioaccessibility in Bakery Products. *Antioxidants* **2020**, *9*, 1216. [CrossRef] [PubMed]
14. Stromsnes, K.; Lagzdina, R.; Olaso-Gonzalez, G.; Gimeno-Mallench, L.; Gambini, J. Pharmacological Properties of Polyphenols: Bioavailability, Mechanisms of Action, and Biological Effects in In Vitro Studies, Animal Models, and Humans. *Biomedicines* **2021**, *9*, 1074. [CrossRef]
15. Schaich, K.M. Lipid Antioxidants: More than Just Lipid Radical Quenchers. In *Lipid Oxidation in Food and Biological Systems: A Physical Chemistry Perspective*; Bravo-Diaz, C., Ed.; Springer International Publishing: Cham, Switzerland, 2022; pp. 151–184. [CrossRef]
16. Jimenez-Alvarez, D.; Giuffrida, F.; Golay, P.A.; Cotting, C.; Lardeau, A.; Keely, B.J. Antioxidant activity of oregano, parsley, and olive mill wastewaters in bulk oils and oil-in-water emulsions enriched in fish oil. *J. Agric. Food Chem.* **2008**, *56*, 7151–7159. [CrossRef]
17. Liu, Y.; Qin, Y.; Bai, R.; Zhang, X.; Yuan, L.; Liu, J. Preparation of pH-sensitive and antioxidant packaging films based on κ-carrageenan and mulberry polyphenolic extract. *Int. J. Biol. Macromol.* **2019**, *134*, 993–1001. [CrossRef] [PubMed]
18. Domínguez, R.; Barba, F.J.; Gómez, B.; Putnik, P.; Bursać Kovačević, D.; Pateiro, M.; Santos, E.M.; Lorenzo, J.M. Active packaging films with natural antioxidants to be used in meat industry: A review. *Food Res. Int.* **2018**, *113*, 93–101. [CrossRef]
19. Zhang, Y.; Wang, B.; Lu, F.; Wang, L.; Ding, Y.; Kang, X. Plant-derived antioxidants incorporated into active packaging intended for vegetables and fatty animal products: A review. *Food Addit. Contam. Part A Chem. Anal. Control Expo. Risk Assess.* **2021**, *38*, 1237–1248. [CrossRef]
20. De Leonardis, A.; Macciola, V.; Lembo, G.; Aretini, A.; Nag, A. Studies on oxidative stabilisation of lard by natural antioxidants recovered from olive-oil mill wastewater. *Food Chem.* **2007**, *100*, 998–1004. [CrossRef]
21. Leo, A.J. Octanol/Water Partition Coefficients. In *Encyclopedia of Computational Chemistry*; John Wiley & Sons, Ltd.: Hoboken, NJ, USA, 2002. [CrossRef]
22. Berthod, A.; Carda-Bosch, S. Determination of liquid-liquid partition coefficients by separation methods. *J. Chromatog. A* **2004**, *1037*, 3–14. [CrossRef]
23. Amézqueta, S.; Subirats, X.; Fuguet, E.; Rosés, M.; Ràfols, C. Chapter 6—Octanol-Water Partition Constant. In *Liquid-Phase Extraction*; Poole, C.F., Ed.; Elsevier: Amsterdam, The Netherlands, 2020; pp. 183–208. [CrossRef]
24. Loureiro, D.R.P.; Soares, J.X.; Lopes, D.; Macedo, T.; Yordanova, D.; Jakobtorweihen, S.; Nunes, C.; Reis, S.; Pinto, M.M.M.; Afonso, C.M.M. Accessing lipophilicity of drugs with biomimetic models: A comparative study using liposomes and micelles. *Eur. J. Pharm. Sci.* **2018**, *115*, 369–380. [CrossRef]
25. Jakobtorweihen, S.; Zuniga, A.C.; Ingram, T.; Gerlach, T.; Keil, F.J.; Smirnova, I. Predicting solute partitioning in lipid bilayers: Free energies and partition coefficients from molecular dynamics simulations and COSMOmic. *J. Chem. Phys.* **2014**, *141*, 045102. [CrossRef] [PubMed]
26. Blokhina, S.V.; Ol'khovich, M.V.; Sharapova, A.V.; Proshin, A.N.; Perlovich, G.L. Partition coefficients and thermodynamics of transfer of novel drug-like spiro-derivatives in model biological solutions. *J. Chem. Thermodyn.* **2013**, *61*, 11–17. [CrossRef]
27. Sangster, J. *Octanol-Water Partition coefficients, Fundamentals and Physcial Chemistry*; J. Wiley & Sons: Chichester, UK, 1997.
28. Dini, I. Contribution of Nanoscience Research in Antioxidants Delivery Used in Nutricosmetic Sector. *Antioxidants* **2022**, *11*, 563. [CrossRef]
29. Sezgin-Bayindir, Z.; Losada-Barreiro, S.; Bravo-Díaz, C.; Sova, M.; Kristl, J.; Saso, L. Nanotechnology-Based Drug Delivery to Improve the Therapeutic Benefits of NRF2 Modulators in Cancer Therapy. *Antioxidants* **2021**, *10*, 685. [CrossRef] [PubMed]
30. Losada-Barreiro, S.; Bravo-Díaz, C.; Paiva-Martins, F. Why encapsulate antioxidants in emulsion-based systems, where they are located, and how location affects their efficiency. In *Emulsion-Based Encapsulation of Antioxidants*; Aboudzadeh, M.A., Ed.; Springer Nature: Cham, Switzerland, 2021. [CrossRef]
31. García-Pérez, P.; Losada-Barreiro, S.; Gallego, P.P.; Bravo-Díaz, C. Cyclodextrin-Elicited Bryophyllum Suspension Cultured Cells: Enhancement of the Production of Bioactive Compounds. *Int. J. Mol. Sci.* **2019**, *20*, 5180. [CrossRef] [PubMed]

32. Kaur, G.; Grewal, J.; Jyoti, K.; Jain, U.K.; Chandra, R.; Madan, J. Chapter 15—Oral controlled and sustained drug delivery systems: Concepts, advances, preclinical, and clinical status. In *Drug Targeting and Stimuli Sensitive Drug Delivery Systems*; Grumezescu, A.M., Ed.; William Andrew Publishing: Norwich, NY, USA, 2018; pp. 567–626. [CrossRef]
33. Chamberlin, A.C.; Levitt, D.G.; Cramer, C.J.; Truhlar, D.G. Modeling Free Energies of Solvation in Olive Oil. *Mol. Pharm.* **2008**, *5*, 1064–1079. [CrossRef]
34. Coe, E.L.; Coe, M.H. A hypothesis relating oil: Water partition coefficients and vapor pressures of nonelectrolytes to their penetration rates through biological membranes. *J. Theor. Biol.* **1965**, *8*, 327–343. [CrossRef]
35. Chiou, C.T. Partition coefficients of organic compounds in lipid-water systems and correlations with fish bioconcentration factors. *Environ. Sci. Technol.* **1985**, *19*, 57–62. [CrossRef]
36. Liu, X.; Testa, B.; Fahr, A. Lipophilicity and Its Relationship with Passive Drug Permeation. *Pharm. Res.* **2011**, *28*, 962–977. [CrossRef]
37. Poulin, P.; Schoenlein, K.; Theil, F.P. Prediction of adipose tissue: Plasma partition coefficients for structurally unrelated drugs. *J. Pharm. Sci.* **2001**, *90*, 436–447. [CrossRef]
38. Krinsky, N.I. Membrane antioxidants. *Ann. N.Y. Acad. Sci.* **1988**, *551*, 17–32. [CrossRef] [PubMed]
39. Tomasi, J.; Persico, M. Molecular Interactions in Solution: An Overview of Methods Based on Continuous Distributions of the Solvent. *Chem. Rev.* **1994**, *94*, 2027–2094. [CrossRef]
40. Palmer, D.S.; Llinàs, A.; Morao, I.; Day, G.M.; Goodman, J.M.; Glen, R.C.; Mitchell, J.B.O. Predicting Intrinsic Aqueous Solubility by a Thermodynamic Cycle. *Mol. Pharm.* **2008**, *5*, 266–279. [CrossRef] [PubMed]
41. Costa, M.; Paiva-Martins, F.; Bravo-Díaz, C.; Losada-Barreiro, S. Control of Lipid Oxidation in Oil-in Water Emulsions: Effects of Antioxidant Partitioning and Surfactant Concentration. In *Lipid Oxidation in Food and Biological Systems: A Physical Chemistry Perspective*; Bravo-Diaz, C., Ed.; Springer International Publishing: Cham, Switzerland, 2022; pp. 201–216. [CrossRef]
42. Romsted, L.S.; Bravo-Díaz, C. Determining Antioxidant Distributions in Intact Emulsions by Kinetic Methods: Application of Pseudophase Models. In *Lipid Oxidation in Food and Biological Systems: A Physical Chemistry Perspective*; Bravo-Diaz, C., Ed.; Springer International Publishing: Cham, Switzerland, 2022; pp. 31–48. [CrossRef]
43. Pérez-Palacios, T.; Estévez, M. Lipid Oxidation in Meat Systems: Updated Means of Detection and Innovative Antioxidant Strategies. In *Lipid Oxidation in Food and Biological Systems: A Physical Chemistry Perspective*; Bravo-Diaz, C., Ed.; Springer International Publishing: Cham, Switzerland, 2022; pp. 93–111. [CrossRef]
44. Costa, M.; Losada-Barreiro, S.; Bravo-Díaz, C.; Paiva-Martins, F. Effects of Emulsion Droplet Size on the Distribution and Efficiency of Antioxidants. In *Lipid Oxidation in Food and Biological Systems: A Physical Chemistry Perspective*; Bravo-Diaz, C., Ed.; Springer International Publishing: Cham, Switzerland, 2022; pp. 217–235. [CrossRef]
45. Bravo-Díaz, C. Advances in the control of lipid peroxidation in oil-in-water emulsions: Kinetic approaches. *Crit. Rev. Food Sci. Nutr.* **2022**, *1*, 1–33. [CrossRef]
46. Wu, H.; Richards, M.P. Lipid oxidation and antioxidant delivery systems in muscle food. *Compr. Rev. Food Sci. Food Saf.* **2022**, *21*, 1275–1299. [CrossRef]
47. Yang, W.; Yue, H.; Lu, G.; Wang, W.; Deng, Y.; Ma, G.; Wei, W. Advances in Delivering Oxidative Modulators for Disease Therapy. *Research* **2022**, *2022*, 9897464. [CrossRef]
48. Nic, M.; Jirat, J.; Kosata, J.; Jenkins, A.; McNaught, A. *Compendium of Chemical Terminology. Gold Book, IUPAC, (Web 2.0 Version)*; IUPAC: Research Triangle Park, NC, USA, 2019. [CrossRef]
49. Leo, A.; Hansch, C.; Elkins, D. Partition Coefficients And Their Uses. *Chem. Rev.* **1971**, *71*, 525–616. [CrossRef]
50. Chen, C.-S.; Lin, S.-T. Prediction of pH Effect on the Octanol–Water Partition Coefficient of Ionizable Pharmaceuticals. *Ind. Eng. Chem. Res.* **2016**, *55*, 9284–9294. [CrossRef]
51. Bannan, C.C.; Calabró, G.; Kyu, D.Y.; Mobley, D.L. Calculating Partition Coefficients of Small Molecules in Octanol/Water and Cyclohexane/Water. *J. Chem. Theory Comput.* **2016**, *12*, 4015–4024. [CrossRef]
52. Sarkar, A.; Kellogg, G.E. Hydrophobicity–shake flasks, protein folding and drug discovery. *Curr. Top. Med. Chem.* **2010**, *10*, 67–83. [CrossRef]
53. Leckband, D.; Israelachvili, J. Intermolecular forces in biology. *Q. Rev. Biophys.* **2001**, *34*, 105–267. [CrossRef]
54. Atkins, P.; de Paula, J. *Physical Chemistry*; Oxford University Press: Oxford, UK, 2022.
55. Israelachvili, J. *Intermolecular and Surface Forces*, 3rd ed.; Elsevier: Amsterdam, The Netherlands, 2011. [CrossRef]
56. Leo, A.J.; Hansch, C. Linear free energy relations between partitioning solvent systems. *J. Org. Chem.* **1971**, *36*, 1539–1544. [CrossRef]
57. Schwarzenbach, R.P.; Gschwend, P.M.; Imboden, D.M. Partitioning: Molecular Interactions and Thermodynamics. In *Environmental Organic Chemistry*; Schwarzenbach, R.P., Gschwend, P.M., Imboden, D.M., Eds.; J. Wiley & Sons: Hoboken, NJ, USA, 2002; pp. 57–96. [CrossRef]
58. Hansch, C.; Leo, A.; Hoelkman, D. *Exploring QSAR: Fundamentals and Applications in Chemistry and Biology*; American Chemical Society: Washington, DC, USA, 1995.
59. Zamora, W.J.; Curutchet, C.; Campanera, J.M.; Luque, F.J. Prediction of pH-Dependent Hydrophobic Profiles of Small Molecules from Miertus–Scrocco–Tomasi Continuum Solvation Calculations. *J. Phys. Chem. B* **2017**, *121*, 9868–9880. [CrossRef]

60. Kempińska, D.; Chmiel, T.; Kot-Wasik, A.; Mróz, A.; Mazerska, Z.; Namieśnik, J. State of the art and prospects of methods for determination of lipophilicity of chemical compounds. *TrAC Trends Anal. Chem.* **2019**, *113*, 54–73. [CrossRef]
61. Montalbán, M.G.; Collado-González, M.; Trigo, R.M.; Fg, D.; Víllora, G. Experimental Measurements of Octanol-Water Partition Coefficients of Ionic Liquids. *J. Adv. Chem. Eng.* **2015**, *5*, 2. [CrossRef]
62. Berthod, A.; García Álvarez-Coque, M.C. *Micellar Liquid Chromatography*; Chromatographic Science Series; Marcel Dekker: New York, NY, USA, 2000.
63. Liang, C.; Lian, H.-z. Recent advances in lipophilicity measurement by reversed-phase high-performance liquid chromatography. *TrAC Trends Anal. Chem.* **2015**, *68*, 28–36. [CrossRef]
64. Kubik, Ł.; Struck-Lewicka, W.; Kaliszan, R.; Wiczling, P. Simultaneous determination of hydrophobicity and dissociation constant for a large set of compounds by gradient reverse phase high performance liquid chromatography–mass spectrometry technique. *J. Chromatogr. A* **2015**, *1416*, 31–37. [CrossRef]
65. Kaliszan, R. QSRR: Quantitative Structure-(Chromatographic) Retention Relationships. *Chem. Rev.* **2007**, *107*, 3212–3246. [CrossRef]
66. Tetko, I.V.; Yan, A.; Gasteiger, J. Prediction of Physicochemical Properties of Compounds. In *Applied Chemoinformatics*; Wiley: Hoboken, NJ, USA, 2018; pp. 53–81. [CrossRef]
67. Freiría-Gándara, J.; Losada-Barreiro, S.; Paiva-Martins, F.; Bravo-Díaz, C. Differential Partitioning of Bioantioxidants in Edible Oil–Water and Octanol–Water Systems: Linear Free Energy Relationships. *J. Chem. Eng. Data* **2018**, *63*, 2999–3007. [CrossRef]
68. Costa, M.; Losada-Barreiro, S.; Paiva-Martins, F.; Bravo-Díaz, C. Polyphenolic Antioxidants in Lipid Emulsions: Partitioning Effects and Interfacial Phenomena. *Foods* **2021**, *10*, 539.
69. Molyneux, P. Octanol/water partition coefficients Kow: A critical examination of the value of the methylene group contribution to logKow for homologous series of organic compounds. *Fluid Phase Equilibria* **2014**, *368*, 120–141. [CrossRef]
70. Cappelli, C.I.; Benfenati, E.; Cester, J. Evaluation of QSAR models for predicting the partition coefficient (logP) of chemicals under the REACH regulation. *Environ. Res.* **2015**, *143*, 26–32. [CrossRef]
71. Klamt, A.; Eckert, F.; Arlt, W. COSMO-RS: An alternative to simulation for calculating thermodynamic properties of liquid mixtures. *Annu. Rev. Chem. Biomol. Eng.* **2010**, *1*, 101–122. [CrossRef] [PubMed]
72. Mannhold, R.; Rekker, R.F.; Sonntag, C.; ter Laak, A.M.; Dross, K.; Polymeropoulos, E.E. Comparative Evaluation of the Predictive Power of Calculation Procedures for Molecular Lipophilicity. *J. Pharm. Sci.* **1995**, *84*, 1410–1419. [CrossRef] [PubMed]
73. Beezer, A.E.; Gooch, C.A.; Hunter, W.H.; Volpe, P.L.O. A thermodynamic analysis of the Collander equation and establishment of a reference solvent for use in drug partitioning studies. *J. Pharm. Pharmacol.* **1987**, *39*, 774–779. [CrossRef]
74. Molinspiration. Available online: https://www.molinspiration.com/ (accessed on 23 February 2023).
75. Goss, K.-U.; Schwarzenbach, R.P. Linear Free Energy Relationships Used To Evaluate Equilibrium Partitioning of Organic Compounds. *Environ. Sci. Technol.* **2001**, *35*, 1–9. [CrossRef]
76. Farooq, S.; Abdullah; Zhang, H.; Weiss, J. A comprehensive review on polarity, partitioning, and interactions of phenolic antioxidants at oil–water interface of food emulsions. *Nat. Antioxid. Sources Compd.* **2021**, *20*, 4250–4277. [CrossRef]
77. Váňová, J.; Liimatta, L.J.; Česla, P.; Wiedmer, S.K. Determination of distribution constants of antioxidants by electrokinetic chromatography. *Cogent Chem.* **2017**, *3*, 1385173. [CrossRef]
78. Losada-Barreiro, S.; Bravo-Díaz, C.; Romsted, L.S. Distributions of phenolic acid antioxidants between the interfacial and aqueous regions of corn oil emulsions: Effects of pH and emulsifier concentration. *Eur. J. Lipid Sci. Technol.* **2015**, *117*, 1801–1813.
79. Dean, J.A. *Lange 's Handbook of Chemistry*; McGraw-Hill, Inc.: New York, NY, USA, 1992.
80. Fazary, A.E.; Ju, Y. Nonaqueous Solution Studies on the Protonation Equilibria of some Phenolic Acids. *J. Solut. Chem.* **2008**, *37*, 1305–1319. [CrossRef]
81. Ozkorucuklu, S.P.; Beltrán, J.L.; Fonrodona, G.; Barrón, D.; Alsancak, G.; Barbosa, J. Determination of Dissociation Constants of Some Hydroxylated Benzoic and Cinnamic Acids in Water from Mobility and Spectroscopic Data Obtained by CE-DAD. *J. Chem. Eng. Data* **2009**, *54*, 807–811. [CrossRef]
82. Arellano, J.B.; Li, H.; González-Pérez, S.; Gutiérrez, J.; Melø, T.B.; Vacha, F.; Naqvi, K.R. Trolox, a Water-Soluble Analogue of α-Tocopherol, Photoprotects the Surface-Exposed Regions of the Photosystem II Reaction Center in Vitro. Is This Physiologically Relevant? *Biochem.* **2011**, *50*, 8291–8301.
83. Raimúndez-Rodríguez, E.A.; Losada-Barreiro, S.; Bravo-Díaz, C. Enhancing the fraction of antioxidants at the interfaces of oil-in-water emulsions: A kinetic and thermodynamic analysis of their partitioning. *J. Colloid Interface Sci.* **2019**, *555*, 224–233. [CrossRef] [PubMed]
84. Pastoriza-Gallego, M.J.; Sánchez-Paz, V.; Losada-Barreiro, S.; Bravo-Diaz, C.; Gunaseelan, K.; Romsted, L.S. Effects of Temperature and Emulsifier Concentration on α-Tocopherol Distribution in a Stirred, Fluid, Emulsion. Thermodynamics of α-Tocopherol transfer between the Oil and Interfacial Regions. *Langmuir* **2009**, *25*, 2646–2653. [PubMed]
85. McClements, D.J. *Food Emulsions, Principles, Practices and Techniques*; CRC Press: Boca Raton, FL, USA, 2015.
86. Dalgleish, D.G. Food emulsions—Their structures and structure-forming properties. *Food Hydrocoll.* **2006**, *20*, 415–422. [CrossRef]
87. Dar, A.A.; Bravo-Diaz, C.; Nazir, N.; Romsted, L.S. Chemical kinetic and chemical trapping methods: Unique approaches for determining respectively the antioxidant distributions and interfacial molarities of water, counter-anions, and other weakly basic nucleophiles in association colloids. *Curr. Opin. Colloid Interface Sci.* **2017**, *32*, 84–93. [CrossRef]

88. Bravo-Díaz, C.; Romsted, L.S.; Liu, C.; Losada-Barreiro, S.; Pastoriza-Gallego, M.J.; Gao, X.; Gu, Q.; Krishnan, G.; Sánchez-Paz, V.; Zhang, Y.; et al. To Model Chemical Reactivity in Heterogeneous Emulsions, Think Homogeneous Microemulsions. *Langmuir* **2015**, *31*, 8961–8979. [CrossRef]
89. Romsted, L.S.; Bravo-Díaz, C. Modelling chemical reactivity in emulsions. *Curr. Opin. Colloid Interface Sci.* **2013**, *18*, 3–14.
90. Losada-Barreiro, S.; Sánchez-Paz, V.; Bravo-Díaz, C. Transfer of antioxidants at the interfaces of model food emulsions: Distributions and thermodynamic parameters. *Org. Biomol. Chem.* **2008**, *6*, 4004–4011.
91. Almeida, J.; Losada-Barreiro, S.; Costa, M.; Paiva-Martins, F.; Bravo-Díaz, C.; Romsted, L.S. Interfacial Concentrations of Hydroxytyrosol and Its Lipophilic Esters in Intact Olive Oil-in-Water Emulsions: Effects of Antioxidant Hydrophobicity, Surfactant Concentration, and the Oil-to-Water Ratio on the Oxidative Stability of the Emulsions. *J. Agric. Food Chem.* **2016**, *64*, 5274–5283.
92. Costa, M.; Losada-Barreiro, S.; Paiva-Martins, F.; Bravo-Díaz, C.; Romsted, L.S. A direct correlation between the antioxidant efficiencies of caffeic acid and its alkyl esters and their concentrations in the interfacial region of olive oil emulsions. The pseudophase model interpretation of the "cut-off" effect. *Food Chem.* **2015**, *175*, 233–242.
93. Costa, M.; Losada-Barreiro, S.; Bravo-Díaz, C.; Monteiro, L.S.; Paiva-Martins, F. Interfacial Concentrations of Hydroxytyrosol Derivatives in Fish Oil-in-Water Emulsions and Nanoemulsions and Its Influence on Their Lipid Oxidation: Droplet Size Effects. *Foods* **2020**, *9*, 1897. [CrossRef] [PubMed]
94. Costa, M.; Freiría-Gándara, J.; Losada-Barreiro, S.; Paiva-Martins, F.; Bravo-Díaz, C. Effects of droplet size on the interfacial concentrations of antioxidants in fish and olive oil-in-water emulsions and nanoemulsions and on their oxidative stability. *J. Colloid Interface Sci.* **2020**, *562*, 352–362. [CrossRef] [PubMed]
95. Costa, M.; Losada-Barreiro, S.; Bravo-Díaz, C.; Vicente, A.A.; Monteiro, L.S.; Paiva-Martins, F. Influence of AO chain length, droplet size and oil to water ratio on the distribution and on the activity of gallates in fish oil-in-water emulsified systems: Emulsion and nanoemulsion comparison. *Food Chem.* **2020**, *310*, 125716. [CrossRef]
96. Freiría-Gándara, J.; Losada-Barreiro, S.; Paiva-Martins, F.; Bravo-Díaz, C. Enhancement of the antioxidant efficiency of gallic acid derivatives in intact fish oil-in-water emulsions through optimization of their interfacial concentrations. *Food Funct.* **2018**, *9*, 4429–4442.
97. Martinez-Aranda, N.; Losada-Barreiro, N.; Bravo-Díaz, C.; Romsted, L.S. Influence of Temperature on the Distribution of Catechin in Corn oil-in-water Emulsions and some Relevant Thermodynamic Parameters. *Food Biophys.* **2014**, *9*, 380–388. [CrossRef]
98. Losada-Barreiro, S.; Bravo-Díaz, C.; Costa, M.; Paiva-Martins, F. Distribution of catechol in emulsions. *J. Phys. Org. Chem.* **2013**, *in press*. [CrossRef]
99. Costa, M.; Losada-Barreiro, S.; Paiva-Martins, F.; Bravo-Diaz, C. Optimizing the efficiency of antioxidants in emulsions by lipophilization: Tuning interfacial concentrations. *RSC Adv.* **2016**, *6*, 91483–91493.
100. Losada-Barreiro, S.; Bravo Díaz, C.; Paiva Martins, F.; Romsted, L.S. Maxima in antioxidant distributions and efficiencies with increasing hydrophobicity of gallic acid and its alkyl esters. The pseudophase model interpretation of the "Cut-off effect. *J. Agric. Food Chem.* **2013**, *61*, 6533–6543.
101. Narkiewicz-Michalek, J.; Szymula, M.; Losada-Barreiro, S.; Bravo-Diaz, C. Concentration of resveratrol at the oil–water interface of corn oil-in-water emulsions. *Adsorption* **2019**, *25*, 903–911. [CrossRef]
102. Lisete-Torres, P.; Losada-Barreiro, S.; Albuquerque, H.; Sánchez-Paz, V.; Paiva-Martins, F.; Bravo-Díaz, C. Distribution of hydroxytyrosol and hydroxytyrosol acetate in olive oil emulsions and their antioxidant efficiency. *J. Agric. Food Chem.* **2012**, *60*, 7318–7325. [PubMed]
103. Galan, A.; Losada-Barreiro, S.; Bravo-Díaz, C. A Physicochemical Study of the Effects of Acidity on the Distribution and Antioxidant Efficiency of Trolox in Olive Oil-in-Water Emulsions. *ChemPhysChem* **2016**, *17*, 296–304.
104. Meireles, M.; Losada-Barreiro, S.; Costa, M.; Paiva-Martins, F.; Bravo-Díaz, C.; Monteiro, L.S. Control of antioxidant efficiency of chlorogenates in emulsions: Modulation of antioxidant interfacial concentrations. *J. Sci. Food Agric.* **2019**, *99*, 3917–3925. [CrossRef] [PubMed]
105. Mitrus, O.; Żuraw, M.; Losada-Barreiro, S.; Bravo-Díaz, C.; Paiva-Martins, F. Targeting Antioxidants to Interfaces: Control of the Oxidative Stability of Lipid-Based Emulsions. *J. Agric. Food Chem.* **2019**, *67*, 3266–3274. [CrossRef] [PubMed]
106. Costa, M.; Losada-Barreiro, S.; Paiva-Martins, F.; Bravo-Díaz, C. Physical evidence that the variations in the efficiency of homologous series of antioxidants in emulsions are a result of differences in their distribution. *J. Sci. Food Agric.* **2017**, *97*, 564–571. [CrossRef]

*antioxidants*

MDPI

*Article*

# Impact of Phenolic Acid Derivatives on the Oxidative Stability of β-Lactoglobulin-Stabilized Emulsions

**Alina Bock [1,*], Helena Kieserling [1], Ulrike Steinhäuser [2] and Sascha Rohn [1,*]**

1 Department of Food Chemistry and Food Analysis, Institute of Food Technology and Food Chemistry, Technische Universität Berlin, Gustav-Meyer-Allee 25, 13355 Berlin, Germany
2 Department of Food Technology and Food Analysis, Berliner Hochschule für Technik, Luxemburger Straße 10, 13353 Berlin, Germany
* Correspondence: a.bock@tu-berlin.de (A.B.); rohn@tu-berlin.de (S.R.)

**Abstract:** Proteins, such as β-lactoglobulin (β-Lg), are often used to stabilize oil–water-emulsions. By using an additional implementation of phenolic compounds (PC) that might interact with the proteins, the oxidative stability can be further improved. Whether PC have a certain pro-oxidant effect on oxidation processes, while interacting non-covalently (pH-6) or covalently (pH.9) with the interfacial protein-film, is not known. This study aimed to characterize the impact of phenolic acid derivatives (PCDs) on the antioxidant efficacy of the interfacial β-Lg-film, depending on their structural properties and pH-value. Electron paramagnetic resonance (EPR) analyses were performed to assess the radical scavenging in the aqueous and oil phases of the emulsion, and the complexation of transition metals: these are well known to act as pro-oxidants. Finally, in a model linseed oil emulsion, lipid oxidation products were analyzed over storage time in order to characterize the antioxidant efficacy of the interfacial protein-film. The results showed that, at pH.6, PCDs can scavenge hydrophilic radicals and partially scavenge hydrophobic radicals, as well as reduce transition metals. As expected, transition metals are complexed to only a slight degree, leading to an increased lipid oxidation through non-complexed reduced transition metals. At pH.9, there is a strong complexation between PCDs and the transition metals and, therefore, a decreased ability to reduce the transition metals; these do not promote lipid oxidation in the emulsion anymore.

**Keywords:** phenolic acid derivatives; β-lactoglobulin; oil-water-emulsions; pro-oxidant; transition metal complexation; iron reducing power; radical scavenging; oxidation products

## 1. Introduction

Most processed foods are emulsions. However, emulsions need to be regarded as both physically and chemically labile systems. Physically, emulsions tend to coalesce, to flocculate, and finally, to cream, resulting in phase separation [1]. Chemically, emulsified oil droplets have a large surface area; this favors lipid oxidation by oxygen, and being triggered through other pro-oxidant compounds that are able to diffuse through the aqueous phase onto the interface and potentially into the oil [2]. To stabilize emulsions, amphiphilic compounds, such as (small molecule) emulsifiers or proteins, are used. Proteins can unfold and align themselves with an amphiphilic conformation at the interface with their side chains by changing their secondary and tertiary structure [3]. Due to intermolecular protein interactions, a protein film forms at the interface, stabilizing the oil–water interface, and prevents physical instability [4]. Typical proteins used as emulsifiers are whey proteins or a variation of plant proteins, with legume proteins with soy being the most prominent representative [2].

Emulsifiers and proteins can have antioxidant properties and decrease lipid oxidation, and there is also the possibility of adding other antioxidants in order to reach a higher antioxidant efficacy in the interface to stabilize the emulsion. High antioxidant efficacy can be achieved when present free radicals can be scavenged and the chain reaction in the course of lipid oxidation can be slowed down or inhibited. Typical antioxidants, naturally present in foods, include phenolic compounds such as flavonoids or phenolic acids. They are resonance stabilized by extended $\pi$-electron systems, enabling them to split off protons, scavenge radicals, and convert themselves into less reactive radicals [5–7]. The property of self-oxidation and electron donation can also be disadvantageous for oxidative emulsion stability, as transition metals present in food can be reduced to pro-oxidant compounds within a redox cycle mode of action [8,9]. However, some studies showed that accelerated lipid peroxidation is due to the pro-oxidant effect of phenolic compounds [10–12]. The initial electron transfer oxidation of phenolic compounds, through the reduction of transition metals, generates the corresponding semiquinone radical of the phenolic compound. This semiquinone radical proceeds through a second electron transfer reaction with $O_2$ to form an ortho-quinone and a superoxide anion ($O_2^{\bullet-}$). Then, $O_2^{\bullet-}$ reacts with reduced transition metals to form hydrogen peroxide ($H_2O_2$). The Fenton reaction can occur as $H_2O_2$ reacts to form a hydroxyl radical ($OH^{\bullet}$). Reactive oxygen species (ROS), such as $OH^{\bullet}$ and $O_2^{\bullet-}$, are formed and promote lipid oxidation [11,13,14]. Lipid oxidation is initially manifested in the formation of primary oxidation products such as conjugated dienes and hydroperoxides. Then, these react to form secondary oxidation products, such as alkanes, esters, alcohols, ketones, or aldehydes [15]. The secondary oxidation products continue to be chemically reactive compounds that can further react to form tertiary oxidation products. While primary oxidation products are colorless and tasteless, secondary and tertiary oxidation products can lead to off-flavors and, in some cases, be harmful to health [16,17]. The formation of tertiary oxidation products is a sign of an advanced oxidative deterioration [18].

The ability of phenolic compounds to reduce transition metals and, therefore, have a pro-oxidant effect, is contrasted by their ability to complex transition metals. Khokhar et al. showed that phenolic compounds with o-dihydroxyl structures can complex Fe(III) [19]. In this regard, Hider et al. showed that the corresponding catechol structures could bind transition metals in a ratio of 1:1 to 1:3 (transition metal: catechol structure) [20]. This makes it difficult to assess whether phenolic compounds are pro-oxidant through their transition metal-reducing capacity or antioxidant through their transition metal complexation in an emulsion.

While transition metals, due to their charge, are mainly located in the aqueous phase of the emulsion or at the interfacial film, if there is no electrostatic repulsion to the interfacial film, phenolic compounds can partition themselves between the oil, interface and water phase, depending on their structural properties. The structural properties of phenolic compounds mainly result from the molecule size, the polarity, which in turn is affected by polar substituents, such as hydroxyl or glycosyl groups, as well as the pH, depending on charge [21]. Thus, the partitioning behavior of the phenolic compounds further has an impact on their anti- and pro-oxidant properties, because charged transition metals in the aqueous phase of the system would have to be reduced in the aqueous phase in order to have a pro-oxidant effect [8]. In addition, radicals already present in an emulsion can only be scavenged by phenolic compounds, provided that they are located in the same phase of the emulsion.

Another factor that may affect the anti- and pro-oxidant activity of phenolic compounds in the emulsion are the interactions between the phenolic compounds and the interfacial protein(s). Depending on the conditions, such as the pH value, temperature, ionic strength, and the presence of chemical reagents or food components, these interactions can be non-covalent or covalent interactions [21]. The effect of non-covalent and covalent interactions between phenolic compounds and proteins on their antioxidant properties has been mainly studied in aqueous systems [22–24].

In some studies, phenolic compounds and proteins were shown to interact primarily in the aqueous phase and were, then, used for emulsification [25,26]. However, these results presented a mixed picture of the antioxidant effect of the interactions. For example, some studies showed that the oxidative stability of the emulsion was improved [27–29], but in others it decreased the oxidative stability through non-covalent and covalent interactions between the phenolic compounds and the proteins [10,12,21,29]. Little attention has been paid to studies in which the interfacial protein film (having an amphiphilic alignment of the protein side chains in the oil and the water phase) interacts directly (on-site at the interface) with the phenolic compounds and how these interactions affect the oxidative stability of the emulsion.

Consequently, this study aimed to characterize the impact of phenolic acid derivatives (PCDs), depending on their chemical–structural properties and interaction conditions (pH value), on the antioxidant efficacy of the interfacial protein film and the resulting oxidative emulsion stability.

To investigate this issue, four hypotheses were conceived: (1) PCDs with a high polarity and thus, high-water solubility, exist in a dissolved state in the aqueous phase of the emulsion or interact with the non-polar side of the interfacial film. From there, PCDs can scavenge radicals from the aqueous phase. (2) Secondly, it is assumed that low molecular weight PCDs, with a non-polar molecular character and thus, a low water solubility, partially diffuse from the aqueous phase of the emulsion through the interfacial film. There, they can interact with the hydrophobic side of the interfacial film or are (partly) dissolved in the lipid phase. From there, PCDs can scavenge lipid radicals present in the oil. (3) The third hypothesis assumes that, depending on the pH, PCDs can maintain the redox cycle in food by reducing Fe(III) to pro-oxidant Fe(II). At pH 6, PCDs interact predominantly non-covalently with the interfacial protein film. The molecular structure of PCDs are, therefore, not modified through covalent bonds; this, in turn, makes it easier for them to split off electrons and reduce Fe(III). Fe(II), in turn, promotes the further reaction of hydroperoxides to secondary oxidation products within an emulsion. (4) It is assumed that at pH 9, PCDs exist in a polymerized state or covalently bound to the interfacial film, which reduces the ability of electron splitting and thus, the Fe reduction power. In addition, the hydroxyl groups of the PCD deprotonate with an increasing pH, which can complex transition metals as a result. Complexed transition metals are, then, no longer available to the redox cycle and lose their pro-oxidant effect. The further reaction of hydroperoxides to secondary oxidation products within an emulsion is therefore not supported. Therefore, the radical scavenging of PCDs in an emulsion was studied by electron paramagnetic resonance (EPR) spectroscopy at pH 6 and 9 for hydrophilic und hydrophobic radicals. The ability of the system to complex transition metals was also characterized by EPR. In a next step, the oxidative stability of a Fe-containing linseed oil emulsion was checked over a storage time of 21 days by detecting primary, secondary, and tertiary oxidation products via photometric, potentiometric and gas chromatographic methods.

## 2. Materials and Methods

### 2.1. *Phenolic Acid Derivatives—Structural Characteristics*

All chosen phenolic acid derivatives (PCDs) were based on caffeic acid (CA) (Figure 1). Chicoric acid (CHA) and cynarine (CY) contain two caffeic acid units, while rosmarinic acid (RS) and verbascoside (VD) contain one caffeic acid unit and a substituted catechol-group. CY and VD are esterified with polar substituents, such as quinic acid and rhamnose/glucose. All PCDs were purchased from Carl Roth GmbH & Co. KG (>98%; Karlsruhe, Germany).

### 2.2. *Phenolic Acid Derivative Solutions*

PCD solutions were prepared in distilled water with concentrations of 0.895 mM. For dissolving the PCDs, an ultrasonic bath was used for 15 min. To prevent polymerization reactions in the PCD solutions, these were prepared freshly, and pH values were set to 6.0 or 9.0 directly before adding them to the emulsion at room temperature. The pH-value was set with 0.1 M of HCl and 0.1 M of NaOH (>99.9%; analytical grade, Carl Roth GmbH & Co. KG, Karlsruhe, Germany).

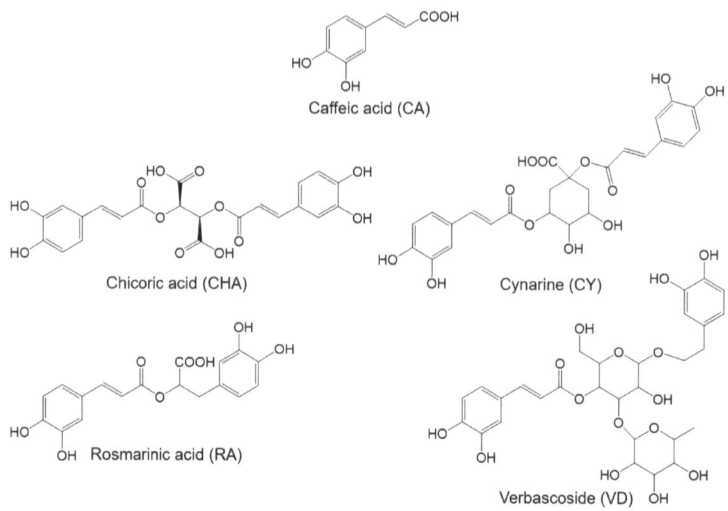


**Figure 1.** Structures of the added phenolic acid derivatives.

### 2.3. *β-Lactoglobulin Solutions*

Native β-lactoglobulin (β-Lg) was purified from a whey protein isolate (Davisco Foods International Inc.; Le Sueur, MN, USA) to 99%, as described by Keppler et al. [30]. An aqueous β-Lg-solution was prepared with distilled water to obtain a β-Lg concentration of 0.1 weight percent (wt.%) in the emulsion. For this purpose, the dissolved β-Lg was stirred at room temperature for 1 h and the pH value was set to 6.0 or 9.0, with 0.1 M of HCl and 0.1M of NaOH (>99.9%; analytical grade, Carl Roth GmbH & Co. KG, Karlsruhe, Germany).

### 2.4. *Oil Purification*

MCT oil (medium-chain-triacylglycerols, >99.9%; ENDIMA GmbH, Illingen, Germany) was used for EPR-measurements. Linseed oil (cold pressed organic linseed oil; Biovenue!, Montpellier, France), which is rich in unsaturated fatty acids and rich in transition metals, such as iron (0.34 mg/100 g), zinc (0.31 mg/100 g), and copper (0.06 mg/100 g), was used [31]. It was used in this study to allow the competitive effects of the antioxidant and the pro-oxidant effects of the PCDs to be investigated in a linseed oil model emulsion. To remove oil-accompanying substances, the oils were purified with Florisil® (100%; MgO × 3.6 SiO2 × 1.53 OH, Carl Roth GmbH & Co. KG, Karlsruhe, Germany) in a ratio of 3:1 (oil: Florisil®). The oil was agitated with a magnetic stirrer for at least 2 h at room temperature and, then, centrifuged at 10,000× *g* for 45 min.

### 2.5. *EPR-Characterization of Copper Complexation in the Emulsion*

To characterize the copper complexation properties of the emulsion, emulsions were prepared in 200 μL micro mixtures containing 0.1% β-Lg and 10% MCT oil in their final concentration. Copper sulfate pentahydrate (>99%; Merck KGaA, Darmstadt, Germany) was added to the samples so that the final sample concentration of Cu(II) was 2.5 mM. The actual emulsification process was carried out in an ultrasonic bath for three minutes. Then, 20 min after emulsification, a PCD was added, and the pH was adjusted to 6 or 9. The PCD concentration was 5 mM in the final sample. EPR measurements were performed with an internal laboratory protocol on a MiniScope MS200 (magnettech GmbH, Berlin, Germany) after an incubation period of 1 h. The following parameters were used for this: B0-field: 330 mT, range: 80 mT, sweep time 30 s, number of passes: 15, modulation: 0.2 mT, microwave (MW) attenuation: 10.0 db and gain 300. The sample collection for the measurement was carried out in disposable capillary pipettes (ringcaps 50 μL, Hirschmann Laborgeräte GmbH, Eberstadt, Germany).

### 2.6. Radical Degradation from the Aqueous Phase of the Emulsion

To characterize the radical degradation kinetic from the aqueous phase of the emulsion, emulsions were prepared in an ultrasonic bath for three minutes containing 0.1% β-Lg and 10% MCT oil in their final concentration. Then, 20 min after emulsification, a PCD (0.895 mM) was added in a molar ratio (β-Lg:PCD) 1:10, the pH was adjusted to 6 or 9, and the emulsion was incubated for one hour. After incubation time, 20 μL of the free hydrophilic radical 4-hydroxy-TEMPO (tempol, 0.054 mM dissolved in distilled water; 97%; Merck KGaA, Darmstadt, Germany) was added to the 180 μL emulsion and dispersed using a vortex mixer. Measurements started immediately afterwards.

The EPR kinetic measurement was performed for 30 min, with a measurement every 60 s, using the following parameters: B0-field: 450 mT, range: 10 mT, sweep time: 30 s, number of passes: 1, modulation: 0.2 mT, MW attenuation: 10.0 db and gain 70.

### 2.7. Radical Degradation from the Lipid Phase of the Emulsion

To characterize the radical degradation kinetics of the lipid phase of the emulsion, the PCD (0.895 mM) and β-Lg (0.2 wt.%) solutions were mixed, the pH-values were set to 6.0 or 9.0, and the samples were incubated for 1 h. After incubation time, 80 μL of the hydrophobic free radical 16-doxyl stearic acid (0.054 mM dissolved in MCT-oil; Merck KGaA, Darmstadt, Germany) was mixed with a 180 μL PCD–β-Lg mixture using a vortexer. After vortexing, emulsions were prepared in an ultrasonic bath for 1 min and measured directly after preparation.

The EPR kinetic measurement was performed for 30 min, with a measurement every 60 s, using the following parameters: B0-field: 350 mT, Range: 50 mT, Sweep time: 30 s, Number of passes: 1, Modulation: 0.2 mT, MW attenuation: 10.0 db and Gain 70. This method was a self-established internal laboratory protocol.

### 2.8. Iron Reducing Power

The reducing power of the aqueous PCD–β-Lg mixtures was studied with a modified method described by Oyaizu (1986) [32]. First, the β-Lg solution (0.1 wt.%) was incubated with the PCD in a molar ration of 1:3 (β-Lg:PCD) at pH 6.0 or pH 9.0 for one hour while stirring the mixture. A volume of 2 mL of the aqueous PCD–β-Lg mixtures, or an ascorbic acid (99.7%; Merck KGaA, Darmstadt, Germany) calibration standard (10–60 mg/L), were mixed with 2 mL of the phosphate buffer (0.2 M, pH 6.6; Merck KGaA, Darmstadt, Germany) and 2 mL of potassium hexacyanoferrate (1%; Merck KGaA, Darmstadt, Germany). The samples were incubated in a water bath at 50 °C for 20 min. Then, 2 mL of trichloroacetic acid (10%; Carl Roth GmbH & Co. KG, Karlsruhe, Germany) were added and the samples were centrifuged at 1,500× $g$ for 10 min. For the photometric measurement, 1.3 mL of the supernatant was mixed with 0.26 mL of iron (III) chloride (0.1 %; neoLab® Migge GmbH, Heidelberg, Germany) and 1.3 mL of distilled water. The photometric measurement was, then, performed exactly 10 min after mixing the reagents in the cuvette at 700 nm [33].

### 2.9. Emulsification and Incubation of the Linseed Oil Emulsion

For the storage tests, an emulsion was prepared containing 10% linseed oil as the final concentration and stabilized with 0.1 wt.% β-Lg. An oil–water stock emulsion was prepared for this purpose. A premix was created by mixing 140 g of 0.1 wt.% β-Lg solution and 70 g of purified linseed oil using a rotor–stator system (Ultra-Turrax® T25 basic, IKA -Werke GmbH & CO. KG, Staufen, Germany) at 13,500 rpm for 60 s. The premix emulsion was homogenized in a high-pressure homogenizer (APV 1000, APV Systems Denmark) at 300 bar with 2 passes. The oil droplet size was measured by static light scattering (LS 13 320 Particle Size Analyzer, Beckman Coulter GmbH, Krefeld, Germany). After emulsification, sodium azide (0.02% in water) was added to prevent microbial growth. The stock emulsion was halved and set to pH 6.0 and 9.0, and split in 8 g samples. Then, 20 min after emulsification, PCD solutions (adjusted to pH 6.0 or 9.0) were added to obtain a β-Lg:PCD molar ratios of 1:10; this was stirred with a glass rod. The storage of the samples took place at room temperature and with sunlight for 21 d.

### *2.10. Determination of Hydroperoxide Values in the Linseed Oil Emulsion*

To extract the oil out of the emulsion, 0.7 g of emulsion was mixed with 2.5 mL of distilled water and 7 mL of an isopropanol/ isooctane (1:1; *v:v*) mixture (>99.9%; Carl Roth GmbH & Co. KG, Karlsruhe, Germany). This mixture was vortexed for 30 s to solve the oil in the organic solvent. For a better water–solvent separation, the samples were centrifuged at 7,500× *g* for 5 min. An aliquot of 2.5 mL of the organic solvent supernatant were evaporated under a nitrogen flow at 40 °C. The leftover oil was weighed, resolved in 30 mL of an acetic acid/chloroform (3:2; *v:v*) mixture (100%/ >98%; VWR International GmbH, Darmstadt, Germany), and transferred to an Erlenmeyer flask. A volume of 0.5 mL of saturated potassium iodide (>99.5%; Th. Geyer GmbH, Renningen, Germany) solution was added to the Erlenmeyer flask. The sample was stirred for 60 s. Then, 30 mL of distilled water was added. This was followed by an automated potentiometric titration (TitroLine® 7000, Pt 61 electrode, Xylem Analytics Germany Sales GmbH & Co. KG, Weilheim, Germany) with sodium thiosulphate (0.001 N; Merck KGaA, Darmstadt, Germany) [18].

### *2.11. Determination of Conjugated Dienes in the Linseed Oil Emulsion*

Conjugated dienes were determined following a modified method, described by Mei et al. (1998). For this purpose, 100 μL of isopropanol/isooctane oil extract was taken from the analysis in Section 2.10 and mixed with 2.7 mL of solvent (methanol/butanol; 2:1; *v:v*; >99.9%; Carl Roth GmbH & Co. KG, Karlsruhe, Germany). This mixture was transferred to a fused silica cuvette and the absorbance was measured at 233 nm on a UV-Vis spectrophotometer (photoLab6600 UV-Vis, Xylem Analytics Germany Sales GmbH & Co. KG, Weilheim, Germany). The increase in the absorbance at 233 nm is characteristic of the conjugated diene formation [34].

### *2.12. GC-Analysis of Secondary Oxidation Products in the Linseed Oil Emulsion*

Secondary lipid oxidation products were quantified after a 21 d storage time by solid phase microextraction (SPME) and gas chromatography (GC), coupled to mass spectrometry (MS). 2-Heptanone (<98%; Merck KGaA, Darmstadt, Germany) was added as an internal standard. The SPME-fibers were put in 20 mL-headspace vials and filled with 8 g of sample (emulsion from Section 2.9). The vials were sealed with a polytetrafluoroethylene (PTFE)-septa. The extraction/adsorption was performed at 60 °C for 10 min. The desorption of the analytes from the SPME fiber takes place at 250 °C for 120 s. The system featured a splitless injection and flow of 2 mL/min with helium. The stationary phase used a medium polar Zebron ZB-1701 capillary column (60 m × 0.32 mm × 0.25 μm) (Phenomenex Inc., Torrance, CA, United States). The temperature program started with a temperature of 40 °C for 5 min, and was, then, raised by 2 °C/min, up to 60 °C. After a 2 min holding time, the temperature was raised at a rate of 10 °C/min, up to 120 °C. In the last step, the temperature was increased at a rate of 40 °C/min, up to 260 °C, and held for 10 min. For detection, electron ionization mass spectrometry was performed at 70 eV [35].

### *2.13. Statistic Analysis*

All measurements were performed in triplicate, except for the qualitative EPR measurements. Following a Shapiro–Wilk test for normal distribution, a one way ANOVA test with a Tukey HSD post hoc ($p < 0.05$) test was conducted with xlstat (Addinsoft SAS, Bordeaux, France) to identify significant differences between the samples. The order of the statistical differences used in this analysis is $a > b > c$. The homogeneity of variances was assumed for this purpose.

## 3. Results and Discussion

The first step was to investigate, by EPR, whether PCDs can directly scavenge radicals at pH 6 and 9 from the oil or water phase of a β-lactoglobulin (β-Lg)-stabilized emulsion. β-Lg was chosen as a model protein, because its interfacial properties have been well studied and, as a whey protein, it is frequently applied in foods [3,36–38]. In the second step, EPR was used to characterize the complexation of transition metals by the interfacial β-Lg film containing PCDs. For this purpose, the extent to which PCD–β-Lg mixtures can reduce transition metals was investigated. In the third step, the impact of the different structured PCDs and the pH value on the formation of primary, secondary and tertiary oxidation products was determined in a ferrous linseed oil emulsion, over a storage period of 21 d.

### 3.1. Radical Scavening of Hydrophilic and Hydrophobic Radicals

The ability of interfacial films to scavenge the free radicals that may start the chain reactions during the lipid oxidation process has been studied in two different ways with the use of a hydrophilic and a hydrophobic radical. Both free radicals in the aqueous phase of the emulsion and in the lipid phase can maintain the chain reaction of lipid oxidation. To imitate a comparable system, a hydrophilic and a hydrophobic free radical were used to investigate in which phase of the emulsion radicals can be scavenged by the interfacial film via EPR.

Figure 2 shows the time-dependent radical scavenging of the hydrophilic free radical tempol by the PCD-crosslinked interfacial films at pH 6 and 9. The emulsions without PCD did not show any radical scavenging, whereas the addition of PCD led to a significant initial decrease in the radical signal at both pH values. At pH 6, the PCD addition could initially decrease the tempol signal by about 30%; this was, therefore, approximately 70% of the initial tempol signal. In the following time course, an asymptotic increase in the tempol signal could be observed for samples containing PCD. For the hydrophobic PCD, CA and CY, this increase was less pronounced; there was approximately 10% of the initial tempol signal in 30 min, while the hydrophilic PCD, RA, CHA, and VD showed an increase of 20% or more in 30 min. These were, therefore, almost at the initial level.

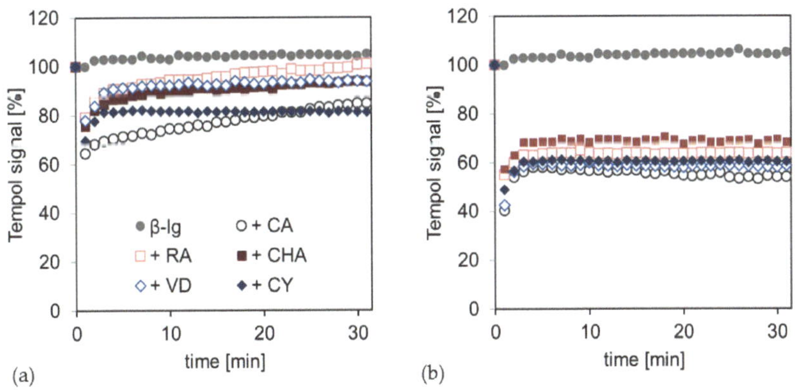


**Figure 2.** Tempol degradation kinetic from the aqueous phase (**a**) of the emulsion at pH 6 and (**b**) at pH 9; here, the β-Lg emulsions are compared without (β-lg) and with PCD (+ PCD) addition. The starting point at t = 0 min represents the output signal of the pure radical. From t = 1 min onwards, the radical degradation or the radical regeneration is shown by an increase or decrease in the original radical signal. The values shown are an individual exemplary series of measurements.

At pH 9, the tempol signal was initially reduced by approximately 50–60% with the addition of PCD, and was thus approximately 40–50%. Comparable to pH 6, an asymptotic increase in the tempol signal of about 20% in 30 min was observed over the time course, but no significant differences between the PCDs were observed.

The pH value, therefore, appears to have a significant impact on the extent to which the PCDs can scavenge free radicals. The ability of the PCDs to deprotonate various phenolic hydroxyl groups increases with rising pH and, therefore, the possibility of radical scavenging, as previously shown for catechins [39] and anthocyanins [40].

The observed increase in the radical signal, following tempol degradation at both pH levels, suggests that tempol radicals, previously scavenged by PCDs, are regenerated under ambient conditions in the emulsion. The scavenged tempol-H turns into the free radical tempol again. This is possible because the addition of radicals in the emulsion methodically deposited oxygen in through vortexing. As Monti et al. have shown, tempol-H is able to be regenerated into the free radical tempol by oxygenation in a very short time [41].

At pH 6, different properties of the PCDs, with regard to the radical scavenging activity, can be identified, depending on their chemical structure. It is striking that PCDs with the highest hydrophobicity and thus, a low water solubility, most strongly prevent tempol-H from regenerating; this is compared to hydrophilic PCDs with a high water solubility. This observation is, in principle, in line with previously described research results, stating that moderately hydrophobic antioxidants are particularly effective in emulsions. This observation is mainly explained by the accumulation of moderately hydrophobic antioxidants at the interface [42,43]. This explanation does not apply to the present study, as the hydrophobicity of the PCDs did not affect radical scavenging per se, but radical regeneration.

At pH 9, no differences in radical scavenging and regeneration, as well as in the deviating hydrophobicity of the PCDs, can be observed due to the chemical–structural differences. This observation is probably due to the polymerization reactions of the PCDs at pH 9; this is because the polymers have more similar chemical properties than the monomers, where individual substituents are more important. This phenomenon has already been noticed in a previous study on physical emulsion stability [44].

Figure 3 shows the time-dependent radical scavenging of the hydrophobic free radical 16-DOXYL-stearic acid by the PCD-crosslinked interfacial films at pH 6 and 9. At pH 6, the pure β-Lg interfacial film, as well as those crosslinked with hydrophilic PCD (RA, CHA, and VD), showed no scavenging of hydrophobic radicals. However, the addition of the hydrophobic poorly water-soluble PCDs (CA and CY) resulted in nearly identical radical scavenging kinetics, where the radical signal was reduced by approximately 20% in 30 min. In contrast, the samples at pH 9 showed no significant tendencies to scavenge the hydrophobic radical from the lipid phase.

At pH 6, the radical scavenging of CA and CY from the lipid phase can be explained by the molecular properties, such as the hydrophobicity, and their following reaction conditions. In previous studies, it has been shown that CA and CY accumulate more profoundly in the lipid phase of an MCT–oil–water system. The other compounds studied (RA, CHA, and VD) were almost completely present in the aqueous phase [45]. In addition, PCDs are protonated and, thus, have a low net charge at pH 6; this allows CA and CY to transition to the lipid phase. On the other hand, the PCDs are negatively charged at pH 9, which, in turn, shifts the partitioning of the PCDs in the direction of the aqueous phase due to an increasing polarity through the charge [46]. As hypothesized, at pH 6, CA and CY may, therefore, be present freely in the lipid phase or interact with the interfacial film coming from the lipid phase, where they can scavenge hydrophobic lipid radicals. Due to the anionic charge of PCDs at pH 9, PCDs attach to the interfacial film, starting from the aqueous phase or are free in the aqueous phase, so that they are sterically hindered and separated from the hydrophobic lipid radical and cannot scavenge it.

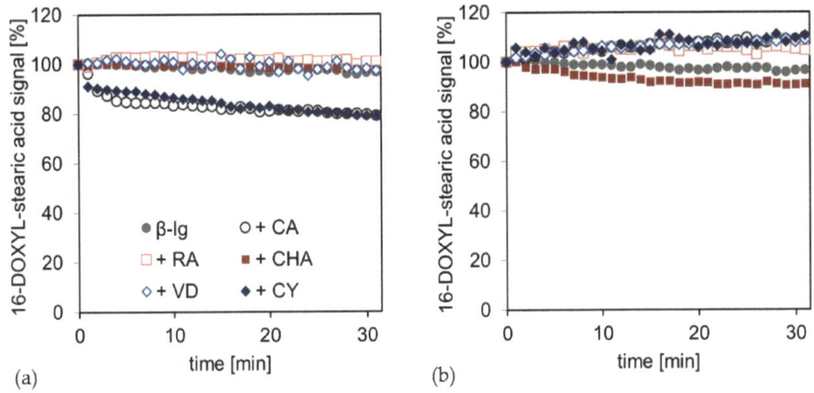


**Figure 3.** 16-DOXYL-stearic acid degradation kinetic from the lipid phase (**a**) of the emulsion at pH 6 and (**b**) at pH 9; Here, the β-Lg emulsions are compared without (β-lg) and with PCD (+PCD) addition. The starting point at t = 0 min represents the output signal of the pure radical. From t = 1 min onwards, the radical degradation or radical regeneration is shown by an increase or decrease in the original radical signal.

In summary, PCDs experience a higher scavenging of radicals in the aqueous phase of the emulsion at pH 9 than at pH 6, because of the hydroxyl groups being deprotonated and the follow-up reaction releasing protons with free radicals. However, at pH 9, electrostatic repulsion, polymerization reactions of the PCDs, and covalent binding to the interfacial protein film can occur. Consequently, the PCDs are hindered from scavenging hydrophobic radicals from the lipid phase, because they are sterically prevented from penetrating the interfacial film. On the other hand, PCDs have a low net charge at pH 6 and are (small) monomers, so that they can penetrate the interfacial film and saturate radicals in the lipid phase, depending on their polarity.

### 3.2. Complexation of Transition Metals

The complexation of transition metals by β-Lg-PCD mixtures was investigated with an EPR measurement at pH 6 and 9, in order to assess the impact of structural effects and pH on the formation of an interfacial complex between the transition metal Cu(II) and the interfacial β-Lg-PCD mixture.

Figure 4 shows the EPR spectra of the interfacial films without PCDs, and with CA and RA, in the presence of Cu(II) at pH 6 and 9. The basic Cu(II) signal presents itself as a wave in the range of 320 mT at pH 6. In the presence of CA and RA, a change in the basic signal in the range of 335 mT was observed, suggesting a partial complexation of Cu(II) by PCDs. At pH 9, the Cu(II) signal was no longer measurable due to the pH value. At pH 9, the dissolved Cu(II) was precipitated as $Cu(OH)_2$, which is not EPR active. However, in the presence of CA and RA, the Cu(II) was complexed and thus held in solution. This was shown by a copper complex signal at around 340 mT.

Goodman et al. (2012) were able to demonstrate a pH-dependent complex formation with the formation of different complexes for tea polyphenols and Cu(II). They observed a slight weakening of the Cu(II) signal at pH 6 for ellagic acid and Cu(II); this shows a clear shift in the EPR spectrum with an increase in the pH to 10, due to the formation of a complex [47]. Pirker et al. (2012) described that, at weakly acidic reaction conditions, silent EPR species form, showing almost no change in the spectrum; these are, therefore, difficult to characterize [48]. Ferreira Severino et al. (2011) suggested that such silent EPR species involve the formation of di- or polymeric complexes in which the Cu(II) is coordinated via carboxyl groups [49]. These observations are consistent with the EPR spectra of the interfacial film at pH 6 in the present study (Figure 4a). The spectra of the emulsions with CA and RA showed only a slight deviation from the Cu(II) spectrum. Thus, it can be assumed that, at pH 6, the carboxyl groups of CA and RA coordinate the Cu(II) present. For alkaline pH values, Pirker et al. showed that other complexes are formed that predominantly mononuclearily bind the Cu(II). The analyses of rotational correlation times, performed by Pirker et al., also showed an increasing molecular mass at alkaline pH values. This led these to assume that di- or polymeric PCDs complexed the Cu(II) [48]. Due to the deprotonation of the hydroxyl groups under alkaline conditions, PCDs can form Cu(II) complexes on the deprotonated phenolic hydroxyl groups [50]. Therefore, complexes with different chemical structures are formed, depending on the pH value; this results in other binding sites, which can be shown through clear differences in the EPR spectra. As hypothesized, the deprotonation at pH 9 leads to the formation of strong Cu(II)-PCD complexes, while at pH 6, only weaker complexes via the carboxyl group are assumed to be formed.

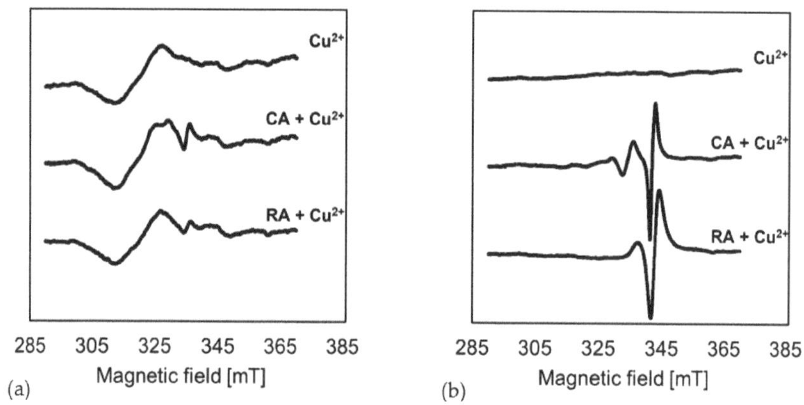


**Figure 4.** The EPR spectra show the change in the Cu(II) signal in the presence of the β-Lg interfacial film of an emulsion without PCD addition (top) and with the addition of CA (middle) and RA (bottom) at (**a**) pH 6 and (**b**) pH 9. The spectra shown are an individual exemplary series of measurements.

### 3.3. *Iron Reducing Power of β-Lg-PCD Mixtures*

The Fe(III)-reducing power of β-Lg-PCD mixtures was investigated by a photometric measurement at pH 6 and 9; this was performed in order to assess the impact of structural effects and pH on pro-oxidant efficacy by maintaining the redox cycle, due to electron donation. The results, calculated as ascorbic acid equivalents (AAE), are shown in Figure 5. For both pH conditions, the pure β-Lg solution showed a low iron-reducing power with an AAE of approximately 15 mg/L. The addition of CA increased the iron-reducing power to approximately 60 mg/L AAE at pH 6 and 45 mg/L AAE at pH 9. The higher molecular weight of the PCD (RA, CHA, VD and CY) led to a further increase in the value to 80–90 mg/L AAE at pH 6, with CHA showing the highest iron-reduction capacity. At pH 9, the increase in the iron-reducing power, due to PCD addition, was less pronounced than at pH 6. The RA, VD, and CY were approximately 70 mg/L AAE; only CHA increased to 90 mg/L AAE.

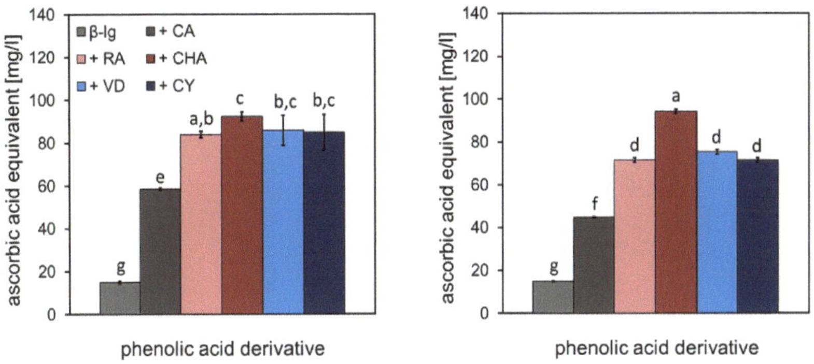


(a) (b)

**Figure 5.** Fe(III)-reducing power calculated as ascorbic acid equivalents (**a**) of β-Lg-PCD mixtures at pH 6 and (**b**) at pH 9; the letters (a–g) are describing the statistical homogeneous groups without significant differences ($p > 0.05$) for (**a**,**b**).

At both pH conditions, there was a trend in molecules with a low molecular weight, such as CA, donating fewer electrons; this reduced transition metals, such as iron, rather than molecules with a higher molecular weight, such as RA, CHA, VD, and CY. This is because molecules with a higher molecular weight can have more conjugated $\pi$-electrons and are therefore more resonance stabilized, donating more electrons for Fe(III) reduction. Yen et al. (2002) described the relationship between the ability to complex metals and donate electrons, and hence the metal reducing power of ascorbic acid and gallic acid [51]. These authors concluded that metal complexation simultaneously decreased the reduction of transition metals and, therefore, their chemical reactivity, especially in lipid oxidation [34,51,52]. The decrease in iron reducing power is, therefore, consistent with the EPR measurements of the present study (Figure 4), showing a significant metal complexation at pH 9, which was not the case at pH 6. In addition, there is the possibility that PCDs, covalently bound to the protein, or even polymerized PCDs at pH 9, can only donate protons to a limited extent and can, therefore, reduce less Fe(III).

In summary, PCDs formed various complexes with transition metals. It is suspected that at pH 6, partially weak complexes are formed with the carboxyl groups of the PCDs and at pH 9, stronger complexes were formed via the deprotonated hydroxyl groups of the catechol structure of the PCDs. The strong complexation of transition metals at pH 9 contributes to the fact that complexed transition metals no longer take part in the redox cycle, thus inhibiting the reduction of transition metals to pro-oxidant compounds.

### 3.4. Formation of Oxidation Products during Storage

In order to consider the antioxidant and pro-oxidant properties of the PCD–β-Lg interfacial films, a storage experiment was performed with a linseed oil emulsion. In the first step, primary oxidation products during storage were investigated. These were initially conjugated dienes (CD) (Figure 6a,b), which in turn react further to form hydroperoxides (Figure 6c,d). At pH 6 and 9, an increase in conjugated dienes in the emulsion had already occurred within the first three days. At pH 6, the pure β-Lg emulsion and the samples with CHA and CY showed a stronger increase in CD than the samples with CA, RA, and VD. After completion of the storage test, however, the CD contents converge again. Even at pH 9, the CD content increased faster with the pure β-Lg emulsion, as well as with CHA, than with the other PCD; there was a similar final level at the end of the storage time.

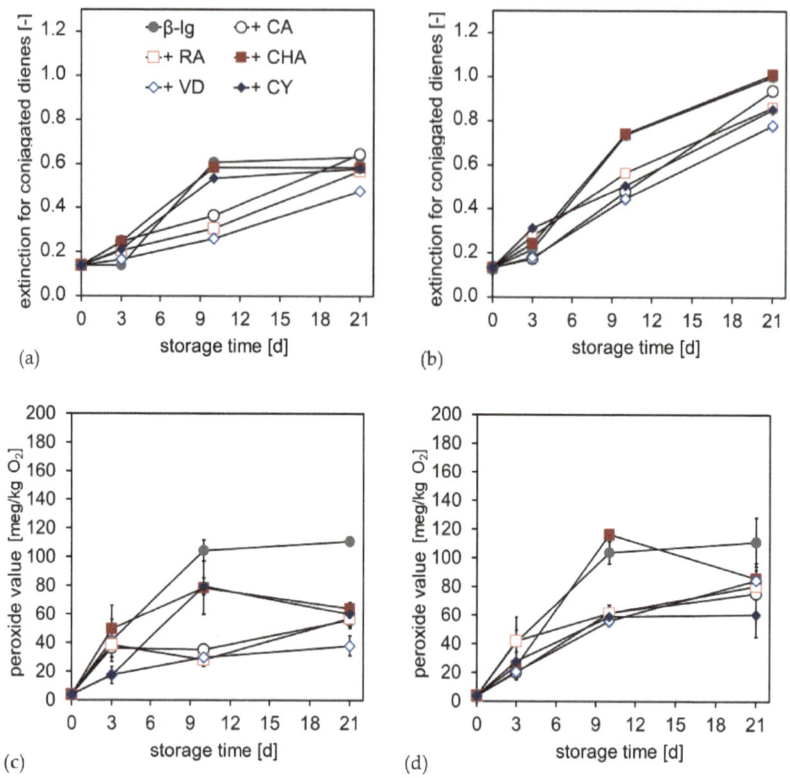


**Figure 6.** Extinction of the emulsion oil extract at 233 nm as an indicator for the formation of conjugated dienes within the emulsion (**a**) at pH 6 and (**b**) at pH 9; peroxide values for emulsions stabilized by different PCD and stored for 21 days (**c**) at pH 6 and (**d**) at pH 9.

With the formation of CD, an increase in the peroxide value (POV) could also be estimated for pH 6 and 9. Considered over the storage time, the POVs were highest for the pure β-Lg emulsion. At pH 6, CHA and CY again showed a faster increase in POV than CA, RA, and VD. However, over the storage time, POV decreased again, so that at the end of the storage time, POV for emulsions with PCDs were between 40 and 60 meq/kg $O_2$, while the emulsion without PCDs was approximately 110 meq/kg of $O_2$. At pH 9, there was a significant increase in the POV of the emulsion with CHA, at about 120 meq/kg of $O_2$ on day 9. The other PCDs were comparatively low on day 9, at 60 meq/kg of $O_2$. The final level, at the end of the storage period, was higher at pH 9 than at pH 6, at about 60 to 80 meq/kg of $O_2$. The pure β-Lg emulsion was again at approximately 110 meq/kg of $O_2$. Thus, for both pH conditions, a stagnation or a decrease in POV could be observed in the second half of the storage time, with the POV of the emulsions with PCDs, generally, significantly lower than without PCDs.

The increase in CD during the storage time showed that the double bonds of polyunsaturated fatty acids were shifted due to oxidation processes, resulting in conjugated dienes. These dienes react further in the course of chain propagation to form hydroperoxides [15,53]. The increase in CD indicated that lipid oxidation progress occurs in all samples, reaching a plateau only for the emulsion at pH 6, without PCD sor with CHA or CY. Within this plateau of CHA and CY, the degradation of hydroperoxides had already occurred. For these compounds, this indicated that secondary oxidation products were increasingly formed [15]. It was notable that CHA and CY were the compounds with the largest $\pi$-electron systems and, therefore, showed the greatest resonance stabilization. In addition, at pH 9, the emulsions with CHA and CY indicated a trend in forming a plateau, in addition to a degradation of the hydroperoxides; meanwhile, an increase in hydroperoxide formation was still observed with the other samples.

At the end of the storage time, the emulsions were analyzed for secondary and tertiary oxidation products. Among other compounds, 2,4-heptadienal and 3,5-octadien-2-one (Figure 7a,b), as well as 2-hexenal and its reaction product ethylfuran (Figure 7c,d), were found to be the main components in terms of concentration.

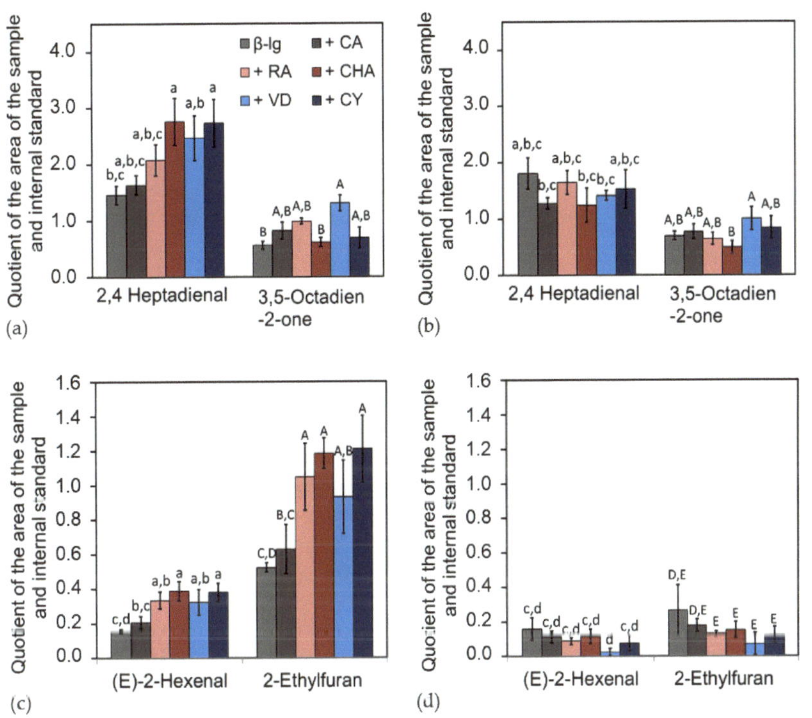


**Figure 7.** Oxidation products of the linseed oil emulsion after 21 d of storage time, shown as quotient of the internal standard: Quantitative main components of the secondary oxidation products of linolenic acid (**a**) at pH 6 and (**b**) at pH 9; the letters (a–c and A–B) are describing the statistical homogeneous groups without significant differences ($p$ >0.05) for (**a**,**b**); and the further reaction of a secondary oxidation product ((E)-2-hexenal) to a tertiary oxidation product (2-ethylfuran) (**c**) at pH 6 and (**d**) at pH 9; the letters (a–d and A-E) are describing the statistical homogeneous groups without significant differences ($p$ >0.05) for (**c**,**d**).

For all samples, significantly more oxidation products were formed during the storage of emulsions at pH 6 than at pH 9. The emulsions with PCDs at pH 6 showed a significant increase in oxidation products, compared to the emulsion without PCDs. There were shifts in the composition of the oxidation products, e.g., CHA and CY promoted the formation of 2,4-heptadienal, while less 3,5-octadien-2-one was formed. In addition, CA led to a less pronounced increase in oxidation products than the PCD RA, CHA, VD, and CY with high molecular weights. For the emulsions at pH 9, no significant differences were observed in the formation of the oxidation products, with and without PCDs. The most significant difference between the emulsion at pH 9 and pH 6 could be seen in the formation of the 2-hexenal. In emulsions at pH 6 with PCDs, significantly more 2-hexenal was formed compared to pH 9, which, in turn, reacted further to a higher extent of 2-

ethylfuran; therefore, approximately five times the amount of ethylfuran was measured at pH 6, compared to pH 9.

The formed primary, secondary, and tertiary oxidation products showed that the pH of the emulsion had the most intense impact on the lipid oxidation. The molecular weight and structural properties of the PCDs equally had an impact on lipid oxidation products.

However, at the end of the storage period, significantly more secondary and tertiary oxidation products were formed in the emulsions at pH 6 with PCDs than when without the addition of PCDs. For the emulsion at pH 9, no significant differences in the formation of secondary and tertiary oxidation products could be determined, in comparison to with or without PCD. Thus, the reaction conditions at pH 6 induced a pro-oxidant effect on the emulsion. This pro-oxidant effect may be due to the following causes: As is obvious from Figures 4 and 5, PCDs have a high potential at pH 6 to reduce Fe(III), but a low ability to complex transition metals. Reduced transition metals, such as Fe(II), are known to react intensively with hydroperoxides, leading to radical formation during autoxidation. Thus, for the iron-containing linseed oil emulsion, it is suspected that the hydroperoxides, formed in the emulsion at pH 6, decompose into secondary oxidation products faster than the hydroperoxides formed in the emulsions at pH 9 do. Therefore, the POV of the emulsion at pH 6 is lower than at pH 9.

For pH 9, it was shown in Figure 5 that PCDs had a significantly low ability to reduce Fe(III). In addition, Figure 4 showed the high ability of PCDs to complex transition metals. The experiments described by Iwahashi et al. indicated that Fe(III) forms a complex with caffeic acid, which cannot be reduced under the present conditions (pH 7.4 in this study) [54]. The redox cycle of the transition metals would thus be interrupted, and the further reaction of the hydroperoxides to secondary oxidation products would not be promoted.

The high content of tertiary oxidation products in the emulsions at pH 6 with PCDs indicates a progressive lipid oxidation, compared to the emulsion at pH 9. According to Grebenteuch et al., the formation of tertiary oxidation products is an indicator for a more advanced lipid oxidation; this should be considered to be differentiated to secondary oxidation products in the context of the evaluation [18].

Hence, the use of PCD seems to promote not only the formation of secondary oxidation products, but also their further reaction to tertiary oxidation products. 2-Ethylfuran, as a tertiary oxidation product, can be formed from a secondary oxidation product, such as the unsaturated aldehyde 2-hexenal. Adams et al. showed, for these compounds, that this reaction can be further triggered in the presence of amino acids, peptides, and proteins [55]. Thus, it is likely that the interfacial protein also has a catalytic effect on the formation of 2-ethylfuran. Effectively, it is not possible to differentiate whether the increased 2-ethylfuran formation is also promoted by PCDs or whether the increased formation of secondary oxidation products at pH 6, in the presence of the interfacial protein film, is the cause of the 2-ethylfuran formation.

In summary, the addition of PCDs to the emulsions at pH 6 had a pro-oxidant effect. The further reaction of primary oxidation products to secondary and possibly tertiary oxidation products is promoted by the presence of PCDs in the emulsion at pH 6. In contrast, there are no significant differences at pH 9. The pro-oxidant behavior of the PCDs could be attributed to their ability to reduce transition metals, making them pro-oxidantly effective. At pH 9, this observation occurs hardly at all, as transition metals can be complexed via the hydroxyl groups of the PCDs and lose their pro-oxidant effect when being complexed, as they can no longer be reduced.

## 4. Conclusions

In this study, the pH and polarity of PCDs were found to have a significant effect on their antioxidant/ pro-oxidant activity at the interfacial protein film.

As hypothesized, the polarity and, therefore, the partitioning behavior of the PCDs, determined whether radicals present could be scavenged from the continuous aqueous phase or the disperse lipid phase of the emulsion. More hydrophobic PCDs could only penetrate the interfacial protein film and scavenge radicals in the lipid phase when they were not electrostatically repelled from it. Polymerized PCDs, as is likely at pH 9, are sterically hindered from penetrating the interfacial protein film and, therefore, do not enter the oil phase. Monomers present in protonated PCDs at pH 6 showed the ability to reduce transition metals to pro-oxidant-active oxidation states via their capability to split off electrons, as hypothesized. They also showed a low complexation ability for transition metals, which likely occurs via the carboxyl groups of the PCDs. Consequently, PCDs have a pro-oxidant effect in an emulsion. Polymeric deprotonated PCDs at pH 9 reduced the transition metals present to a lesser extent than at pH 6, as hypothesized. At pH 9, the hydroxyl groups of the catechol structures of the PCDs were partially deprotonated and formed stable complexes with the transition metals over them. The transition metals could, then, no longer be reduced and, accordingly, they could no longer have a pro-oxidant effect on the lipid phase.

The results suggest that the technological use of PCDs as an antioxidant should definitely be performed, considering the food prevailing conditions. For example, a pro-oxidant effect of PCDs can be expected at low pH values and in the presence of transition metals. Alternatively, provided that PCDs occur naturally in food, it is conceivable to selectively induce covalent bonds and polymerization reactions to control their pro-oxidant properties.

In the follow-up research, it would be interesting to look at how the system behaves if, instead of varying the pH, oxidative conditions can be created by other means. The use of enzymes or electrochemical potentials would be conceivable here. Thus, the impact of electrostatic effects, due to the different protonation at pH 6 and 9, could be minimized, and the results could be related more concretely to the predominant bond types.

**Author Contributions:** Conceptualization, H.K. and S.R.; Investigation, A.B.; Methodology, A.B.; Project administration, U.S. and S.R.; Resources, U.S. and S.R.; Supervision, H.K., U.S. and S.R.; Validation, A.B., H.K. and S.R.; Visualization, A.B.; Writing—original draft, A.B.; Writing—review & editing, H.K. and S.R. All authors have read and agreed to the published version of the manuscript.

**Funding:** This research received no external funding.

**Institutional Review Board Statement:** Not applicable.

**Informed Consent Statement:** Not applicable.

**Data Availability Statement:** The data are contained within this article.

**Acknowledgments:** The authors would especially like to thank Simon Meurer, Tuan Anh Nguyen, and Nathanael Enghardt, who carried out some of the practical laboratory experiments at the University of Applied Sciences, Berlin, Germany. In addition, we thank Carsten Schwandt for his continuous technical support. We acknowledge support from the Open Access Publication Fund of the TU Berlin.

**Conflicts of Interest:** The authors declare no conflict of interest.

## References

1. Damodaran, S. Protein Stabilization of Emulsions and Foams. *Food Sci.* **2005**, *70*, 54–66. [CrossRef]
2. Padial-Domínguez, M.; Espejo-Carpio, F.J.; García-Moreno, P.J.; Jacobsen, C.; Guadix, E.M. Protein Derived Emulsifiers with Antioxidant Activity for Stabilization of Omega-3 Emulsions. *Food Chem.* **2020**, *329*, 127148. [CrossRef] [PubMed]
3. Zhai, J.L.; Day, L.; Aguilar, M.I.; Wooster, T.J. Protein Folding at Emulsion Oil/Water Interfaces. *Curr. Opin. Colloid Interface Sci.* **2013**, *18*, 257–271. [CrossRef]
4. Dickinson, E. Protein-Stabilized Emulsions. *J. Food Eng.* **1994**, *22*, 59–74. [CrossRef]

5. Brewer, M.S. Natural Antioxidants: Sources, Compounds, Mechanisms of Action, and Potential Applications. *Compr. Rev. Food Sci. Food Saf.* **2011**, *10*, 221–247. [CrossRef]
6. Logan, A.; Fagan, P. Strategies to Prevent Oxidative Deterioration in Oil-in-Water Emulsion Systems Canola-Based Phenolic Applications. In *Lipid Oxidation*; AOCS Press: Urbana, Illinois, USA, 2013.
7. Costa, M.; Losada-Barreiro, S.; Paiva-Martins, F.; Bravo-Díaz, C. Polyphenolic Antioxidants in Lipid Emulsions: Partitioning Effects and Interfacial Phenomena. *Foods* **2021**, *10*, 539. [CrossRef] [PubMed]
8. Zhou, L.; Elias, R.J. Understanding Antioxidant and Prooxidant Mechanisms of Phenolics in Food Lipids. In *Lipid Oxidation*; AOCS Press: Urbana, IL, USA, 2013.
9. Decker, E.A.; McClements, D.J.; Bourlieu-Lacanal, C.; Durand, E.; Figueroa-Espinoza, M.C.; Lecomte, J.; Villeneuve, P. Hurdles in Predicting Antioxidant Efficacy in Oil-in-Water Emulsions. *Trends Food Sci. Technol.* **2017**, *67*, 183–194. [CrossRef]
10. Zhou, L.; Elias, R.J. Antioxidant and Pro-Oxidant Activity of (-)-Epigallocatechin-3-Gallate in Food Emulsions: Influence of PH and Phenolic Concentration. *Food Chem.* **2013**, *138*, 1503–1509. [CrossRef]
11. Iwasaki, Y.; Hirasawa, T.; Maruyama, Y.; Ishii, Y.; Ito, R.; Saito, K.; Umemura, T.; Nishikawa, A.; Nakazawa, H. Effect of Interaction between Phenolic Compounds and Copper Ion on Antioxidant and Pro-Oxidant Activities. *Toxicol. In Vitro* **2011**, *25*, 1320–1327. [CrossRef]
12. Tian, L.; Zhang, S.; Yi, J.; Zhu, Z.; Cui, L.; Decker, E.A.; McClements, D.J. Factors Impacting the Antioxidant/Prooxidant Activity of Tea Polyphenols on Lipids and Proteins in Oil-in-Water Emulsions. *LWT-Food Sci. Technol.* **2022**, *156*, 113024. [CrossRef]
13. Iwasaki, Y.; Manabe, R.; Kimoto, M.; Fukuda, M.; Mase, N.; Miyazawa, M.; Hosokawa, K.; Kamei, J. Copper-Induced Interactions of Caffeic Acid and Sinapic Acid to Generate New Compounds in Artificial Biological Fluid Conditions. *Antioxidant* **2022**, *11*, 1307. [CrossRef] [PubMed]
14. Fan, G.J.; Jin, X.L.; Qian, Y.P.; Wang, Q.; Yang, R.T.; Dai, F.; Tang, J.J.; Shang, Y.J.; Cheng, L.X.; Yang, J.; et al. Hydroxycinnamic Acids as DNA-Cleaving Agents in the Presence of Cu II Ions: Mechanism, Structure-Activity Relationship, and Biological Implications. *Chem.-A Eur. J.* **2009**, *15*, 12889–12899. [CrossRef] [PubMed]
15. Moon, J.K.; Shibamoto, T. Antioxidant Assays for Plant and Food Components. *J. Agric. Food Chem.* **2009**, *57*, 1655–1666. [CrossRef] [PubMed]
16. Kubow, S. Routes of Formation and Toxic Consequences of Lipid Oxidation Products in Foods. *Free Radic. Biol. Med.* **1992**, *12*, 63–81. [CrossRef]
17. Kubow, S. Toxicity of Dietary Lipid Peroxidation Products. *Trends Food Sci. Technol.* **1990**, *1*, 67–71. [CrossRef]
18. Grebenteuch, S.; Kroh, L.W.; Drusch, S.; Rohn, S. Formation of Secondary and Tertiary Volatile Compounds Resulting from the Lipid Oxidation of Rapeseed Oil. *Foods* **2021**, *10*, 2417. [CrossRef]
19. Khokhar, S.; Owusu Apenten, R.K. Iron Binding Characteristics of Phenolic Compounds: Some Tentative Structure-Activity Relations. *Food Chem.* **2003**, *81*, 133–140. [CrossRef]
20. Hider, R.C.; Liu, Z.D.; Khodr, H.H. Polyphenol Metal Complexes. *Methods Enzymol.* **2001**, *335*, 190–203. [CrossRef]
21. Zhang, Q.; Cheng, Z.; Wang, Y.; Fu, L. Dietary Protein-Phenolic Interactions: Characterization, Biochemical-Physiological Consequences, and Potential Food Applications. *Crit. Rev. Food Sci. Nutr.* **2021**, *61*, 3589–3615. [CrossRef]
22. Stojadinovic, M.; Radosavljevic, J.; Ognjenovic, J.; Vesic, J.; Prodic, I.; Stanic-Vucinic, D.; Cirkovic Velickovic, T. Binding Affinity between Dietary Polyphenols and β-Lactoglobulin Negatively Correlates with the Protein Susceptibility to Digestion and Total Antioxidant Activity of Complexes Formed. *Food Chem.* **2013**, *136*, 1263–1271. [CrossRef]
23. Kanakis, C.D.; Hasni, I.; Bourassa, P.; Tarantilis, P.A.; Polissiou, M.G.; Tajmir-Riahi, H.A. Milk β-Lactoglobulin Complexes with Tea Polyphenols. *Food Chem.* **2011**, *127*, 1046–1055. [CrossRef] [PubMed]
24. Wang, X.; Zhang, J.; Lei, F.; Liang, C.; Yuan, F.; Gao, Y. Covalent Complexation and Functional Evaluation of (-)-Epigallocatechin Gallate and α-Lactalbumin. *Food Chem.* **2014**, *150*, 341–347. [CrossRef] [PubMed]
25. von Staszewski, M.; Pizones Ruiz-Henestrosa, V.M.; Pilosof, A.M.R. Green Tea Polyphenols-β-Lactoglobulin Nanocomplexes: Interfacial Behavior, Emulsification and Oxidation Stability of Fish Oil. *Food Hydrocoll.* **2014**, *35*, 505–511. [CrossRef]
26. Rodríguez, S.D.; von Staszewski, M.; Pilosof, A.M.R. Green Tea Polyphenols-Whey Proteins Nanoparticles: Bulk, Interfacial and Foaming Behavior. *Food Hydrocoll.* **2015**, *50*, 108–115. [CrossRef]
27. Jia, X.; Zhao, M.; Xia, N.; Teng, J.; Jia, C.; Wei, B.; Huang, L.; Chen, D. Interaction between Plant Phenolics and Rice Protein Improved Oxidative Stabilities of Emulsion. *J. Cereal Sci.* **2019**, *89*, 102818. [CrossRef]
28. Sui, X.; Sun, H.; Qi, B.; Zhang, M.; Li, Y.; Jiang, L. Functional and Conformational Changes to Soy Proteins Accompanying Anthocyanins: Focus on Covalent and Non-Covalent Interactions. *Food Chem.* **2018**, *245*, 871–878. [CrossRef] [PubMed]
29. Tian, L.; Kejing, Y.; Zhang, S.; Yi, J.; Zhu, Z.; Decker, E.A.; McClements, D.J. Impact of Tea Polyphenols on the Stability of Oil-in-Water Emulsions Coated by Whey Proteins. *Food Chem.* **2021**, *343*, 128448. [CrossRef]
30. Keppler, J.K.; Koudelka, T.; Palani, K.; Stuhldreier, M.C.; Temps, F.; Tholey, A.; Schwarz, K. Characterization of the Covalent Binding of Allyl Isothiocyanate to β-Lactoglobulin by Fluorescence Quenching, Equilibrium Measurement, and Mass Spectrometry. *J. Biomol. Struct. Dyn.* **2014**, *32*, 1103–1117. [CrossRef]
31. Szyczewski, P.; Frankowski, M.; Zioła-Frankowska, A.; Siepak, J.; Szyczewski, T.; Piotrowski, P. A Comparative Study of the Content of Heavy Metals in Oils: Linseed Oil, Rapeseed Oil and Soybean Oil in Technological Production Processes. *Arch. Environ. Prot.* **2016**, *42*, 37–40. [CrossRef]

32. Oyaizu, M. Studies on Products of Browning Reactions: Antioxidative Activities of Browning Products of Browning Reaction Prepared from Glucosamine. *Jpn. J. Nutr. Diet.* **1986**, *44*, 307–315. [CrossRef]
33. Tamm, F.; Härter, C.; Brodkorb, A.; Drusch, S. Functional and Antioxidant Properties of Whey Protein Hydrolysate/Pectin Complexes in Emulsions and Spray-Dried Microcapsules. *LWT-Food Sci. Technol.* **2016**, *73*, 524–527. [CrossRef]
34. Mei, L.; McClements, D.J.; Decker, E.A. Lipid Oxidation in Emulsions as Affected by Charge Status of Antioxidants and Emulsion Droplets. *J. Agric. Food Chem.* **1999**, *47*, 2267–2273. [CrossRef] [PubMed]
35. Böttcher, S.; Steinhäuser, U.; Drusch, S. Off-Flavour Masking of Secondary Lipid Oxidation Products by Pea Dextrin. *Food Chem.* **2015**, *169*, 492–498. [CrossRef] [PubMed]
36. Zhai, J.; Wooster, T.J.; Hoffmann, S.V.; Lee, T.H.; Augustin, M.A.; Aguilar, M.I. Structural Rearrangement of β-Lactoglobulin at Different Oil-Water Interfaces and Its Effect on Emulsion Stability. *Langmuir* **2011**, *27*, 9227–9236. [CrossRef] [PubMed]
37. Zhai, J.; Miles, A.J.; Pattenden, L.K.; Lee, T.-H.; Augustin, M.A.; Wallace, B.A; Aguilar, M.-I.; Wooster, T.J. Changes in Beta-Lactoglobulin Conformation at the Oil/Water Interface of Emulsions Studied by Synchrotron Radiation Circular Dichroism Spectroscopy. *Biomacromolecules* **2010**, *11*, 2136–2142. [CrossRef] [PubMed]
38. Engelhardt, K.; Lexis, M.; Gochev, G.; Konnerth, C.; Miller, R.; Willenbacher, N.; Peukert, W.; Braunschweig, B. PH Effects on the Molecular Structure of β-Lactoglobulin Modified Air-Water Interfaces and Its Impact on Foam Rheology. *Langmuir* **2013**, *29*, 11646–11655. [CrossRef] [PubMed]
39. Muzolf, M.; Szymusiak, H.; Gliszczynska-Swiglo, A.; Rietjens, I.M.C.M.; Tyrakowska, B. PH-Dependent Radical Scavenging Capacity of Green Tea Catechins. *J. Agric. Food Chem.* **2008**, *56*, 816–823. [CrossRef] [PubMed]
40. Borkowski, T.; Szymusiak, H.; Gliszczyńska-Świgło, A.; Rietjens, I.M.C.M.; Tyrakowska, B. Radical Scavenging Capacity of Wine Anthocyanins Is Strongly PH-Dependent. *J. Agric. Food Chem.* **2005**, *53*, 5526–5534. [CrossRef]
41. Monti, E.; Cova, D.; Guido, E.; Morelli, R.; Oliva, C. Protective Effect of the Nitroxide Tempol against the Cardiotoxicity of Adriamycin. *Free Radic. Biol. Med.* **1996**, *21*, 463–470. [CrossRef]
42. Farooq, S.; Abdullah; Zhang, H.; Weiss, J. A Comprehensive Review on Polarity, Partitioning, and Interactions of Phenolic Antioxidants at Oil–Water Interface of Food Emulsions. *Compr. Rev. Food Sci. Food Saf.* **2021**, *20*, 4250–4277. [CrossRef]
43. Costa, M.; Losada-Barreiro, S.; Paiva-Martins, F.; Bravo-Díaz, C.; Romsted, L.S. A Direct Correlation between the Antioxidant Efficiencies of Caffeic Acid and Its Alkyl Esters and Their Concentrations in the Interfacial Region of Olive Oil Emulsions. the Pseudophase Model Interpretation of the "Cut-off" Effect. *Food Chem.* **2015**, *175*, 233–242. [CrossRef]
44. Bock, A.; Kieserling, H.; Rohn, S.; Steinhäuser, U.; Drusch, S. Impact of Phenolic Acid Derivatives on β-Lactoglobulin Stabilized Oil- Water-Interfaces. *Food Biophys.* **2022**, *17*, 508–522. [CrossRef]
45. Bock, A.; Steinhäuser, U.; Drusch, S. Partitioning Behavior and Interfacial Activity of Phenolic Acid Derivatives and Their Impact on β-Lactoglobulin at the Oil-Water Interface. *Food Biophys.* **2021**, *16*, 191–202. [CrossRef]
46. Bravo-Díaz, C.; Romsted, L.S.; Liu, C.; Losada-Barreiro, S.; Pastoriza-Gallego, M.J.; Gao, X.; Gu, Q.; Krishnan, G.; Sánchez-Paz, V.; Zhang, Y.; et al. To Model Chemical Reactivity in Heterogeneous Emulsions, Think Homogeneous Microemulsions. *Langmuir* **2015**, *31*, 8961–8979. [CrossRef] [PubMed]
47. Goodman, B.A.; Severino, J.F.; Pirker, K.F. Reactions of Green and Black Teas with Cu(Ii). *Food Funct.* **2012**, *3*, 399–409. [CrossRef] [PubMed]
48. Pirker, K.F.; Baratto, M.C.; Basosi, R.; Goodman, B.A. Influence of PH on the Speciation of Copper(II) in Reactions with the Green Tea Polyphenols, Epigallocatechin Gallate and Gallic Acid. *J. Inorg. Biochem.* **2012**, *112*, 10–16. [CrossRef]
49. Severino, J.F.; Goodman, B.A.; Reichenauer, T.G.; Pirker, K.F. Is There a Redox Reaction between Cu(II) and Gallic Acid? *Free Radic. Res.* **2011**, *45*, 123–132. [CrossRef]
50. Ong, H.R.; Khan, M.M.R.; Ramli, R.; Du, Y.; Xi, S.; Yunus, R.M. Facile Synthesis of Copper Nanoparticles in Glycerol at Room Temperature: Formation Mechanism. *RSC Adv.* **2015**, *5*, 24544–24549. [CrossRef]
51. Yen, G.-C.; Duh, P.-Der; Tsai, H.-L. Antioxidant and Pro-Oxidant Properties of Ascorbic Acid and Gallic Acid. *Food Chem.* **2002**, *79*, 307–313. [CrossRef]
52. Dunford, H.B. Free Radicals in Iron-Containing Systems. *Free Radic. Biol. Med.* **1987**, *3*, 405–421. [CrossRef]
53. Farhoosh, R.; Moosavi, S.M.R. Evaluating the Performance of Peroxide and Conjugated Diene Values in Monitoring Quality of Used Frying Oils. *J. Agric. Sci. Technol.* **2009**, *11*, 173–179.
54. Iwahashi, H.; Ishii, T.; Sugata, R.; Kido, R. The Effects of Caffeic Acid and Its Related Catechols on Hydroxyl Radical Formation by 3-Hydroxyanthranilic Acid, Ferric Chloride, and Hydrogen Peroxide. *Arch. Biochem. Biophys.* **1990**, *276*, 242–247. [CrossRef] [PubMed]
55. Adams, A.; Bouckaert, C.; Van Lancker, F.; De Meulenaer, B.; De Kimpe, N. Amino Acid Catalysis of 2-Alkylfuran Formation from Lipid Oxidation-Derived α,β-Unsaturated Aldehydes. *J. Agric. Food Chem.* **2011**, *59*, 11058–11062. [CrossRef] [PubMed]

antioxidants

MDPI

*Article*

# Physical and Oxidative Stability of Emulsions Stabilized with Fractionated Potato Protein Hydrolysates Obtained from Starch Production Side Stream

**Betül Yesiltas [1,*], Pedro J. García-Moreno [2], Rasmus K. Mikkelsen [1], Simon Gregersen Echers [3], Dennis K. Hansen [4], Mathias Greve-Poulsen [5], Grethe Hyldig [1], Egon B. Hansen [1] and Charlotte Jacobsen [1,*]**

[1] National Food Institute, Technical University of Denmark, 2800 Kgs. Lyngby, Denmark; rkrmik@food.dtu.dk (R.K.M.); grhy@food.dtu.dk (G.H.); egbh@food.dtu.dk (E.B.H.)
[2] Department of Chemical Engineering, University of Granada, 18012 Granada, Spain; pjgarcia@ugr.es
[3] Department of Chemistry and Bioscience, Aalborg University, 2450 Copenhagen, Denmark; sgr@bio.aau.dk
[4] Lihme Protein Solutions, 2750 Ballerup, Denmark; d.hansen@lps-dk.com
[5] KMC Amba, 7330 Brande, Denmark; mgp@kmc.dk
* Correspondence: betye@food.dtu.dk (B.Y.); chja@food.dtu.dk (C.J.)

**Abstract:** This work studies the emulsifying and antioxidant properties of potato protein hydrolysates (PPHs) fractions obtained through enzymatic hydrolysis of potato protein using trypsin followed by ultrafiltration. Unfractionated (PPH1) and fractionated (PPH2 as >10 kDa, PPH3 as 10–5 kDa, PPH4 as 5–0.8 kDa, and PPH5 as <0.8 kDa) protein hydrolysates were evaluated. Pendant drop tensiometry and dilatational rheology were applied for determining the ability of PPHs to reduce interfacial tension and affect the viscoelasticity of the interfacial films at the oil–water interface. Peptides >10 kDa showed the highest ability to decrease oil–water interfacial tension. All PPH fractions predominantly provided elastic, weak, and easily stretchable interfaces. PPH2 provided a more rigid interfacial layer than the other hydrolysates. Radical scavenging and metal chelating activities of PPHs were also tested and the highest activities were provided by the unfractionated hydrolysate and the fractions with peptides >5 kDa. Furthermore, the ability of PPHs to form physically and oxidatively stable 5% fish oil-in-water emulsions (pH 7) was investigated during 8-day storage at 20 °C. Our results generally show that the fractions with peptides >5 kDa provided the highest physicochemical stability, followed by the fraction with peptides between 5 and 0.8 kDa. Lastly, promising sensory results with mostly mild attributes were obtained even at protein concentration levels that are higher than needed to obtain functional properties. The more prominent attributes (e.g., bitterness and astringency) were within an acceptable range for PPH3 and PPH4.

**Keywords:** protein hydrolysate; ultrafiltration; emulsifying peptides; oil-in-water emulsions; antioxidant activity; sensory profile

## 1. Introduction

Following the growing demand for sustainable food ingredients sourced from natural and sustainable origins, utilization of industrial side streams gained traction as they are often abundant in nutrients and contain bioactive and techno-functional compounds [1]. By harnessing these side streams, the food industry can address the demand while adhering to clean label criteria and considering consumer acceptance [2]. By re-incorporating side streams into food production, the industry can optimize resource utilization, minimize waste, and foster a more sustainable future, where the entire lifecycle of food is maximized. As an emerging application of alternative and natural food ingredients, plant-based protein hydrolysates were investigated for their functional properties in model oil-in-water emulsion systems to gather insight into their properties in real food emulsions [3–8]. Oxidatively prone oils, such as fish oil rich in omega-3 polyunsaturated fatty acids, are commonly used in these model systems to increase the rate of oxidation due to the bis-allylic carbons in their molecular structure. Peptides may potentially provide protection against lipid oxidation due to a combination of both surface activity and antioxidant properties. The former is attributed to their amphiphilic character and ability to interact with/adsorb the interface between oil and water in emulsions [9,10]. The antioxidant activity of peptides was suggested to relate to the content of specific amino acids that have the ability to scavenge free radicals (e.g., Cys, Met, and Try) or chelate metals (e.g., His, Glu, and Asp) [11,12].

Despite a modest protein content (1–2%), the magnitude of proteins obtained from the starch industry (~240,000 t/year) makes potatoes a highly relevant source of plant-based protein [13]. The identification and release of abundant proteins/peptides from plant-based sources gained significant attention from the food industry in recent years [9,14]. To meet the increasing demand for sustainable protein-based ingredients from alternative sources, the development of approaches combining advanced mass spectrometry-based proteomics and bioinformatics tools allowed for more time and cost-efficient research [15,16]. Using this approach, we previously identified several potent emulsifier peptides from abundant potato proteins [14,17]. While recent years showed tremendous progress in the fundamental understanding of peptide emulsification, most studies contributing to this were naturally performed on simple model systems and with individually isolated/synthesized peptides [9,10,18,19]. However, the use of purified peptides is not economically viable in the food industry, and thus, it is essential to further investigate and understand how hydrolysates from alternative protein sources may affect the physicochemical stability of oil-in-water emulsions. For instance, trypsin was determined the best candidate for targeted release of validated emulsifier peptides from potato protein and a tryptic hydrolysate was found to show superior emulsifying properties compared to the native protein and a range of hydrolysates obtained through conventional trial-and-error hydrolysis [20].

Previous studies reported that the size of the peptides contained in whey protein hydrolysates affects the emulsification ability; therefore, the degree of hydrolysis should be controlled [21–23]. In another study, sweet potato protein hydrolysates were obtained by Alcalase under high hydrostatic pressure, which significantly increased the degree of hydrolysis and the <3 kDa fraction content, thereby improving antioxidant activity [7]. Indeed, the size fractionation of the hydrolysates is an important factor to investigate both emulsifying and antioxidant activities in emulsion systems.

Thus, this study aimed to investigate the emulsifying and antioxidant activities of a potato protein hydrolysate obtained by trypsin and its fractions obtained by ultrafiltration. First, the ability of the unfractionated hydrolysate and fractionated PPHs to reduce interfacial tension, as well as their viscoelasticity of the interfacial layer, was evaluated. Moreover, the radical scavenging and metal chelating activities of the PPHs were measured. In addition, we investigated the ability of PPHs to stabilize low-fat oil-in-water emulsions and control lipid oxidation by following emulsions' physical and oxidative stability during 8 days of storage. Finally, the sensory properties of the PPHs were evaluated in a solution with a trained panel to map their sensory profile.

## 2. Materials and Methods

### 2.1. Materials

Potato protein hydrolysate (PPH1) was provided by KMC AmbA (Brande, Denmark). PPH1 was produced by the hydrolysis of denatured potato protein (5 *w*/*v*%) in water using 0.05 *w*/*v*% Trypsin (Pancreatic Trypsin Novo, 6.0 S, 6.0 AU/g) (Novozymes A/S, Bagsværd, Denmark) at pH 8, 50 °C for 24 h. The ratio between protein powder and Trypsin was 100:1 (*w*:*w*). Hydrolysis was carried out in a free-fall pH mode and the change in the pH was tracked every hour. When the pH decreased to 6.6 after 7 h, it was adjusted to 7 and the final pH was noted as 7.45 after 24 h. The enzymatic hydrolysis was followed by enzyme inhibition, centrifugation (2759× *g* for 15 min) (Hermle Z 513 K, Wehingen, Germany), and freezing of the supernatant (PPH1) at −18 °C. The degree of hydrolysis of PPH1 was measured and reported as 3.2%. PPH1 was further fractionated into >10 kDa (PPH2),

5–10 kDa (PPH3), 0.8–5 kDa (PPH4), and <0.8 kDa (PPH5) at Lihme Protein Solutions using sequential ultrafiltration with the following membranes: 10 kDa (K02-E010-05-S, 1.25 $m^2$, Repligen, Waltham, MA, USA), 5 kDa (K02-E005-05-S, 1.25 $m^2$, Repligen, Waltham, MA, USA), and 0.8 kDa (NFG-2B, 1812F, 0.37 $m^2$, Synder, Vacaville, CA, USA). The pump used was KMPi with 8 L/min crossflow and 1 Bar differential pressure (Repligen, Waltham, MA, USA). The protein content of the unfractionated (PPH1) and fractionated (PPH2-PPH5) potato protein hydrolysates was determined by DUMAS method and the results are 2.37, 3.08, 1.28, 1.14, and 0.27 *w/v*% for PPH1, PPH2, PPH3, PPH4, and PPH5, respectively. Medium-chain triglycerides (MCT) oil (WITARIX® MCT 60/40) was provided by OI Oleo GmbH (Hamburg, Germany), which was used in interfacial tension and dilatational rheology measurements. Cod liver oil was provided by Vesteraalen's (Nordland, Norway), which was used in emulsions that were subjected to storage experiments. The peroxide value (PV) of the fish oil was 0.23 ± 0.00 meq. $O_2$/kg oil. The fatty acid composition (%, *w/w*) of the fish oil was as follows: C14:0 (3.7), C16:0 (9.0), C16:1n-7 (9.6), C18:0 (1.9), C18:1n-9 (15.8), C18:1n-7 (4.5), C18:2n-6 (2.4), C18:3n-3 (1.0), C20:1n-9 (13.7), C20:5n-3 (8.4), C22:1n-11 (5.6), and C22:6n-3 (10.7), which was determined using GC analysis of fatty acid methyl esters (FAME) (AOCS official methods Ce2-66 and 1b-89). The tocopherol content of the fish oil was also analyzed and found as follows: α-tocopherol, 195 ± 3 μg/g oil; β-tocopherol, 5 ± 0 μg/g oil; γ-tocopherol, 116 ± 2 μg/g oil; and δ-tocopherol, 43 ± 1 μg/g oil (AOCS Official Method Ce 8–89).

### 2.2. Methods

#### 2.2.1. Characterization of the Fractionated PPHs

##### Molecular Weight Distribution of Hydrolysate and Fractions

PPHs were analyzed by size exclusion chromatography (FPLC) using an Äkta Pure chromatography system (GE Healthcare, Brondby, Denmark). A Tricorn 10/300 column was packed with Superdex 30 prep grade (GE Healthcare, Brondby, Denmark). The mobile phase used was 0.050 M potassium phosphate buffer, 0.100 M sodium chloride pH 7.5, at a flow rate of 0.500 mL/min. The absorbance was measured at the 280 nm wavelength.

##### Interfacial Tension and Interfacial Dilatational Rheology

The dilatational surface properties were analyzed using a drop tensiometer (OCA 25, Dataphysics Instruments GmbH, Filderstadt, Germany) at 20 °C. A pendant droplet of the protein fractions with a surface area of 59 $mm^2$ was formed at the tip of the needle (diameter 1.83 mm) in MCT oil (Witarix MCT 60/40, IOI Oleo GmbH, Hamburg, Germany) contained in a quartz glass cuvette. The aqueous solutions (pH 7) were prepared at 0.1 g protein/L. The drop contour was monitored and fitted with the Young–Laplace–Gauss equation to obtain the interfacial tension. The change in interfacial tension was monitored for 1 h, followed by a frequency and amplitude sweep. The frequency sweep was performed with frequencies of 0.01, 0.0325, 0.055, 0.0775, and 0.1 Hz at an amplitude of 2.5%. The amplitude sweep was performed with amplitudes of 1, 2, 3, 4, and 5% at a frequency of 0.01 Hz. The droplet was subjected to five oscillatory cycles at every amplitude or frequency. Via oscillatory dilatational deformation experiments (compression and expansion), the complex surface dilatational modulus was calculated from the first harmonic of the Fourier spectrum of the oscillatory interfacial tension signal. The complex surface dilatational modulus is equal to $E^*_d = E'_d + iE''_d$, where $E'_d$ is the dilatational storage modulus (elastic modulus), and $E''_d$ is the dilatational loss modulus (viscous modulus). The loss tangent is defined as $\tan\Phi = (E'_d/E''_d)$ [10]. The measurements were conducted in triplicate.

##### In Vitro Antioxidant Activity

The radical scavenging activity of predicted antioxidant peptides was measured using the DPPH radical scavenging activity method of Yang et al. [24] with some modifications. PPHs were dissolved in water and different concentrations (0.2–50 mM) of the PPH solutions were obtained. Equal parts (100 μL) of 0.1 mM ethanolic DPPH solution and a peptide solution were mixed. The mixture was transferred to the microtiter plate, kept in darkness at room temperature for 30 min, and the absorbance was measured at 515 nm using a spectrophotometer. The scavenging effect was calculated as an inhibition percentage following Equation (1):

$$\text{DPPH inhibition (\%)} = (1 - (As - A0)/Ab) \times 100, \quad (1)$$

where As is the sample absorbance (PPHs + DPPH), Ao is the sample blank absorbance (no DPPH), and Ab is the reagent blank absorbance (no PPHs). Results were calculated for 50% inhibition concentration (IC50) and presented in mg/mL. butylated hydroxy toluene (BHT) solution was included as a positive control. All measurements were performed in triplicate.

The iron chelating activity of the PPHs was measured as described in Farvin et al. [25] with some modifications. PPHs were dissolved in water and different concentrations (0.2–50 mM) of PPH solutions were obtained. Concisely, 100 μL of the antioxidant sample was transferred to the microtiter plate and 110 μL of water was added. To start the reaction, 20 μL of 0.5 mM ferrous chloride was added and the plate was shaken. After 3 min, 20 μL of 2.5 mM ferrozine was added and the plate was shaken again. After 10 min in the dark at room temperature, the absorbance was measured at 562 nm. Ethylenediaminetetraacetic acid (EDTA) was employed as a positive control. The chelating capacity and IC50 values were calculated in the same way as for the DPPH inhibition described above.

#### 2.2.2. Production and Physicochemical Stability of Oil-in-Water Emulsions Stabilized with the PPHs

##### Emulsion Production

A total of 250 g of fish oil-in-water emulsions (5 wt%) were stabilized with 0.2 wt.% protein of PPHs. PPH solutions were made in distilled water and kept in a stirrer in a refrigerator in darkness overnight to allow full hydration. The pH of the PPH solutions was checked and adjusted to 7 when necessary, using 1 M NaOH and 1 M HCl. Pre-homogenization was performed using ultra turrax (Ystral, Ballrechten-Dottingen, Germany) for 3 min at 16,000 rpm. Fish oil was added into the aqueous phase within the first minute of mixing. Secondary homogenization was performed using a microfluidizer (M110L Microfluidics, Newton, MA, USA) equipped with a ceramic interaction chamber (CIXC, F20Y, internal dimension 75 μm) at 9 kpsi pressure for three passes. Sodium azide (0.05%) and 50 μM of $FeSO_4$ were added to the emulsion obtained after microfluidizer homogenization and it was stirred with a spoon. The final pH of the emulsion was measured. Emulsions were stored for 8 days at 20 °C in darkness. Samples were collected for physical characterization and oxidative stability analyses during storage.

##### Physical Stability of Emulsions

(1) Creaming index

Creaming index was followed throughout the storage on days 0, 1, 2, 4, and 8. The creaming index was calculated based on the equation below:

$$\text{creaming index (\%)} = (b/a) \times 100,$$

where a is the total height of the emulsion in the tube, and b is the height of the clear aqueous phase at the bottom of the tube.

(2) Emulsion stability

Emulsions were placed in the Turbiscan cabinet on the same day they were produced and scanned over days (Formulaction, Toulouse, France). The scanning program is shown in Table S1 in the Supplementary Material. Based on the scans, the Turbiscan stability index (TSI) is calculated by the computer. This calculation is based on an integrated algorithm that sums up the evolution of transmittance and backscattering light at every position (h) and based on the scan-to-scan difference over total sample height (h).

$$TSI(t) = \frac{1}{N_h} \sum_{t_i=1}^{t_{max}} \sum_{z_i=z_{min}}^{z_{max}} |BST(t_i, z_i) - BST(t_{i-1}, z_i)|$$

where $N_h$ is the number of height position, t is the time, and $z_{min}$ and $z_{max}$ are the lower and upper height limit. BST is the signal that is taken into account (BS when T < 0.2%, otherwise T). The Turbiscan stability index (TSI) values above 3 are considered unstable while values below 1 are stable according to Formulaction's webpage. Values in between are in the beginning of destabilization, but mostly non-visual.

(3) Zeta potential

The zeta potential was measured using a Zetasizer Nano ZS (Malvern Instruments, Ltd., Worcestershire, UK) on day 1. Samples were prepared by taking 80 µL emulsion and diluting in 40 mL distilled water and mixing by a vortex. A DTS-1070 disposable folded capillary cell (Malvern Instruments, Ltd., Malvern, UK) was used for loading the samples. Measurements from the same sample were carried out in triplicate using a zeta potential range of (−) 100 to (+) 50 mV, and samples were analyzed with 100 runs at 25 °C.

(4) Droplet size

The droplet size distribution of the emulsions was measured using the Mastersizer 2000 (Malvern Instruments, Ltd., Worcestershire, UK) based on the laser diffraction technique on days 0 and 8 of the storage experiment. Emulsions were diluted in recirculating water set at 3000 rpm until an obscuration of approximately 12–15% was reached. For particle and dispersant, the refractive indices of sunflower oil (1.469) and water (1.330) were used, respectively. Measurements of droplet size were carried out in duplicates or triplicates and they were given as the surface-weighted (D[3,2]) and volume-weighted (D[4,3]) mean diameters.

Oxidative Stability of Emulsions

Oxidative stability analyses were carried out on the samples collected on days 0, 1, 2, 5, and 8 during the storage experiment.

(1) Peroxide value

Lipids were extracted using the method described by Bligh and Dyer [26] using a reduced amount of chloroform/methanol (1:1, *w/w*). Two extractions were made from each emulsion sample. Peroxide value (PV) was determined on lipid extracts using the colorimetric ferric thiocyanate method at 500 nm according to Shantha and Decker [27] with a spectrophotometer (Shimadzu UV-1280, Holm&Halby, Brøndby, Denmark). Measurements were performed in duplicate.

(2) Tocopherol consumption

Tocopherols were determined using an Agilent 1100 series HPLC (Agilent Technologies, Palo Alto, CA, USA), equipped with a fluorescence detector. Lipid extracts (2 g) from the above-mentioned Bligh and Dyer extraction were weighed and evaporated under nitrogen. Then, lipids were dissolved in 10 mL n-heptane. Heptane solutions (1 mL) were taken into separate vials before injection of an aliquot on a Spherisorb S5 W column (250 × 4.6 mm) (Phase Separation Ltd., Deeside, UK). Elution was performed with an isocratic mixture of n-heptane and 2-propanol (100:0.4, *v:v*) at a flow of 1 mL/min injection volume of 20 µL in a column (Waters Spherisorb 3 µm Silica, 4.6 mm I.D. × 150 mm), preceded by a guard column (Waters Spherisorb, 5 µm Silica, 4.6 mm I.D. ×10 mm). A fluorescence detector with excitation wavelength at 290 nm and emission wavelength at 330 nm was used according to the AOCS method [28]. A tocopherol standard mix including α, β, γ, and δ-tocopherol standards was used (Calbiochem 613424). The external quantitative standard was used for calculations. Measurements were performed in duplicate.

(3) Volatile compound formation using dynamic headspace GC-MS

Four grams of emulsion, 5 mL distilled water, and 30 mg of an internal standard (4-methyl-1-pentanol) were placed in a purge bottle. The purge bottle was subjected to a water bath at 45 °C for 30 min under purging with nitrogen (150 mL/min) and the volatile compounds were trapped in Tenax GR tubes. Afterward, the tubes were put into an Automatic Thermal Desorber (ATD-350, Perkin Elmer, Norwalk, CN), which was connected to gas chromatography (GC Agilent 6890 N, Palo Alto, CA, USA; Column: DB-1701, 30 m × 0.25 mm × 1.0 μm). The oven program had an initial temperature of 45 °C for 5 min, increasing with 1.5 °C/min until 55 °C, with 2.5 °C/min until 90 °C, and with 12.0 °C/min until 220 °C, where the temperature was kept for 4 min. A mass spectrometer (MS Agilent 5973, Agilent Technologies, USA; electron ionization mode, 70 eV; mass to charge ratio scan between 30 and 250) was used to separate the individual volatile compounds and these compounds were identified by MS library searches (Wiley 138 K, John Wiley and Sons, Hewlett-Packard, New York, NY, USA). The following standards were selected for quantification: 2-ethyl furan, 1-penten-3-one, pentanal, 1-penten-3-ol, 1-pentanol, hexanal, (E)-2-hexenal, heptanal, (E)-2-heptenal, (Z)-4-heptenal, octanal, (E,E)-2,4-hexadienal, (E,E)-2,4-heptadienal, (E)-2-nonenal, and (E,E)-2,4-decadienal. An ethanolic solution of the standards was prepared and diluted into nine levels (0.1–250 μg/mL) for the calibration curve. Each dilution (30 mg) was added into an emulsion produced in the same way as in one of the emulsions (PPH2), which allows a similar release for standards, as it was for volatile compounds formed in all emulsions. Analysis was run in triplicate.

#### 2.2.3. Sensory Analysis of the PPHs

The sensory evaluation was performed by a tested and trained objective sensory panel at DTU Food. The panel was specifically trained in objective descriptive analysis and the sensory evaluations were performed in a sensory lab with separated booths under normal daylight and at ambient temperature according to the standards and guidelines for the design and construction of sensory lab to ISO 13300-25492 (2006); ISO 3972 (2011) and ISO 3972 (2011)/Cor 1 (2012); 8586 (2012); and ISO 8589, 2007; NMKL Procedure No. 6, 2023 (Table S3, Supplementary Material).

The first sessions were used to develop a vocabulary to describe the sensory characteristics of the samples. The next three sessions were used to train the sensory panel in measuring and scoring the intensity of the selected sensory attributes. For the training, a set of samples with the different attributes and intensities of the attributes was used to ensure that all experimental conditions were represented. The scale was an unstructured 15 cm line scale with an anchor point at 1.5 cm and 13.5 cm. The samples were poured in plastic cups with lids and stored at 20 °C for one hour before serving. All samples were served in three replicates and in random order with three digital numbers and the assessors were served neutral flat bread and water to clean their mouths between samples.

The final vocabulary was defined in four categories as listed here: appearance: transparent, color (light/dark); odor: sweet, floral-like, sourish; flavor: sweet, earthy, potato, salty, green, cold tea, umami, bitter; texture: sticky, astringency, and viscous.

#### 2.2.4. Statistical Analysis

Mean value and standard deviations were introduced to Statgraphics 18 (Statistical Graphics Corp., Rockville, MD, USA) for the data analysis. Multiple sample comparison was performed to identify the significant differences between samples at certain sampling days and between sampling days for each sample during storage using Tukey as a post hoc test at $p < 0.05$ significance level.

## 3. Results and Discussion

### *3.1. Characterization of the Unfractionated Hydrolysate and Their Fractions Obtained by Ultrafiltration*

#### 3.1.1. Molecular Weight Distribution

The retention volume of the peptides by size exclusion increases from PPH2, PPH3, PPH4, to PPH 5, owing to decreasing peptide sizes (Figure 1). This is in accordance with the decreasing molecular weight cut-off of the membranes that they were retained on during the sequential filtration (10 kDa, 5 kDa, and 0.8 kDa permeate, respectively). PPH1, unfractionated PPH, spans the complete range seen in the fractionated samples. Larger molecules, with a retention volume of less than 11 mL, were exclusively retained in the PPH2 sample.

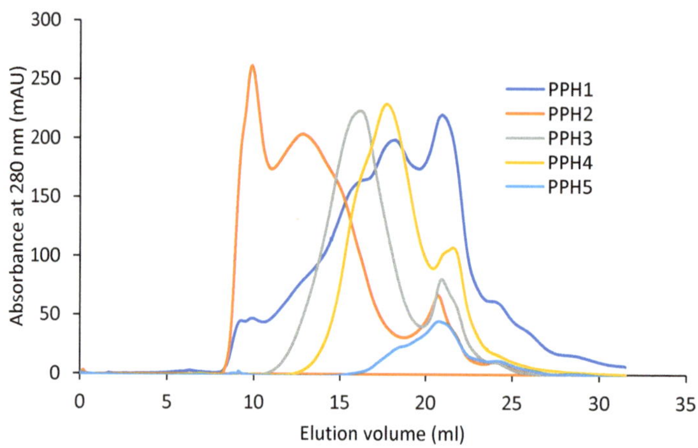


**Figure 1.** Fast protein liquid chromatography results from potato protein hydrolysates (PPHs). The absorbance was measured at a 280 nm wavelength. PPH1 to PPH5 denotes: unfractionated PPH, >10 kDa, 5–10 kDa, 0.8–5 kDa, and <0.8 kDa, respectively.

In contrast, lower-mass peptides (i.e., a retention volume in the range 20–25 mL) seem to be distributed across the different fractions, indicating that lower-mass peptides reside in the retentate post ultracentrifugation. To facilitate more efficient fractionation by size, samples may be passed through the filters with multiple passes by washing in a diafiltration-like manner. However, this would come at the expense of sample dilution and impose substantial additional costs if adapting the process to full-scale production.

#### 3.1.2. Interfacial Tension and Interfacial Dilatational Rheology

Emulsifying peptides may reduce interfacial tension by (i) diffusing from the bulk aqueous phase to the interface, (ii) adsorbing at the oil/water interface, and (iii) rearranging their structure at the interface to project their hydrophobic residues to the oil phase [16]. Figure 2a shows the reduction in the interfacial tension of the potato protein hydrolysates studied (PPH1) and their fractions obtained by ultrafiltration (PPH2-PPH5). It was observed that the bulk hydrolysate (PPH1) notoriously reduced the interfacial tension during the first 10 min, leveling off at a value of 20.9 $\pm$ 0.4 mN/m after 60 min. An identical trend to that observed for the bulk hydrolysate was also found for the hydrolysate fraction PPH4 (5–0.8 kDa). On the other hand, the hydrolysate fractions PPH3 (10–5 kDa) and PPH5 (<0.8 kDa) reduced interfacial tension to a significantly ($p < 0.05$) lower extent when compared to PPH1, reaching a value of 22.9 $\pm$ 0.3 mN/m after 60 min. Interestingly, the hydrolysate fraction PPH2 (>10 kDa) led to an initial significantly ($p < 0.05$) faster decrease in interfacial tension when compared to PPH1, leveling off after 25 min and reaching a value of 17.7 $\pm$ 0.4 mN/m after 60 min. This finding is in agreement with previous studies reporting that limited protein hydrolysis (degree of hydrolysis, DH, of 3–6%), which results in larger peptides, favored the emulsifying properties of the obtained hydrolysates [20–23,29]. In fact, PPH2 presents larger peptides than the rest of the fractions with higher potential to present large hydrophobic patches within their structure. These large hydrophobic patches are key to the diffusion and later adsorption of the peptides at the oil/water interface [14,29]. In any case, it should be noted that the interfacial tension values obtained for the potato protein hydrolysate, or its fractions obtained by ultrafiltration were higher than those previously reported for potato emulsifying peptides [10,14]. This finding is ascribed to the complex nature of a hydrolysate and its fractions, where not all peptides present have amphiphilic properties.

Interfacial dilatational rheology allows the characterization of interfacial layers, providing information on their viscoelasticity, structure, and intermolecular connections [30]. The complex dilatational modulus ($E^*_d = E'_d + iE''_d$) determines the resistance of the interfacial layer to external perturbations of its equilibrium state [31]. The elastic part ($E'_d$) of the complex dilatational modulus indicates the recoverable energy stored in the interface, whereas the viscous part ($E''_d$) represents the energy lost through relaxation processes [32]. The hydrolysate and all the ultrafiltration fractions formed predominantly elastic interfacial layers at the tested frequencies and amplitudes, having values of $E'_d$ markedly larger than $E''_d$ as well as low values of phase angles ($\Phi \leq 26°$). This finding denotes that all the peptide-based interfacial layers evaluated showed a gel-like structure with peptide–peptide interactions. The complex dilatational modulus observed for the interfaces studied ranged from $3.5 \pm 0.9$ to $33.2 \pm 1.5$ mN/m (Figure 2b,c), which is in line with previously reported values for whey and fish protein hydrolysates [21,31]. All the interfacial layers were amplitude-independent in the range 1–5%, denoting that they were within the linear viscoelastic regime (LVE). Moreover, the lack of change in the interfacial moduli with increasing amplitude (Figure 2b) suggests weak and easily stretchable interfaces that did not suffer a disruption in their microstructure in this amplitude range [10]. Figure 2c shows that the elasticity of all the interfaces, but particularly PPH2, PPH1, and PPH4, increased when increasing frequency. This is due to the shorter time needed to adapt to the deformation via the diffusional exchange of peptides with the aqueous phase or relaxation processes within the network [31]. Interestingly, Figure 2b,c shows that PPH2 formed the most rigid interfacial layer (e.g., highest $E^*_d$), followed by the unfractionated hydrolysate (PPH1) and PPH3-PPH4, with PPH5 presenting the lowest $E^*_d$ (Figure 2b,c). Indeed, PPH2 (>10 kDa) presented large peptides and unhydrolyzed proteins that re-arrange at the interface, facilitating intermolecular connections that result in an interface with higher elasticity and strength [21]. On the contrary, low molecular weight peptides present in PPH5 (<0.8 kDa) were not able to interconnect at the interface, leading to the weakest interface.

(a)

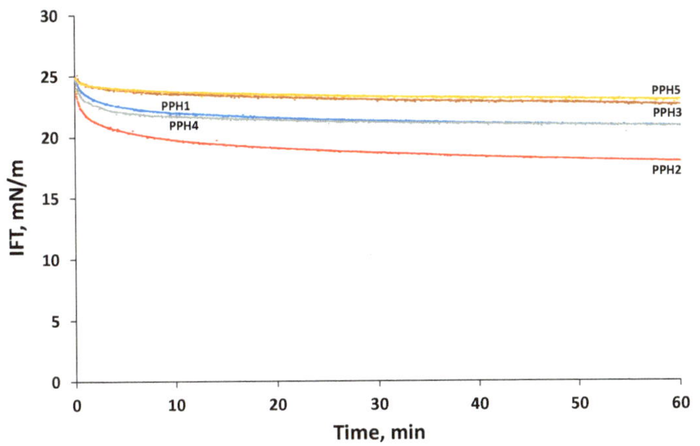


**Figure 2.** *Cont.*

(b)

(c)

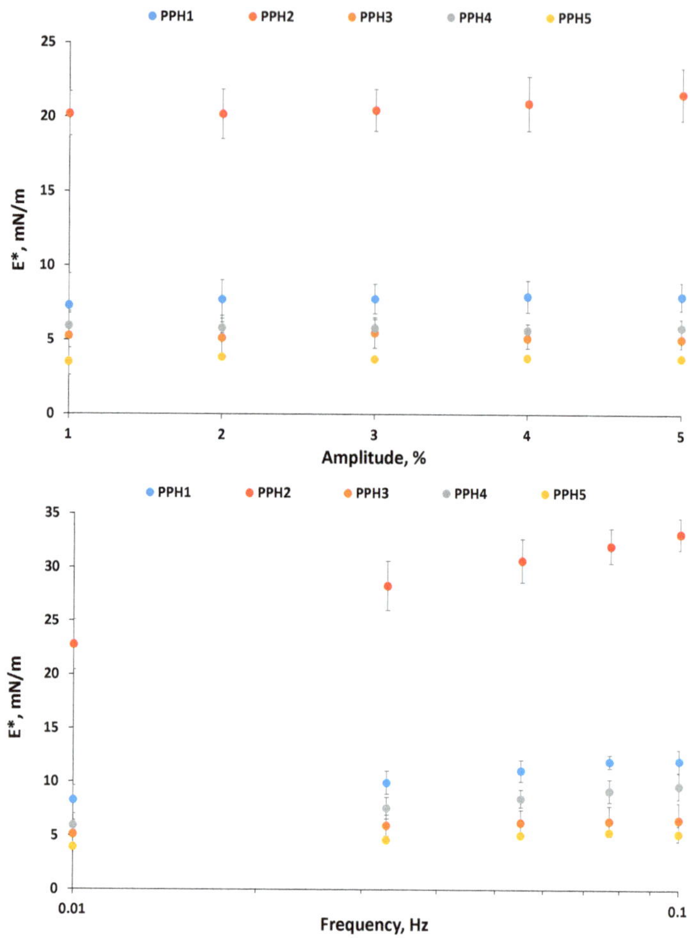


**Figure 2.** Interfacial properties of PPHs: (**a**) Interfacial tension as a function of time, (**b**) complex surface dilatational modulus ($E^*$) as a function of amplitude (1–5%, frequency: 0.01 Hz), and (**c**) complex surface dilatational modulus ($E^*$) as a function of frequency (0.005 to 0.1 Hz, amplitude: 2.5%). Measurements were carried out for 0.1 g protein/L PPHs aqueous solutions (pH 7) at 20 °C using MCT oil. The bare MCT oil–water interfacial tension was 26 mN/m. PPH1 to PPH5 denotes: unfractionated PPH, >10 kDa, 5–10 kDa, 0.8–5 kDa, and <0.8 kDa, respectively.

### 3.1.3. In Vitro Antioxidant Activity

Based on the DPPH radical scavenging activity results, PPH2 exhibited a significantly lower IC50 value compared to PPH1, PPH3, and PPH4 (Figure 3). PPH5 was found to be the least effective to scavenge radicals. On the other hand, based on the ferrozine assay, PPH3 was better at chelating metals followed by PPH2, PPH1, PPH5, and PPH4 (Figure 3). This is a different trend compared to the radical scavenging activity of the same PPHs, which could be attributed to the impact of peptide size on their ability to act as antioxidants via different mechanisms. In a previous study, peptide size was reported to affect the potency of peptides to retard lipid oxidation, where shorter peptides are generally attributed to higher scavenging and lipid peroxidation inhibition activities compared to long-chain peptides [16]. Ren et al. outlined that peptide chains containing <3 kDa peptides were more effective scavengers of hydroxyl radicals and inhibitors of lipid peroxidation when compared to chains containing >3 kDa peptides [33]. Therefore, it can be inferred that shorter peptide chains have higher DPPH radical scavenging activity than longer peptide chains. However, our results show that the smallest fraction (as for PPH5) had the highest IC50 value, indicating the lowest radical scavenging activity, which suggests that the smaller peptides (e.g., <0.8 kDa) should be reconsidered and tested for their inferior activity compared to the fractions with a larger size, but still considered short. Indeed, PPH4, containing peptides between 0.8 and 5 kDa, needed to be used at a lower concentration than PPH5 to provide 50% DPPH inhibition, which presents them as a more promising fraction, followed by PPH2 (>10 kDa). On the other hand, it should be borne in mind that the inorganic salt content present in the different fractions could impact the DPPH color-quenching activity and interfere with the assay [34].

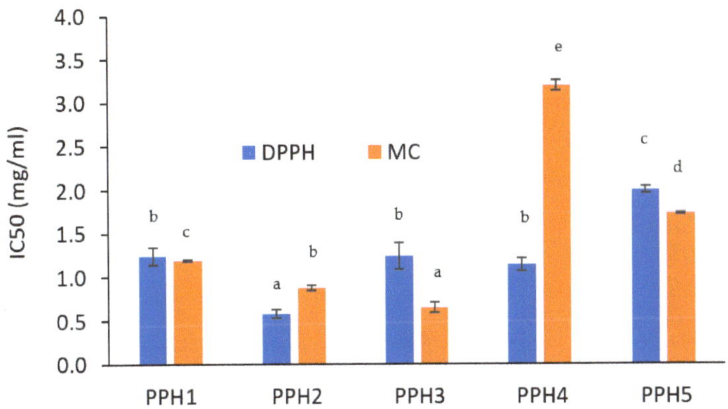


**Figure 3.** Half maximal inhibitory concentration (IC50) of PPHs is shown for DPPH radical scavenging (blue) and ferrozine metal chelating (MC) (orange) activities. Letters a–e indicate significant difference between samples ($p < 0.05$). PPH1 to PPH5 denotes: unfractionated PPH, >10 kDa, 5–10 kDa, 0.8–5 kDa, and <0.8 kDa, respectively.

Metal chelation activity of PPH was also attributed to smaller peptides in a previous study, generally having a size of less than 1 kDa [35]. However, this was not the case for the PPH5 (<0.8 kDa) in our study. PPHs higher than 5 kDa had lower IC50 concentrations, indicating their better metal chelation activity. Indeed, the size of the peptides cannot be considered as the only factor to control the antioxidant activity. There are other factors, such as peptide charge-to-mass ratio, conformation, amino acid composition, as well as the presence of AA residues with carboxyl or imidazole groups as addressed by García-Moreno et al. [14]. Nevertheless, the antioxidant activity of emulsifying hydrolysates should be further tested in model emulsion systems where oxidation is evaluated by advanced techniques such as GC-MS.

### 3.2. *Physicochemical Stability of Fish Oil-in-Water Emulsions Stabilized with PPHs*

#### 3.2.1. Physical Stability of Emulsions

##### Emulsion Stability

According to the creaming results, samples are ranked from the most physically stable to the least in the following order: PPH3 > PPH2 > PPH4 > PPH1 > PPH5 (Figure 4a). PPH5 emulsion was the least stable, which is expected, as this fraction contains the smallest peptides that are not large enough to adsorb at the oil–water interface, as also shown by the interfacial tension measurements (Figure 2a). Low molecular peptides tend to have weak interaction with each other and cannot form a strong interfacial layer as stated in Section 3.1.3. Similar to the creaming results, the Turbiscan stability index (TSI) values indicate that the emulsions from the most stable to the least stable can be listed as the following: PPH3, PPH2, PPH4, PPH5, and PPH1 (Figure 4b). The only difference was that the PPH1, unfractionated hydrolysate, was the least stable in terms of not only creaming, but also other instability phenomena, such as sedimentation, flocculation, and phase separation. This could be attributed to the inferior surface activity of peptides <0.8 kDa contained in PPH1.

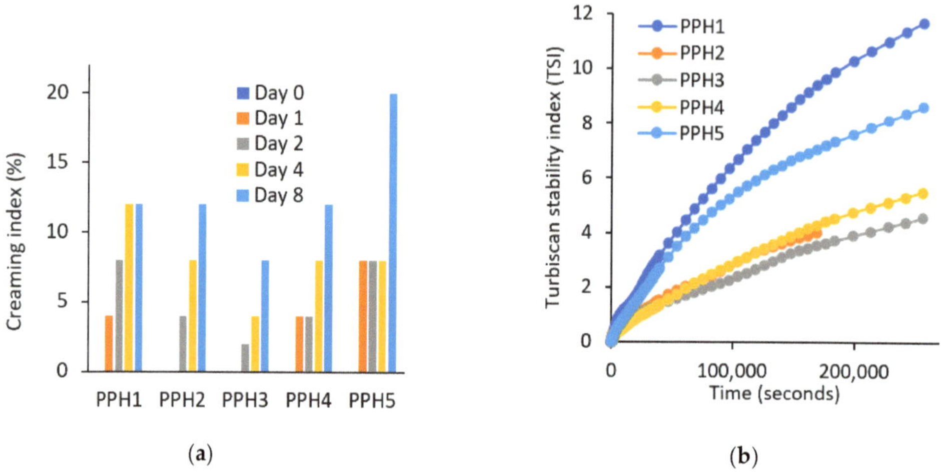


**Figure 4.** Physical stability of emulsions produced with PPHs; (**a**) creaming (%) and (**b**) Turbiscan stability index. PPH1 to PPH5 denotes: unfractionated PPH, >10 kDa, 5–10 kDa, 0.8–5 kDa, and <0.8 kDa, respectively.

Overall, based on the two stability indicators, the creaming index and TSI discussed, emulsions stabilized with PPH2, PPH3, and PPH4 (fractions between 10 and 0.8 kDa) were more physically stable compared to PPH1 (unfractionated) and PPH5 (<0.8 kDa). As concluded in Section 3.1.2, PPH2 led to the highest reduction in interfacial tension and to the interfacial film with the highest viscoelasticity, denoting the ability to form small oil droplets and a strong interface that provides resistance against physical destabilization phenomena.

Zeta Potential and Droplet Size

Zeta potential and droplet size results are shown in Table 1. The zeta potential value of the PPH5 emulsion was significantly lower compared to the other PPH fractions, indicating that the repulsion between droplets could be lower; however, the net value is still higher than 30 mV, which is considered to provide sufficient repulsion to prevent droplet merging [36]. The rest of the PPHs (PPH1 to PPH4) had a net zeta potential of around −60 mV, which is also in line with the findings in a previous study, in which 5% oil-in-water emulsion was stabilized with a synthetic peptide (namely, g1) derived from potato protein, reporting a value of −62 mV at pH 7 [16]. In another study, zeta potential values were reported as 36.3 ± 0.06, −46.8 ± 0.72 and −44.0 ± 0.60 mV for the 5% fish oil-in-water emulsions stabilized with blue witting hydrolysate (pH 2), soy hydrolysate (pH 8), and whey protein hydrolysate (pH 8), respectively [37]. The absolute zeta potential values in the current study were higher (Table 1) compared to the emulsions containing the different hydrolysates from fish, soy, and milk sources. Another study reported on the concentration effect of the protein emulsifier where the maximum zeta potential was obtained for whey protein-stabilized emulsions (pH 7.8) at the concentrations of 0.5 wt% and 1 wt% with −25 mV and slightly declined to −21 and −19 mV at the concentrations of 2 and 4 wt%, respectively [21]. The concentration of the protein hydrolysate also has an impact on the magnitude of the zeta potential values.

**Table 1.** Surface-weighted mean diameter (D[3,2]), volume-weighted mean diameter (D[4,3]), and zeta potential of potato protein hydrolysates. PPH1 to PPH5 denotes: unfractionated PPH, >10 kDa, 5–10 kDa, 0.8–5 kDa, and <0.8 kDa, respectively.

| Sample Code | ξ-Potential (mV) Day 1 | D[3,2] (μm) Day 1 | D[3,2] (μm) Day 8 | D[4,3] (μm) Day 1 | D[4,3] (μm) Day 8 |
|---|---|---|---|---|---|
| PPH1 | $-59.6 \pm 1.3$ [a] | $0.217 \pm 0.002$ [d] | $0.327 \pm 0.000$ [d] | $3.30 \pm 0.04$ [c] | $4.77 \pm 0.15$ [c] |
| PPH2 | $-61.0 \pm 2.0$ [a] | $0.143 \pm 0.000$ [b] | $0.188 \pm 0.008$ [b] | $0.33 \pm 0.01$ [a] | $2.71 \pm 1.85$ [b] |
| PPH3 | $-60.2 \pm 1.4$ [a] | $0.134 \pm 0.001$ [a] | $0.159 \pm 0.001$ [a] | $0.26 \pm 0.00$ [a] | $0.65 \pm 0.00$ [a] |
| PPH4 | $-60.7 \pm 2.2$ [a] | $0.193 \pm 0.003$ [c] | $0.243 \pm 0.000$ [c] | $1.73 \pm 0.21$ [b] | $2.02 \pm 0.20$ [ab] |
| PPH5 | $-53.2 \pm 0.3$ [b] | $0.334 \pm 0.006$ [e] | $0.367 \pm 0.002$ [e] | $4.49 \pm 0.88$ [d] | $6.73 \pm 0.28$ [d] |

[a–e] indicates significant differences between samples at a certain sampling day ($p < 0.05$).

Emulsions stabilized with PPH3 provided the formation of the smallest droplets compared to the rest of the PPHs based on D[3,2], and this was followed by PPH2 and PPH4 (Table 1). Based on the D[4,3] results, the smallest droplets were obtained for PPH3 and PPH2, followed by PPH4 (Table 1). Overall, emulsions stabilized with PPH fractions with peptides larger than 5 kDa (PPH3 and PPH2) provided smaller droplets (Table 1) and better physical stability (Figure 4) on day 1 of the storage experiment compared to the emulsions stabilized with unfractionated PPH (PPH1) and peptides smaller than 5 kDa. This could be due to better coverage of the oil–water interface with the larger peptides compared to the smaller ones. Indeed, in a previous study, it was reported for whey protein hydrolysate-stabilized emulsions that the peptides larger than 5 kDa were more predominant at the oil–water interface compared to smaller peptides (<5 kDa) [21]. Similarly, results obtained from emulsions stabilized with fava bean hydrolysates at two different levels of degrees of hydrolysis (8.4 and 15.6%) showed that emulsions with the lower degree of hydrolysis yielding larger peptides provided more favorable emulsifying activity [8].

In the previously mentioned study on the synthetic potato-derived peptide, g1, authors reported similar droplet sizes obtained for both the D[3,2] and D[4,3] as 151 nm and 250 nm, respectively [16]. However, the synthetic peptide, g1, was much more stable throughout the storage, with no significant increase in both droplet size values. This could be due to there being less complexity when having one compound for the synthetic peptide compared to a hydrolysate, where a plethora of peptides are present and compete for adsorption at the oil–water interface and form a stable layer. This could be attributed to the differences in the adsorption behavior and self-assembling (e.g., based on their secondary structure and amphipathic properties) of the different peptides in the hydrolysates at the oil–water interface [14,38].

3.2.2. Oxidative Stability of Emulsions

Emulsions containing PPH2 provided the highest protection against the formation of hydroperoxides throughout the 8 days of storage as indicated by the peroxide value (Figure 5a). The lag phase was the shortest for the emulsions containing PPH1 or PPH4. PV for these emulsions already started to increase significantly after day 4. Overall, if we look at the status of oxidation on the last day of the storage (day 8), PPH2 had the lowest concentration of hydroperoxides, followed by PPH3 and PPH1, whereas emulsions containing PPH4 and PPH5 were the least stable ones (Figure 5a). Similarly, the consumption of α-tocopherol, which is a natural antioxidant in fish oil, in these emulsions followed the opposite order, where emulsions PPH4 and PPH5 depleted significantly earlier than emulsions containing PPH1, PPH2, and PPH3 (Figure 5b). Indeed, the consumption of α-tocopherol followed the same trend as the formation of peroxide values. Moreover, emulsions with PPH4 or PPH5 had significantly lower levels of α-tocopherols already at day 0, indicating that the oxidation was happening already during the emulsification process.

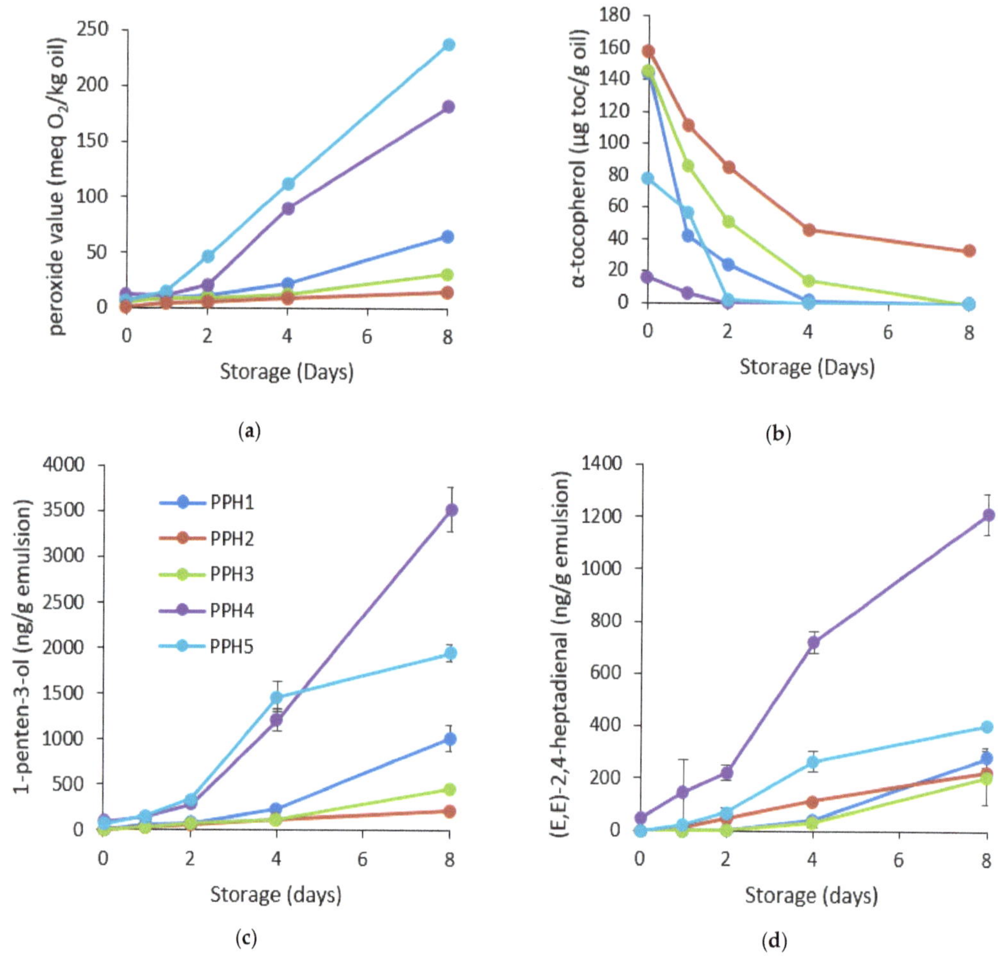


**Figure 5.** Oxidative stability of the emulsions containing PPHs: (**a**) formation of hydroperoxides; (**b**) consumption of α-tocopherols; (**c**) development of 1-penten-3-ol; and (**d**) development of (*E*,*E*)-2,4-heptadienal during 8 days of storage at room temperature in darkness. The PV values lower than 50 have less than a 10% average coefficient of variation, and for the values above 50, the average coefficient of variation is less than 5%. For tocopherols, the average coefficient of variation is less than 5%. PPH1 to PPH5 denotes: unfractionated PPH, >10 kDa, 5–10 kDa, 0.8–5 kDa, and <0.8 kDa, respectively.

Among the formed secondary oxidation volatile compounds, 1-penten-3-ol and (*E*,*E*)-2,4-heptadienal were selected as omega-3-derived polyunsaturated fatty acids developed to the highest concentrations during the 8 days of storage. The trend between these 2 volatile compounds was similar with some slight differences in the lag phases of the different emulsions. Formation of 1-penten-3-ol indicated that emulsions containing PPH1, PPH2, and PPH4 had a lag phase of 4 days, whereas PPH3 and PPH5 had a lag phase of 2 days (Figure 5c, Table S2). On the last day of storage, the concentration of 1-penten-3-ol was significantly lower ($p < 0.05$) for PPH2 and PPH3, followed by PPH1, PPH5 and PPH4. Similarly, the formation of (*E*-*E*)-2,4-heptadienal was significantly slower ($p < 0.05$) for the emulsions produced with PPH1 and PPH3 starting after day 4 (Figure 5d, Table S2). The emulsion containing PPH4 had significantly ($p < 0.05$) higher amounts of

(*E-E*)-2,4-heptadienal after day 2, followed by PPH2 and PPH5 starting after day 1. For this specific volatile compound, the emulsion containing PPH4 was the most oxidized at the end of the storage by around 6 folds in comparison to emulsions containing PPH 3 and PPH2. In a previous study, where the synthetic emulsifier peptides were used for stabilizing 5% fish oil-in-water emulsions, the volatile compounds *(E,E)*-2,4-heptadienal, *(E,E)*-2,4-decadienal, 2-ethyl furan, and 1-penten-3-ol had the highest concentration during 10-day storage, indicating a similar profile of secondary lipid oxidation markers in fish oil emulsions [16]. The concentrations of the volatile compounds developed for emulsions containing synthetic peptides were comparable to the current study, where the emulsions contained PPHs. When all the detected volatiles were screened, the emulsions containing PPH2, PPH3, and PPH1 were the most oxidatively stable, followed by the emulsions containing PPH5 and PPH4 (Figure 5 and Figure S1). Fish oil-in-water emulsions stabilized with sardine or small-spotted catshark hydrolysates also reported high concentrations of 1-penten-3-ol and (E,E)-2,4-heptadienal at day 7, and the levels were comparable with PPH2 and PPH3 in the current study [23].

Overall, based on the formation of hydroperoxides and secondary volatile oxidation products and consumption of tocopherols, PPH2 (>10 kDa) had the lowest oxidation levels, followed by PPH3 (10–5 kDa), indicating that the PPH fractions higher than 5 kDa were the most effective for protection against lipid oxidation and performed better in terms of providing oxidative stability, whereas PPH4, as the fraction 5–0.8 kDa, provided the lowest oxidative stability among others, followed by PPH5, the fraction lower than 0.8 kDa. These results are also in line with the in vitro antioxidant activity of the same PPHs. The DPPH radical scavenging activity of the PPH2 (>10 kDa) had higher activity than the PPH3 and PPH4, which contained peptides between 0.8 and 10 kDa, and the lowest activity was obtained for the peptides <0.8 kDa (PPH5). Similarly, the metal chelating activity of the peptides between 5 and 10 kDa (PPH3) was significantly higher followed by the peptides >10 kDa (PPH2), <0.8 kDa (PPH5), and 0.8–5 kDa (PPH4), respectively. The better oxidative stability of the emulsions containing larger peptides of potatoes can potentially be attributed to the antioxidant activity of unadsorbed peptides in the bulk aqueous phase of the emulsions. Specifically, the negatively charged metal chelator peptides at pH 7 may inhibit oxidation reactions by attracting and chelating positively charged metal ions in the aqueous phase. Moreover, peptides efficiently scavenge radicals at the oil–water interface where lipid oxidation is initiated due to having prooxidants in close proximity to lipids [23,39].

The physical stability of emulsions is greatly influenced by the interfacial tension and interfacial dilatational rheology, which are critical factors impacting lipid oxidation [40]. According to Sections 3.1.2 and 3.2.1 of the study, when emulsifying peptides in the PPHs are introduced into the emulsion, they reduce interfacial tension by adsorbing at the interface between oil and water and reorganizing their structure. Peptides with a molecular weight larger than 5 kDa, such as PPH2 and PPH3, exhibit superior interfacial properties. They effectively decrease interfacial tension and/or display higher viscoelasticity. Consequently, these peptides promote the formation of stable emulsions with smaller droplet sizes, indicating the presence of a strong interface that resists physical destabilization. The physical stability of the emulsion is closely linked to its oxidative stability. Emulsions stabilized with PPH2 and PPH3 demonstrate higher physical stability and smaller droplet sizes, which can contribute to improved oxidative stability. The larger peptides offer better coverage of the oil–water interface, potentially reducing lipid oxidation reactions at the interface by providing both chemical protection through radical scavenging or metal chelating activities and a physical barrier by forming a tightly packed interface. This is likely due to the more complex arrangements of the hydrolysates at the oil–water interface, which prevent interaction with iron. However, this may differ for emulsions stabilized with pure synthetic peptides, as previously reported [10]. Additionally, iron diffusion through the interface is minimal since its solubility in water is higher [41]. Thus, the key is to prevent the interaction between iron in the aqueous phase and polar lipid hydroperoxides at the oil–water interface. Consequently, emulsions stabilized with PPHs larger than 5 kDa, particularly PPH2 and PPH3, exhibit enhanced physical stability, smaller droplet sizes, and potentially improved oxidative stability compared to those stabilized with smaller peptides (PPH4 and PPH5) or unfractionated PPH (PPH1).

In summary, emulsions stabilized with PPHs larger than 5 kDa demonstrated enhanced physical and oxidative stability, together with their antioxidant activity, PPH fractions are promising candidates for applications requiring stable emulsions with enhanced resistance to lipid oxidation.

### 3.3. Sensory Analysis on the PPHs

Size-fractionated potato protein hydrolysates were measured with objective sensory analysis as hydrolysate solutions with a tested and trained panel. Results are shown in a spider plot in Figure 6. For PPH3 and PPH4, the more prominent attributes were astringency, bitter and cold tea, which were not in a very high intensity and may be in an acceptable range when used in a food product. It should be noted that PPH2 had a higher bitter flavor and astringent taste compared to PPH3, PPH4, and PPH5.

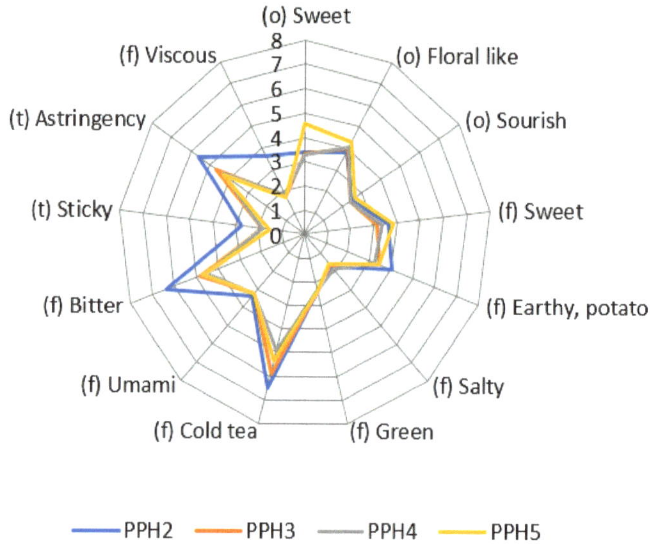


**Figure 6.** Sensory analysis of the size fractionated PPH solutions. The numbers from 0 to 8 indicate the increasing intensity of the different attributes. PPH2 to PPH5 denotes: >10 kDa, 5–10 kDa, 0.8–5 kDa, and <0.8 kDa, respectively.

Peptides obtained through protein hydrolysis often possess a bitter taste, which hampers their usefulness in food products [42]. To address this, researchers investigated the impact of different proteolytic treatments on the bitterness of pea protein hydrolysates. The study revealed that among the tested hydrolysates, Alcalase hydrolysate exhibited the highest level of bitterness, while papain and α-chymotrypsin hydrolysates demonstrated the lowest bitterness [43]. Consequently, it is crucial to assess the bitterness intensity and select appropriate fractions of protein hydrolysates for incorporation into food products. Notably, PPH3 and PPH4 are better suited than PPH2 for foods where the final product's taste could potentially be affected by bitterness. However, it is important to conduct consumer studies to evaluate the acceptance of food products containing protein hydrolysates, as this largely depends on the specific food matrix used. Moreover, masking flavors may be necessary for many plant proteins and protein hydrolysates to achieve desirable sensory attributes [44–47]. Thus, additional research is needed to determine effective strategies for mitigating the bitter taste and ensuring the acceptability of food products incorporating protein hydrolysates.

## 4. Conclusions

Emulsions produced with unfractionated and size-fractionated potato protein hydrolysates (PPHs) obtained by targeted hydrolysis with trypsin provided physiochemically stable emulsions. Different fractions showed differences in their in vitro antioxidant activity in terms of radical scavenging and metal chelating activities, where PPH2 (>10 kDa) had the best radical scavenging activity and PPH3 (5–10 kDa) had the best metal chelating activity. PPH2 (>10 kDa) had significantly stronger peptide–peptide interactions leading to the formation of more rigid and stable oil–water interfacial layers. Overall, emulsions containing PPHs higher than 5 kDa resulted in higher physical and oxidative stability during 8-day storage, indicating that the size fractionation influences techno-functional properties. This is crucial in the utilization of these ingredients as the fractions can be selected based on their best performance in the required functional property as antioxidants, emulsifiers, or both. Promising sensory results were also obtained at a higher protein concentration than needed to obtain expected functional properties.

**Supplementary Materials:** The following supporting information can be downloaded at: https://www.mdpi.com/article/10.3390/antiox12081622/s1, Table S1: Scanning program for the Turbiscan; Table S2: Statistical analysis for the (a) formation of hydroperoxides; (b) consumption of α-tocopherols; (c) development of 1-penten-3-ol and (d) development of (*E*,*E*)-2,4-heptadienal; Figure S1: The development of volatile compounds during 8 days of storage; Table S3: References for sensory analysis.

**Author Contributions:** Conceptualization, B.Y., P.J.G.-M. and C.J.; methodology, B.Y., P.J.G.-M., D.K.H., M.G.-P., G.H. and C.J.; validation, B.Y., P.J.G.-M., R.K.M. and G.H.; formal analysis, B.Y., P.J.G.-M., R.K.M., D.K.H., M.G.-P. and G.H.; investigation, B.Y., P.J.G.-M., R.K.M., D.K.H., M.G.-P. and G.H.; data curation, B.Y., P.J.G.-M., R.K.M. and G.H.; writing—original draft preparation, B.Y. and P.J.G.-M.; writing—review and editing, B.Y., P.J.G.-M., R.K.M., S.G.E., D.K.H., M.G.-P., G.H. and C.J.; visualization, B.Y., P.J.G.-M. and G.H.; supervision, B.Y. and C.J.; project administration, E.B.H. and C.J.; funding acquisition, E.B.H. and C.J. All authors have read and agreed to the published version of the manuscript.

**Funding:** This research was funded by Innovation Fund Denmark, grant number 7045-00021B (PROVIDE Project).

**Data Availability Statement:** Data is available on request.

**Acknowledgments:** We thank Lis Berner and Rie Sørensen for their skillful work in the lab with the physicochemical characterization of the emulsions and sensory analysis, respectively.

**Conflicts of Interest:** The authors declare no conflict of interest. The funders had no role in the design of the study; in the collection, analyses, or interpretation of data; in the writing of the manuscript; or in the decision to publish the results.

## References

1. Verni, M.; Casanova, F. The Potential of Food By-Products Bioprocessing, Bioactive Compounds Extraction and Functional Ingredients Utilization. *Foods* **2022**, *11*, 4092. [CrossRef]
2. Asioli, D.; Aschemann-Witzel, J.; Caputo, V.; Vecchio, R.; Annunziata, A.; Næs, T.; Varela, P. Making Sense of the "Clean Label" Trends: A Review of Consumer Food Choice Behavior and Discussion of Industry Implications. *Food Res. Int.* **2017**, *99*, 58–71. [CrossRef]
3. Yan, J.; Yang, Z.; Qiao, X.; Kong, Z.; Dai, L.; Wu, J.; Xu, X.; McClements, D.J. Interfacial Characteristics and in Vitro Digestion of Emulsion Coated by Single or Mixed Natural Emulsifiers: Lecithin and/or Rice Glutelin Hydrolysates. *J. Sci. Food Agric.* **2022**, *102*, 2990–2999. [CrossRef]
4. Cheng, Y.; Xiong, Y.L.; Chen, J. Antioxidant and Emulsifying Properties of Potato Protein Hydrolysate in Soybean Oil-in-Water Emulsions. *Food Chem.* **2010**, *120*, 101–108. [CrossRef]
5. Cheng, Y.; Chen, J.; Xiong, Y.L. Interfacial Adsorption of Peptides in Oil-in-Water Emulsions Costabilized by Tween 20 and Antioxidative Potato Peptides. *J. Agric. Food Chem.* **2014**, *62*, 11575–11581. [CrossRef]

6. Padial-Domínguez, M.; Espejo-Carpio, F.J.; Pérez-Gálvez, R.; Guadix, A.; Guadix, E.M. Optimization of the Emulsifying Properties of Food Protein Hydrolysates for the Production of Fish Oil-in-Water Emulsions. *Foods* **2020**, *9*, 636. [CrossRef] [PubMed]
7. Zhang, M.; Mu, T.-H. Identification and Characterization of Antioxidant Peptides from Sweet Potato Protein Hydrolysates by Alcalase under High Hydrostatic Pressure. *Innov. Food Sci. Emerg. Technol.* **2017**, *43*, 92–101. [CrossRef]
8. Felix, M.; Cermeño, M.; FitzGerald, R.J. Structure and in Vitro Bioactive Properties of O/W Emulsions Generated with Fava Bean Protein Hydrolysates. *Food Res. Int.* **2021**, *150*, 110780. [CrossRef]
9. García-Moreno, P.J.; Yesiltas, B.; Gregersen Echers, S.; Marcatili, P.; Overgaard, M.T.; Hansen, E.B.; Jacobsen, C. Recent Advances in the Production of Emulsifying Peptides with the Aid of Proteomics and Bioinformatics. *Curr. Opin. Food Sci.* **2023**, *51*, 101039. [CrossRef]
10. García-Moreno, P.J.; Yang, J.; Gregersen, S.; Jones, N.C.; Berton-Carabin, C.C.; Sagis, L.M.C.; Hoffmann, S.V.; Marcatili, P.; Overgaard, M.T.; Hansen, E.B.; et al. The Structure, Viscoelasticity and Charge of Potato Peptides Adsorbed at the Oil-Water Interface Determine the Physicochemical Stability of Fish Oil-in-Water Emulsions. *Food Hydrocoll.* **2021**, *115*, 106605. [CrossRef]
11. Elias, R.J.; McClements, D.J.; Decker, E.A. Antioxidant Activity of Cysteine, Tryptophan, and Methionine Residues in Continuous Phase β-Lactoglobulin in Oil-in-Water Emulsions. *J. Agric. Food Chem.* **2005**, *53*, 10248–10253. [CrossRef] [PubMed]
12. Elias, R.J.; Kellerby, S.S.; Decker, E.A. Antioxidant Activity of Proteins and Peptides. *Crit. Rev. Food Sci. Nutr.* **2008**, *48*, 430–441. [CrossRef] [PubMed]
13. Alting, A.C.; Pouvreau, L.; Giuseppin, M.L.F.; van Nieuwenhuijzen, N.H. Potato Proteins. In *Handbook of Food Proteins*; Elsevier: Amsterdam, The Netherlands, 2011.
14. García-Moreno, P.J.; Gregersen, S.; Nedamani, E.R.; Olsen, T.H.; Marcatili, P.; Overgaard, M.T.; Andersen, M.L.; Hansen, E.B.; Jacobsen, C. Identification of Emulsifier Potato Peptides by Bioinformatics: Application to Omega-3 Delivery Emulsions and Release from Potato Industry Side Streams. *Sci. Rep.* **2020**, *10*, 690. [CrossRef] [PubMed]
15. Gregersen Echers, S.; Abdul-Khalek, N.; Mikkelsen, R.K.; Holdt, S.L.; Jacobsen, C.; Hansen, E.B.; Olsen, T.H.; Sejberg, J.J.P.; Overgaard, M.T. Is Gigartina a Potential Source of Food Protein and Functional Peptide-Based Ingredients? Evaluating an Industrial, Pilot-Scale Extract by Proteomics and Bioinformatics. *Future Foods* **2022**, *6*, 100189. [CrossRef]
16. Yesiltas, B.; Soria Caindec, A.M.; García-Moreno, P.J.; Echers, S.G.; Olsen, T.H.; Jones, N.C.; Hoffmann, S.V.; Marcatili, P.; Overgaard, M.T.; Hansen, E.B.; et al. Physical and Oxidative Stability of Fish Oil-in-Water Emulsions Stabilized with Emulsifier Peptides Derived from Seaweed, Methanotrophic Bacteria and Potato Proteins. *Colloids Surf. A Physicochem. Eng. Asp.* **2023**, *663*, 131069. [CrossRef]
17. García-Moreno, P.J.; Jacobsen, C.; Marcatili, P.; Gregersen, S.; Overgaard, M.T.; Andersen, M.L.; Sørensen, A.-D.M.; Hansen, E.B. Emulsifying Peptides from Potato Protein Predicted by Bioinformatics: Stabilization of Fish Oil-in-Water Emulsions. *Food Hydrocoll.* **2020**, *101*, 105529. [CrossRef]
18. Yesiltas, B.; García-Moreno, P.J.; Gregersen, S.; Olsen, T.H.; Jones, N.C.; Hoffmann, S.V.; Marcatili, P.; Overgaard, M.T.; Hansen, E.B.; Jacobsen, C. Antioxidant Peptides Derived from Potato, Seaweed, Microbial and Spinach Proteins: Oxidative Stability of 5% Fish Oil-in-Water Emulsions. *Food Chem.* **2022**, *385*, 132699. [CrossRef]
19. Yesiltas, B.; Gregersen, S.; Lægsgaard, L.; Brinch, M.L.; Olsen, T.H.; Marcatili, P.; Overgaard, M.T.; Hansen, E.B.; Jacobsen, C.; García-Moreno, P.J. Emulsifier Peptides Derived from Seaweed, Methanotrophic Bacteria, and Potato Proteins Identified by Quantitative Proteomics and Bioinformatics. *Food Chem.* **2021**, *362*, 130217. [CrossRef]
20. Gregersen Echers, S.; Jafarpour, A.; Yesiltas, B.; García-Moreno, P.J.; Greve-Poulsen, M.; Hansen, D.K.; Jacobsen, C.; Overgaard, M.T.; Hansen, E.B. Targeted Hydrolysis of Native Potato Protein: A Novel Workflow for Obtaining Hydrolysates with Improved Interfacial Properties. *Food Hydrocoll.* **2023**, *137*, 108299. [CrossRef]
21. Schröder, A.; Berton-Carabin, C.; Venema, P.; Cornacchia, L. Interfacial Properties of Whey Protein and Whey Protein Hydrolysates and Their Influence on O/W Emulsion Stability. *Food Hydrocoll.* **2017**, *73*, 129–140. [CrossRef]
22. Morales-Medina, R.; Tamm, F.; Guadix, A.M.; Guadix, E.M.; Drusch, S. Functional and Antioxidant Properties of Hydrolysates of Sardine (*S. pilchardus*) and Horse Mackerel (*T. mediterraneus*) for the Microencapsulation of Fish Oil by Spray-Drying. *Food Chem.* **2016**, *194*, 1208–1216. [CrossRef] [PubMed]
23. García-Moreno, P.J.; Guadix, A.; Guadix, E.M.; Jacobsen, C. Physical and Oxidative Stability of Fish Oil-in-Water Emulsions Stabilized with Fish Protein Hydrolysates. *Food Chem.* **2016**, *203*, 124–135. [CrossRef] [PubMed]
24. Yang, J.; Guo, J.; Yuan, J. In Vitro Antioxidant Properties of Rutin. *LWT—Food Sci. Technol.* **2008**, *41*, 1060–1066. [CrossRef]
25. Sabeena Farvin, K.H.; Baron, C.P.; Nielsen, N.S.; Jacobsen, C. Antioxidant Activity of Yoghurt Peptides: Part 1-in Vitro Assays and Evaluation in ω-3 Enriched Milk. *Food Chem.* **2010**, *123*, 1081–1089. [CrossRef]
26. Bligh, E.G.; Dyer, W.J. A Rapid method of total lipid extraction and purification. *Can. J. Biochem. Physiol.* **1959**, *37*, 911–917. [CrossRef]
27. Shantha, N.C.; Decker, E.A. Rapid, Sensitive, Iron-Based Spectrophotometric Methods for Determination of Peroxide Values of Food Lipids. *J. AOAC Int.* **1994**, *77*, 421–424. [CrossRef]
28. AOCS Official Method Ce 8-89. *Determination of Tocopherols and Tocotrienols in Vegetable Oils and Fats by HPLC*; American Oil Chemists' Society: Champaign, IL, USA, 1998.
29. Tamm, F.; Gies, K.; Diekmann, S.; Serfert, Y.; Strunskus, T.; Brodkorb, A.; Drusch, S. Whey Protein Hydrolysates Reduce Autoxidation in Microencapsulated Long Chain Polyunsaturated Fatty Acids. *Eur. J. Lipid Sci. Technol.* **2015**, *117*, 1960–1970. [CrossRef]

30. Carrera Sánchez, C.; Rodríguez Patino, J.M. Contribution of the Engineering of Tailored Interfaces to the Formulation of Novel Food Colloids. *Food Hydrocoll.* **2021**, *119*, 106838. [CrossRef]
31. Ruiz-Álvarez, J.M.; del Castillo-Santaella, T.; Maldonado-Valderrama, J.; Guadix, A.; Guadix, E.M.; García-Moreno, P.J. PH Influences the Interfacial Properties of Blue Whiting (*M. poutassou*) and Whey Protein Hydrolysates Determining the Physical Stability of Fish Oil-in-Water Emulsions. *Food Hydrocoll.* **2022**, *122*, 107075. [CrossRef]
32. Tamm, F.; Drusch, S. Impact of Enzymatic Hydrolysis on the Interfacial Rheology of Whey Protein/Pectin Interfacial Layers at the Oil/Water-Interface. *Food Hydrocoll.* **2017**, *63*, 8–18. [CrossRef]
33. Ren, J.; Zhao, M.; Shi, J.; Wang, J.; Jiang, Y.; Cui, C.; Kakuda, Y.; Xue, S.J. Optimization of Antioxidant Peptide Production from Grass Carp Sarcoplasmic Protein Using Response Surface Methodology. *LWT—Food Sci. Technol.* **2008**, *41*, 1624–1632. [CrossRef]
34. Al-Dabbas, M.M.; Al-Ismail, K.; Kitahara, K.; Chishaki, N.; Hashinaga, F.; Suganuma, T.; Tadera, K. The Effects of Different Inorganic Salts, Buffer Systems, and Desalting of Varthemia Crude Water Extract on DPPH Radical Scavenging Activity. *Food Chem.* **2007**, *104*, 734–739. [CrossRef]
35. Cheng, Y.; Chen, J.; Xiong, Y.L. Chromatographic Separation and Tandem MS Identification of Active Peptides in Potato Protein Hydrolysate That Inhibit Autoxidation of Soybean Oil-in-Water Emulsions. *J. Agric. Food Chem.* **2010**, *58*, 8825–8832. [CrossRef] [PubMed]
36. McClements, D.J. *Food Emulsions*; CRC Press: Boca Raton, FL, USA, 2015; ISBN 9780429154034.
37. Padial-Domínguez, M.; Espejo-Carpio, F.J.; García-Moreno, P.J.; Jacobsen, C.; Guadix, E.M. Protein Derived Emulsifiers with Antioxidant Activity for Stabilization of Omega-3 Emulsions. *Food Chem.* **2020**, *329*, 127148. [CrossRef] [PubMed]
38. Ricardo, F.; Pradilla, D.; Cruz, J.C.; Alvarez, O. Emerging Emulsifiers: Conceptual Basis for the Identification and Rational Design of Peptides with Surface Activity. *Int. J. Mol. Sci.* **2021**, *22*, 4615. [CrossRef]
39. Berton-Carabin, C.C.; Ropers, M.-H.; Genot, C. Lipid Oxidation in Oil-in-Water Emulsions: Involvement of the Interfacial Layer. *Compr. Rev. Food Sci. Food Saf.* **2014**, *13*, 945–977. [CrossRef]
40. Berton-Carabin, C.C.; Sagis, L.; Schroën, K. Formation, Structure, and Functionality of Interfacial Layers in Food Emulsions. *Annu. Rev. Food Sci. Technol.* **2018**, *9*, 551–587. [CrossRef] [PubMed]
41. Homma, R.; Johnson, D.R.; McClements, D.J.; Decker, E.A. Influence of Iron Solubility and Charged Surface-Active Compounds on Lipid Oxidation in Fatty Acid Ethyl Esters Containing Association Colloids. *Food Chem.* **2016**, *199*, 862–869. [CrossRef]
42. O'Sullivan, D.; Nongonierma, A.B.; FitzGerald, R.J. Bitterness in Sodium Caseinate Hydrolysates: Role of Enzyme Preparation and Degree of Hydrolysis. *J. Sci. Food Agric.* **2017**, *97*, 4652–4655. [CrossRef]
43. Humiski, L.M.; Aluko, R.E. Physicochemical and Bitterness Properties of Enzymatic Pea Protein Hydrolysates. *J. Food Sci.* **2007**, *72*, S605–S611. [CrossRef]
44. Mittermeier-Kleßinger, V.K.; Hofmann, T.; Dawid, C. Mitigating Off-Flavors of Plant-Based Proteins. *J. Agric. Food Chem.* **2021**, *69*, 9202–9207. [CrossRef] [PubMed]
45. Wang, Y.; Tuccillo, F.; Lampi, A.; Knaapila, A.; Pulkkinen, M.; Kariluoto, S.; Coda, R.; Edelmann, M.; Jouppila, K.; Sandell, M.; et al. Flavor Challenges in Extruded Plant-based Meat Alternatives: A Review. *Compr. Rev. Food Sci. Food Saf.* **2022**, *21*, 2898–2929. [CrossRef] [PubMed]
46. Maehashi, K.; Arai, S. Taste Evaluation for Peptides in Protein Hydrolysates from Soybean and Other Plants. In *Analysis of Taste and Aroma*; Springer: Berlin/Heidelberg, Germany, 2002; pp. 47–68.
47. Li, H.; Li, L.-F.; Zhang, Z.-J.; Wu, C.-J.; Yu, S.-J. Sensory Evaluation, Chemical Structures, and Threshold Concentrations of Bitter-Tasting Compounds in Common Foodstuffs Derived from Plants and Maillard Reaction: A Review. *Crit. Rev. Food Sci. Nutr.* **2023**, *63*, 2277–2317. [CrossRef] [PubMed]

*antioxidants*

MDPI

*Article*

# Red Raspberry Seed Oil Low Energy Nanoemulsions: Influence of Surfactants, Antioxidants, and Temperature on Oxidative Stability

**Ana Gledovic [1,*], Aleksandra Janosevic-Lezaic [2], Slobodanka Tamburic [3] and Snezana Savic [1,*]**

1 Department of Pharmaceutical Technology and Cosmetology, Faculty of Pharmacy, University of Belgrade, 11121 Belgrade, Serbia
2 Department of Physical Chemistry and Instrumental Methods, Faculty of Pharmacy, University of Belgrade, 11121 Belgrade, Serbia
3 Cosmetic Science Research Group, London College of Fashion, University of the Arts London, London WC1V 7EY, UK
* Correspondence: ana.gledovic@pharmacy.bg.ac.rs (A.G.); snezana.savic@pharmacy.bg.ac.rs (S.S.)

**Abstract:** The aim of this study was to assess and improve the oxidative stability of red raspberry seed oil–RO, a potential topical ingredient derived from food industry by-products, on its own and when incorporated in low energy nanoemulsion (NE). The RO's oxidative stability was assessed at 5, 25, and 40 °C during one month of storage and expressed in: peroxide value, p-anisidine, and thiobarbituric reactive substances—TBARS value, while for NEs, lipid hydroperoxides and TBARS values were monitored. Both synthetic (butylated hydroxytoluene—BHT and ethylenediaminetetraacetic acid—EDTA), and natural (oregano essential oil—ORE and oak fruit extract—OAK) antioxidants were used. Pure RO and RO with BHT or ORE were stable at 5 °C and 25 °C, but at 40 °C BHT showed only moderate protection, while ORE was prooxidant. NEs prepared with new biodegradable polyglycerol esters-based surfactants, with droplet sizes of < 50 nm and narrow size distribution, showed improved physicochemical stability at room temperature, and especially at 40 °C, compared to NEs with polysorbate 80, which required the addition of antioxidants to preserve their stability. Natural antioxidants ORE and OAK were compatible with all NEs; therefore, their use is proposed as an effective alternative to synthetic antioxidants.

**Keywords:** nanoemulsification; polyglycerol ester; polysorbate 80; natural; oregano oil; oak extract

## 1. Introduction

Nowadays, growing popularity of natural products is evident in all health-related industries due to consumer preferences and ecological concerns, as well as marketing trends [1,2]. In particular, ingredients obtained from food industry waste during fruit and vegetable processing are being increasingly investigated in pharmaceutical and cosmetic industry, aiming to create safe products with added value [1–3]. One such by-product is red raspberry (*Rubus idaeus* L.) seed oil—RO, a "specialty oil" with potentially important dietary and functional properties. RO is recognized as a rich source of polyunsaturated fatty acids (PUFAs): C18:2 ω6 linoleic acid and C18:3 ω3 α-linolenic acid, with ω6: ω3 ratio from 1.5:1 to 2.7:1, which is considered a favorable ratio for the prevention of cardiovascular and carcinogenic diseases [4,5]. Moreover, there is a growing body of evidence that oral consumption of PUFAs, as well topical application of PUFA-loaded products, improve skin barrier function and vitality, leading to alleviation of various skin disorders accompanied with inflammation (i.e., photoaging, atopic dermatitis or psoriasis) [6,7].

Previous studies have shown that RO exhibits antioxidant activity [4,8–10], due to the presence of minor bioactive components such as carotenoids, tocopherols, and phenolic compounds [11,12]. However, despite these notable antioxidant properties, the shelf life of plant oils with high PUFA content remains limited (usually 6 to 12 months, if stored protected from light and air), due to their poor oxidative stability, especially at elevated temperatures [13,14]. Since natural antioxidants themselves exhibit instability, it would be useful to explore the use of innovative delivery systems as a strategy to ensure their stabilization, biocompatibility, and efficacy. In that regard, nanotechnological approach, especially nanoemulsions, has shown much promise recently [15–17].

In particular, nanoemulsions of the oil-in-water type (O/W NEs), constituted from oil phase dispersed in water phase in the form of small nanodroplets (usually from 20 to 200 nm) are suitable for encapsulation of lipophilic actives (such as RO). They provide various benefits (improved physical stability, prolonged or controlled release, pleasant visual appearance due to increased transparency) compared to other emulsion systems. Another benefit is a possibility to prepare NE systems by employing low energy methods, i.e., without high temperature or high-shear devices, which preserves quality of the natural and other sensitive ingredients [18,19]. The literature data regarding the lipid oxidation in classical emulsions are conflicting, since it was reported that lipid oxidation can be increased compared to bulk oil [20], but the opposite behavior was also described [21]. Due to the smaller size of nanodroplets compared to classical emulsions, the larger available oil–water interfacial surface area sometimes leads to even more pronounced oxidative instability [22,23], but the same controversy on the subject is evident in the reported results regarding NEs [24–26]. To avoid such problems, additional fine-tuning of the system composition may be required, based on elucidation of the role that surfactants and additional antioxidants play in the antioxidant stability of NE formulations [27,28].

Therefore, it was of interest to investigate the possibility of improving oxidative stability of RO in low energy O/W NEs prepared with novel polyglycerol ester-based mixture in comparison to polysorbate 80 as a standard small molecule surfactant in nanoformulations. This is a continuation of our previous work, where such systems were successfully developed and characterized, showing preliminary stability at room temperature and notable antioxidant properties [10,29,30]. Additionally, it was of interest to explore the feasibility of using natural antioxidants in NEs, specifically oregano seed oil (lipophilic) and oak extract (hydrophilic) instead of synthetic ones, since they are sometimes linked to adverse effects on human health [24,31,32]. Therefore, the aim of this study was to provide new insights into the antioxidant stability of nanoemulsions with natural ingredients, in order to contribute to the development of stable, safe, and efficient topical products.

## 2. Materials and Methods

### 2.1. Materials

Nonionic surfactants: Polysorbate® 80 (polyoxyethylene-20 sorbitan monooleate) produced by Fagron, Trikala, Greece and polyglycerol ester-based mixture prepared from Tego Care® PL4 (polyglyceryl 4-laurate) and Tego Solve® 61 (polyglyceryl-6 caprylate, polyglyceryl-3 cocoate, polyglyceryl-4 caprate, polyglyceryl-6 ricinoleate at 1:1:1:1 ratio). Mutual ratio of Tego Care® PL4 to Tego Solve® 61 was 6:4; these surfactants were produced by Evonik Nutrition and Care GmbH, Essen, Germany. Preservative mixture with additional cosurfactant properties was Euxyl PE® 9010 (phenoxyethanol, ethylhexylglycerin), produced by Schülke GmbH, Norderstedt, Germany.

Oil phase components: Pelemol® OPG (ethylhexyl pelargonate) was received as a free sample from Phoenix Chemical, Inc. Branchburg, NJ, USA; red raspberry (*Rubus idaeus*) seed oil—RO, cold pressed, unrefined oil contained 80% PUFAs out of 92% total unsaturated faffy acids (C18:2 ω6 linoleic —58.05%, C18:3 ω3 linolenic —22%, C18:1 ω9 oleic acid—12%, according to manufacturer specification) was produced by Aromaaz International, New Delhi, India.

Water phase components: ultra-purified water was obtained with GenPure apparatus (TKA Wasseranfbereitungssysteme GmbH, Neiderelbert, Germany). The polyglycerol ester-based NEs also contained Glycerol Solvagreen® (≥98%, anhydrous, Ph. Eur. grade), produced by Carl Roth GmbH, Karlsruhe, Germany; their pH value was adjusted with freshly prepared 0.1 M sodium hydroxide aqueous solution.

Antioxidants: synthetic butylated hydroxytoluene (BHT) was produced by Sigma-Aldrich Co., St. Lois, MO, USA and ethylenediaminetetraacetic acid disodium salt dihydrate (EDTA) was produced by Lach-ner, Ltd., Neratovice, Czech Republic; both were obtained from local distributors. Two natural antioxidants were also used: Oregano (*Origanum vulgare*) essential oil—ORE containing 80% of carvacrol (produced by Aromaaz International, New Delhi, India, for domestic brand Eterra/company Terra Co, Novi Sad) and oak fruit (acorn) extract, Phytessence® French oak—OAK (INCI: Aqua, Glycerin, Quercus petraea fruit extract), standardized at 1% polyphenols, received as free sample from Crodarom, Chanac, France. Lipophilic antioxidants (BHT or ORE) were added to the pure RO or to the NE oil phase, while the hydrophilic antioxidants (EDTA or OAK) were added to the NE water phase.

Reagents for determination of primary and secondary oxidation products in RO and nanaoemulsions: barium chloride dihydrate and ammonium thiocyanate were produced by Carl Roth GmbH, Karlsruhe, Germany; iron (II) sulfate heptahydrate, cumene hydroperoxide, thiobarbituric acid, trichloroacetic acid, butylated hydroxytoluene (BHT), 1,1,3,3-tetraethoxypropane (TEP), chloroform, 2-propanol, isooctane and 1-butanol were produced by Sigma-Aldrich Co., St. Lois, MO, USA. Hydrochloric acid was produced by Fischer Scientific, Loughborough, UK. Ethanol and methanol were produced by Honeywell, Muskegon, MI, USA. All solvents and reagents were of analytical grade or higher.

### 2.2. Methods

#### 2.2.1. Nanoemulsion Preparation

The NEs were prepared using the isothermal phase inversion composition (PIC) method suitable for heat-sensitive ingredients, at room temperature, without high-shear homogenization equipment or heating, which is recognized as low energy nanoemulsification due to considerable energy savings compared to the conventional production [18,19]. Briefly, the samples were prepared by stepwise addition of water phase with continuous vortex mixing (at 1300 rpm) to previously prepared surfactant-oil (SO) mixtures (2 min, at 1000 rpm). Formulation optimization of these RO-loaded NEs and the mechanism of nanoemulsification are described in detail in our previously published papers [10,29]. Therefore, for the purpose of the current study, similar formulations were prepared with polysorbate 80 (P80 NEs) or polyglycerol ester-based surfactants (PG NEs), with or without additional antioxidants (BHT—0.02 wt%, EDTA—0.2 wt%, ORE—0.02 wt% or OAK—1 wt%), while the nanoemulsion oil phase composition was kept constant. The antioxidants concentrations were chosen to be suitable for topical application, based on the common practice in cosmetic or pharmaceutical formulations, producer specifications, and the available literature [26–32]. After all NE components were added, a short homogenization of samples followed (for 2 min, at 1300 rpm) to obtain oil-in-water NEs containing 10 wt% surfactant, 10 wt% oil phase, and 80 wt% water phase, where wt% represents percentage weight by weight. The NE composition is presented in Table 1. The samples were kept in 10 mL glass bottles, capped tightly, and stored in the dark place for 30 days at different temperatures (5, 25, and 40 ± 2 °C, i.e., refrigeration, standard room temperature, and slightly elevated temperature, for example during summer, respectively), as a standard procedure to assess influence of the usual storage conditions for topical products and food [26,32].

**Table 1.** Composition of nanoemulsions prepared with different surfactants.

| Formulation Name | Surfactant (10 wt%) | Oil Phase (10 wt%) | Water Phase (80 wt%) |
|---|---|---|---|
| F0 P80 | P80 | 9.5 EP, 0.5 EUX | Water |
| F1 P80 | | 7.5 EP, 2 RO, 0.5 EUX | Water |
| F1 P80 + BHT | | 7.48 EP, 2 RO, 0.5 EUX, 0.02 BHT | Water |
| F1 P80 + ORE | | 7.48 EP, 2 RO, 0.5 EUX, 0.02 ORE | Water |
| F1 P80 + EDTA | | 7.5 EP, 2 RO, 0.5 EUX | Water—79.8, EDTA—0.2 |
| F1 P80 + OAK | | 7.5 EP, 2 RO, 0.5 EUX | Water—79.0, OAK—1.0 |
| F0 + PG | PG | 9.5 EP, 0.5 EUX | Water—76, Glycerol—4 |
| F1 + PG | | 7.5 EP, 2 RO, 0.5 EUX | Water—56, Glycerol—24 |
| F1 PG + BHT | | 7.48 EP, 2 RO, 0.5 EUX, 0.02 BHT | Water—56, Glycerol—24 |
| F1 PG + ORE | | 7.48 EP, 2 RO, 0.5 EUX, 0.02 ORE | Water—56, Glycerol—24 |
| F1 PG + EDTA | | 7.5 EP, 2 RO, 0.5 EUX | Water—55.8, Glycerol—24, EDTA—0.2 |
| F1 PG + OAK | | 7.5 EP, 2 RO, 0.5 EUX | Water—55.8, Glycerol—24, OAK—1.0 |

Table abbreviations: BHT—butylated hydroxytoluene, EDTA—ethylenediaminetetraacetic acid, disodium salt dihydrate, EP—ethylhexyl pelargonate, EUX—preservative/ cosurfactant mixture: phenoxyethanol, ethylhexylglycerin, OAK—oak fruit (acorn) extract, ORE—oregano essential oil, PG—mixture of polyglycerol ester-based surfactants: Tego Care® PL4 and Tego Solve® 61 at 6:4 ratio, P80—polysorbate 80, RO—red raspberry seed oil.

#### 2.2.2. Nanoemulsion Characterization and Physical Stability

##### Particle Size Distribution

The mean droplet size Z-average diameter (Z ave) and droplet size distribution (polydispersity index—PDI) of the NEs were determined by dynamic light scattering (Zetasizer Nano ZS90, Malvern Instruments Ltd., Worcestershire, UK). Before measurement, each NE sample was diluted with ultra-pure water (1:100, *v*/*v*). The samples were measured in triplicate at 25 ± 2 °C, 24 h after production and after one month of storage at different temperatures (storage stability study).

##### Electrical Conductivity and pH Value Measurements

Electrical conductivity of NEs was measured using SENSION+ EC71 apparatus (HACH, Loveland, CO, USA), while pH values were measured with HI2223 pH/ORP meter (Hanna Instruments Inc., Ann Arbor, MI, USA). The measurements were performed on undiluted NE samples, in triplicate, at 25 ± 2 °C according to the stability testing protocol.

### 2.3. *Oxidative Stability Investigations*

#### 2.3.1. Red Raspberry Seed Oil (RO) Oxidative Stability

For the purpose of oxidative stability investigations, lipophilic antioxidants were added to RO in the concentrations intended for skincare products—BHT at 0.2 wt% or 0.6 wt%, and ORE at 0.2 wt% or 1 wt%. The samples of pure RO or RO with added antioxidants (RO-BHT 0.2, RO-BHT 0.6, RO-ORE 0.2 and RO-ORE 1) were stored in tightly capped 5 mL glass bottles protected from light. The oil samples were analyzed initially (24 h after opening and/or addition of antioxidant), and again after 15 and 30 days of storage, at different temperatures (5, 25, or 40 ± 2 °C). The following parameters were analyzed: peroxide value (PV) as a measure of primary oxidation products, and p-anisidine value (PA) and thiobarbituric reactive substances (TBARS) value as measures of secondary oxidation products.

Peroxide Value (PV)

PV was determined according to the official International Dairy Federation method (IDF), a reliable spectrophotometric method suitable for small amounts of oil ($\leq$0.01–0.3 g) which is convenient for fast analysis of numerous samples. The IDF method is based on the ability of lipid hydroperoxides to oxidize Fe(II) to Fe(III) ions, which form a deep red colored complex with thiocyanate anion [33]. The concentration of lipid peroxides can be quantified by measuring the absorbance of this complex at 500 nm after 5 min of reaction time; the calibration curve is usually prepared with Fe(III) solution. The preparation of reagents and analysis of oil samples was performed exactly as reported previously [33], and the calibration curve was prepared by analyzing standard samples containing 2.5–50 μg Fe(III) instead of oil samples. The results were expressed as milliequivalents of peroxide per kilogram of sample, calculated according to the following formula:

$$PV = \frac{(As - Ab) \times s}{55.84 \times m \times 2} \tag{1}$$

where $A_s$ is absorbance of the sample, $A_b$ is absorbance of the blank (containing all reagents without the oil sample), s is slope, (in this experiment, s was 31.18), m is mass in grams of the oil sample, and 55.84 is atomic weight of iron. The division by factor 2 is necessary to express the peroxide value as milliequivalents of peroxide instead of milliequivalents of oxygen, as mentioned in reference [33]. The absorbances were measured via the Evolution 300 UV-VIS spectrophotometer (Thermo Scientific, Loughborough, UK).

p-Anisidine Value (PA)

The method used to determine the PA of oil samples was based on the procedure previously described [32] and according to the AOCS official method Cd 18-90 [34] with slight modifications regarding the reduction of amount of oil and reagents. The reaction is based on the reaction between p-anisidine and aldehydic compounds (mainly 2-alkenals and 2,4-alkadienals) in oils and fats in acidic conditions, resulting in the formation of a yellow-colored compound with absorbance at 350 nm. Briefly, p-anisidine reagent was prepared by dissolving of p-anisidine (0.025 g) in 10 mL of glacial acetic acid. Next, 0.5 g of oil sample was dissolved in isooctane and filled to 25 mL in a volumetric flask. First, the unreacted solution was prepared by mixing 2.5 mL of the oil solution and 0.5 mL of glacial acetic acid in a glass vial (for 10 s, with vortex mixer). The other 2.5 mL of the oil solution was mixed with 0.5 mL of p-anisidine reagent in another glass vial and kept in the dark for 10 min (reacted solution). The absorbance of these solutions and the blank solution (isooctane 2.5 mL mixed with 0.5 mL of p-anisidine reagent) were measured at 350 nm against isooctane, using a UV-VIS spectrophotometer. For the calculation of PA, the following equation was used:

$$PA = 25 \times [1.2 \times (A2 - Ab) - A1] \div m \tag{2}$$

where A1 is the absorbance of unreacted solution, A2 is the absorbance of reacted solution, and Ab is the absorbance of the blank, m is the mass of oil sample. The measurements were performed in triplicate.

Thiobarbituric Reactive Substances (TBARS)

TBARS measurements were carried out following the previously described procedure [35] with slight modifications. In brief, the oil samples (0.1 g, i.e., 120 μL) were mixed (for 10 s, with vortex mixer) with 3.75 mL of the freshly prepared TBA reagent containing 15% $w/v$ trichloroacetic acid and 0.375% $w/v$ thiobarbituric acid, dissolved in 0.25 M

HCl and 600 µL 2% *w*/*v* BHT solution in ethanol, and filled up with ultra-purified water up to 5 mL in screw-capped glass test tubes. The reaction mixtures were then heated in the water bath at 90 ± 5 °C for 15 min, followed by quick cooling in tap water to room temperature and centrifuged for 15 min at 3000 rpm (MPW centrifuge, MPW MED. Instruments, Warszawa, Poland). After 10 min, the absorbance of the supernatant containing pink colored complex formed from TBA and malonaldehyde was measured at 532 nm. The blank sample was prepared with ultra-pure water instead of oil, and its absorbance was subtracted from the measured absorbances of the samples. The TBARS content (expressed as mg malonaldehyde per kilogram of oil (MDA/kg oil) was obtained from the calibration curve prepared with 1,1,3,3-tetraethoxypropane (TEP) as a standard compound.

#### 2.3.2. Nanoemulsion Oxidative Stability

##### Determination of Lipid Hydroperoxides (LH) in Nanoemulsions

The formation of the primary oxidation products in the NE samples (blank NE and RO-loaded NEs with or without additional antioxidants), as well as in the surfactant dispersion in water phase without oil was performed according to the method described by Walker et al. [35] with slight modifications. First, the Fe(II) solution was freshly prepared prior to analysis, by mixing equal amounts of 0.132 M barium chloride and 0.144 M ferrous sulfate solutions in ultra-pure water, followed by centrifugation at 5000 rpm for 5 min to obtain a clear solution. The lipid extraction from NE samples was performed by mixing 0.2 mL of NE with 1 mL of isooctane/isopropanol (3:1 *v*/*v*), followed by vortexing for 10 s, and then centrifugation at 3000 rpm, for 2 min. The top layer (0.2 mL) was then collected and mixed with 2.8 mL of methanol/1-butanol (2:1 *v*/*v*), followed by addition of 15 µL of Fe(II) solution and 15 µL of ammonium thiocyanate solution as activator (3.94 M). The reaction mixtures were kept protected from light, and the absorbance was measured after 20 min against methanol/butanol mixture. The blank sample was prepared with isooctane instead of NE extract, and its absorbance was subtracted from the absorbance of the samples. For the method validation, cumene hydroperoxide was used as standard compound, and the results were expressed as mmol cumene hydroperoxide/L nanoemulsion (mmol CUM/L nanoemulsion).

##### Determination of TBARS in Nanoemulsions

TBARS values of NEs were determined when 500 µL of NE was mixed with 4 mL of the TBA reagent and 500 µL of BHT solution, by following the same procedure as described for the oil samples. The results were expressed as mg MDA/L nanoemulsion.

### 2.4. Statistical Analysis

Results are presented as means and standard deviations (SD) of measurements conducted in triplicate. The influence of added antioxidants on the droplet sizes and PDI values of NEs was investigated by one-way analysis of variance (ANOVA) using OriginPro8.5 (Originlab Corporation, Northampton, MA, USA). Evaluation of the effect of two independent variables (i.e., different antioxidants and different storage temperatures) on the oxidative stability of oils and NEs was performed via the two-way analysis of variance for different time intervals (15 and 30 days), followed by Tukey's post hoc test, while *p*-value of <0.05 was considered statistically significant.

## 3. Results and Discussion

### 3.1. Influence of Red Raspberry Seed Oil, Antioxidants, and Temperature on Nanoemulsion Properties

#### 3.1.1. The Nanoemulsion Droplet Size Distribution

Nanocarriers, in particular O/W NEs produced via low-energy methods are being increasingly proposed as promising carriers for natural and sensitive lipophilic ingredients such as natural oils and extracts, aiming to preserve/improve their stability and efficacy [19,36,37]. As previously described in the literature, the low-energy nanoemulsification (such as the PIC method used in this study) is based on the release of chemical energy when selected ingredients are mixed in a specific way, leading to the formation of small nanodroplets due to surfactant self-assembly [18,19,38]. Therefore, every change in system composition, such as the choice of surfactants, or addition of natural oil or other lipophilic or hydrophilic additives (i.e., antioxidants) can significantly influence NE properties and their stability [18,36].

For the purpose of this study, two different non-ionic surfactants: polysorbate 80—a standard surfactant suitable for various applications [23,39] and polyglycerol ester-based mixture—a novel biocompatible and biodegradable surfactant blend also suitable for food and cosmetics [40,41], were chosen to prepare NEs (Table 1). In our previous studies, NEs with similar composition were thoroughly optimized and characterized [10,29]. However, to the best of our knowledge, the influence of polyglycerol-ester based surfactants on NE physicochemical stability with additional antioxidants at different storage temperatures has not been reported before, neither it was compared to polysorbate 80 as the traditionally used stabilizer in submicron/NE formulations.

As it can be seen from Table 2, the choice of surfactant and addition of RO both had a significant impact on NE properties, while the addition of lipophilic (BHT or ORE) or hydrophilic antioxidants (EDTA or OAK) did not have significant effect. Namely, 24 h after preparation, RO-loaded P80 NEs had droplet sizes ranging from 203.50 nm to 211.43 nm, while F0 P80 without RO and antioxidants had bigger droplets (245.3 nm). As a result, all P80 NEs were milky white with slight bluish tone, indicative of NEs [42]. On the other hand, PG NEs were semi-transparent, due to very fine droplet sizes (Z-ave < 65 nm), also with notable differences between blank NE (F0 PG: 62.32 nm) and RO-loaded NEs (38.63–46.56 nm). Droplet size distribution was very narrow (PDI < 0.15) in both types of NE systems (i.e., 0.116–0.125 for P80 NEs, and 0.051–0.110 for PG NEs, respectively), indicating good preliminary stability [26,43].

**Table 2.** Z-average droplet size (Z-ave), polydispersity index (PDI), pH value, and electrical conductivity (El. cond.) of nanoemulsions stabilized with different surfactants and antioxidants.

| Formulation Name | Z-ave (nm) | PDI | pH | El. Cond. (μS/cm) |
|---|---|---|---|---|
| F0 P80 | 245.30 ± 3.12 | 0.121 ± 0.015 | 5.37 ± 0.006 | 159.33 ± 1.71 |
| F1 P80 | 207.63 ± 1.46 | 0.108 ± 0.017 | 5.45 ± 0.015 | 156.03 ± 0.76 |
| F1 P80 + BHT | 210.03 ± 1.62 | 0.125 ± 0.004 | 5.38 ± 0.021 | 167.37 ± 0.65 |
| F1 P80 + ORE | 203.50 ± 1.15 | 0.108 ± 0.014 | 5.40 ± 0.015 | 155.07 ± 0.49 |
| F1 P80 + EDTA | 206.03 ± 1.39 | 0.084 ± 0.027 | 4.98 ± 0.058 | 744.67 ± 0.58 |
| F1 P80 + OAK | 211.43 ± 1.66 | 0.116 ± 0.020 | 6.16 ± 0.025 | 242.67 ± 2.08 |
| F0 PG | 62.32 ± 0.32 | 0.060 ± 0.014 | 4.16 ± 0.006 | 167.73 ± 0.40 |
| F1 PG | 41.96 ± 0.72 | 0.059 ± 0.018 | 4.19 ± 0.010 | 81.07 ± 1.51 |
| F1 PG + BHT | 38.36 ± 0.73 | 0.110 ± 0.008 | 4.20 ± 0.017 | 75.77 ± 0.68 |
| F1 PG + ORE | 41.40 ± 0.95 | 0.079 ± 0.016 | 4.21 ± 0.026 | 71.27 ± 0.75 |
| F1 PG + EDTA | 46.56 ± 1.01 | 0.051 ± 0.013 | 4.59 ± 0.060 | 346.67 ± 1.53 |
| F1 PG + OAK | 41.57 ± 0.83 | 0.069 ± 0.024 | 4.35 ± 0.056 | 89.85 ± 0.23 |

After 30 days of storage at different temperatures, both systems showed satisfactory preliminary stability according to droplet size measurement, since there was no significant increase in droplet sizes for the majority of samples and PDI values remained without significant changes (Table S1, Supplementary file). However, visual observations (Figure 1) revealed that some changes occurred in P80 NEs that could not be detected via the droplet size measuring instrument due to its limited range (up to 5 microns) and because of the dilution prior to measurements. Although a standard procedure, the dilution could prevent the detection of some common instability phenomena (i.e., flocculation and formation of larger aggregates) [42]. For example, the creaming process obviously occurred in blank P80 formulation at all temperatures, indicating that addition of RO had a stabilizing effect on P80 NEs (Figure 1).

**Figure 1.** Visual appearance of nanoemulsions prepared with polysorbate 80 without raspberry oil on the left (sample 1—F0 P80 NE, where creaming process is visible on the glass vial wall) and with raspberry oil on the right (samples 2—F1 P80 NE, with improved stability) stored at different temperatures: (**a**) 5 °C, (**b**) 25 °C, and (**c**) 40 °C.

In the case of PG NEs, some interesting properties were revealed: for example, at 5 °C most of PG NEs became cloudy, while at 40 °C they became 2-phase systems (most likely a concentrated NE on the top with Z-ave 52.23 nm, PDI 0.18, and with an excess water phase below). However, these transformations of PG NEs were reversible, since after a few hours at RT and slight shaking (for 1 min) all PG NEs spontaneously recovered to their original state, with very similar droplet size distribution as initially (Figure 2a,b, Table S1, Supplementary file).

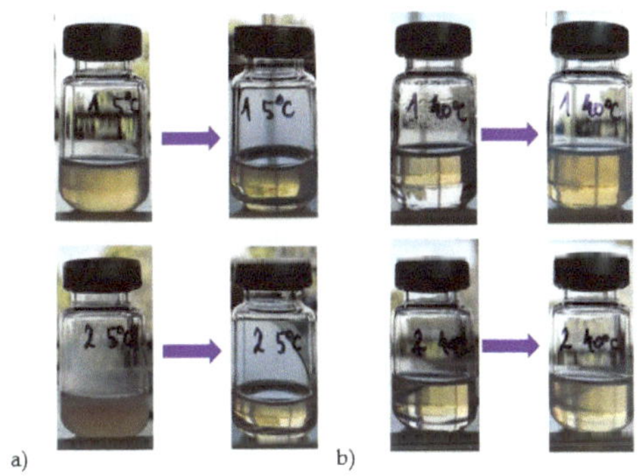

**Figure 2.** Visual appearance of nanoemulsions prepared with polyglycerol ester-based mixture without red raspberry oil on the left (sample 1—F0 PG NE) or with red raspberry oil on the right (sample 2—F1 PG NE) after one month of storage: (**a**) at 5 °C (cloudy samples), (**b**) at 40 °C (2-phase samples). All samples became clear/semi-transparent after one hour at room temperature (RT).

### 3.1.2. The pH Value and Electrical Conductivity

Further differences in stability among the NEs stabilized with different surfactants were manifested as changes in their pH value and electrical conductivity (data shown in Table S2, Supplementary file). It was observed that RO-loaded PG NEs exhibited good stability at all tested temperatures, since there were no major changes in their pH values, having in mind that decline in pH value is an indicator of fatty acid leakage from the nanoemulsion core [23,44]. Some changes in electrical conductivity were observed in most samples: up to 20% (for F1 PG+EDTA and F1 PG+OAK), up to 55% (for samples F1 PG+BHT, F1 PG+ORE), and especially for blank PG NE at all temperatures (up to 71% increase), although this was not reflected in overall stability of the samples. On a contrary, P80 NEs exhibited high instability at 40 °C, with very pronounced decrease of pH values (from 5.37 to 3.39 and from 5.45 to 3.72 for blank F0 P80 NE and F1 P80 NE with RO without added antioxidants, respectively). This can be attributed to low stability of P80 and natural oils with high PUFA content at elevated temperatures, as previously reported [23]. However, all tested antioxidants successfully prevented this NE deterioration, proving that it is necessary to add antioxidant to this type of surfactant.

### 3.2. Oxidative Stability of Red Raspberry Seed Oil

RO used in this study was an organic, cold-pressed, unrefined seed oil, known for its high PUFA content (>80%) and minor bioactive compounds (i.e., carotenoids and tocopherols), with orange color and pleasant herbal scent. In our previous investigations, this particular RO exhibited antioxidant activity per se or in different NE formulations [10,29]. However, similar to other natural oils with high PUFA content, there are concerns regarding the oxidative stability of RO during storage, especially at elevated temperatures, due to the fast formation of oxidation products, with potential hazardous effects on human health. Additionally, deterioration of natural oils leads to poor sensory attributes of products, such as rancid odor, change in color and texture [23,31,45]. Therefore, the first step was to investigate the oxidative quality of pure RO as a raw material, but also with added standard synthetic antioxidant (BHT) or ORE as a natural essential oil with remarkable antioxidant activity [30,46,47].

It is well established that the lower the PV the better is the oil quality, while the PV value above 10 is an indicator of oil oxidation (formation of primary oxidation products, such as lipid hydroperoxides) [31]. In the case of cold-pressed oils, this limit goes up to 15 [13]. It can be seen from Figure 3 that pure RO and RO with all tested antioxidants exhibited satisfactory PV values below 10 during the first 15 days of storage at all tested temperatures. However, at the end of storage stability study, at 40 °C the majority of samples exceeded the limit of PV = 10, but still remained at a relatively satisfactory level (<20). Interestingly, it was also observed that ORE (at 0.2 wt%) acted as prooxidant, with a very prominent rise of PV (to 41.34). In conclusion, there were no significant differences among PV values of pure oil, and RO with added BHT (0.2 and 0.6 wt%), as well as among samples stored at 5 and 25 °C, which remained in the desired PV limits.

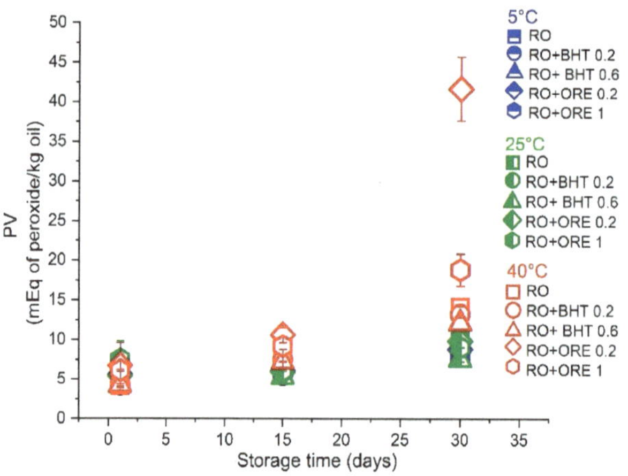


**Figure 3.** Peroxide values of oil samples (pure raspberry oil–RO) and RO with added antioxidants (butylated hydroxytoluene—BHT, 0.2 or 0.6 wt%, and oregano essential oil—ORE, 0.2 or 1 wt%) at 5, 25, and 40 °C during one month of storage. The results are expressed as milliequivalents of peroxide/kg oil, determined according to the IDF method.

The formation of secondary oxidation products, such as malonaldehyde, was monitored via the TBARS test (Figure 4) and the formation of other aldehydes (i.e., 2-alkenals and 2,4-alkadienals) was expressed as PA (Figure 5).

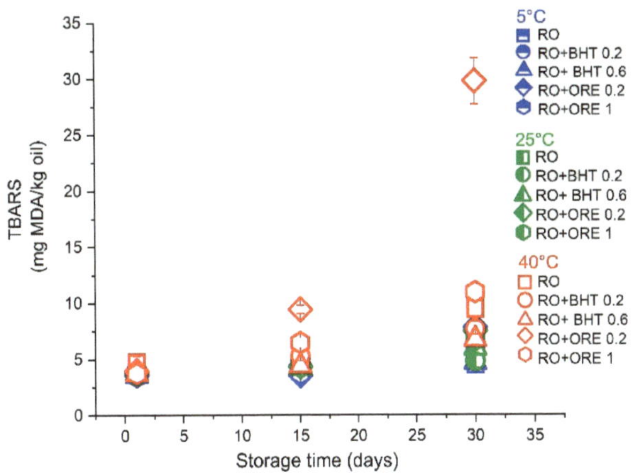


**Figure 4.** TBARS values of oil samples (pure raspberry oil–RO) and RO with added antioxidants (butylated hydroxytoluene—BHT, 0.2 or 0.6 wt%, and oregano essential oil—ORE, 0.2 or 1 wt%) at 5, 25, and 40 °C during one month of storage. The results are expressed as mg of malonaldehyde/kg oil (mg MDA/kg oil).

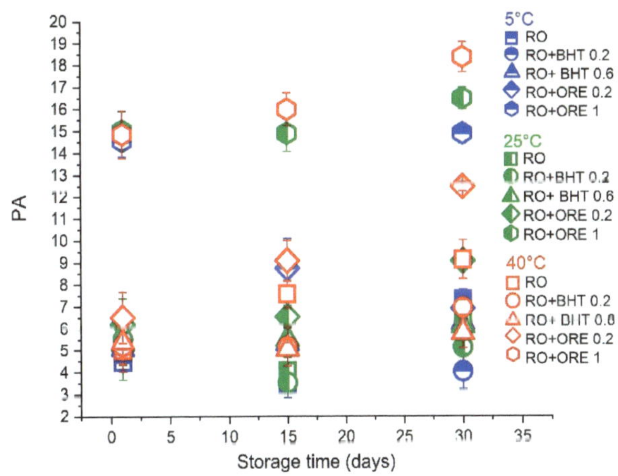


**Figure 5.** p-anisidine value (PA) of oil samples (pure raspberry oil–RO) and RO with added antioxidants (butylated hydroxytoluene—BHT, 0.2 or 0.6 wt%, and oregano essential oil—ORE, 0.2 or 1 wt%) during one month of storage at 5, 25, and 40 °C.

Similar to the results of PV measurements, TBARS values at day 15 at all tested temperatures remained ≤5 mg MDA/kg oil for pure RO and RO-BHT 0.2 and 0.6 wt%, indicating slow degradation of the formed lipid hydroperoxides. However, there was a significant increase of TBARS in RO-ORE samples stored at 40 °C—from 3.96 to 9.40 (for 0.2 wt% ORE) and from 3.74 to 6.47 (for 1 wt% ORE), which was in line with the rise of PV in these samples. Finally, at 30 days of storage, there were no significant differences among samples stored at 5 and 25 °C, neither among samples of pure RO and RO with BHT (0.2 and 0.6 wt%) which had moderately increased TBARS. However, RO-ORE samples at both tested concentrations exhibited notable increase of TBARS.

The results of PA measurements (Figure 5) were also generally in line with the PV and TBARS values, since PA values of RO with or without added BHT remained below the recommended limit of 10 [13] at all tested temperatures after 30 days of storage, and there were no significant differences among samples stored at 5 and 25 °C. Interestingly, RO with

1 wt% ORE showed high PA even on the first day of the study, indicating the possibility that some minor component of this essential oil interacted with the test results.

To conclude, it was found that oxidative stability of pure RO and RO with added antioxidants was satisfactory during one month of storage at 5 and 25 °C, with all investigated parameters (PV, PA, and TBARS) within optimal limits, which can be attributed to antioxidant activity of RO per se [10]. However, at 40 °C the BHT sample (at both tested concentrations) showed only moderate protection against oxidation, while RO-ORE samples showed high instability.

### 3.3. Oxidative Stability of Nanoemulsions with Red Raspberry Seed Oil

As was previously reported, incorporation of oils into classical emulsions (with oil droplets in the micrometer range) and nanoemulsions (preferably up to ~200 nm) can have a controversial outcome on the oil oxidative stability. One of the main reasons for increased instability of oil in emulsion system compared to bulk oil is a large contact area between the oil and the water phase containing prooxidants. Moreover, previous studies showed that structural and physicochemical properties of surfactants, as well as their interactions with antioxidants in formulations, significantly influence oxidative stability of emulsified oils [23,25,26,45,48].

The results of the formation of lipid hydroperoxides—LH as primary oxidation products for the NEs stabilized with polysorbate 80—P80 and polyglycerol ester-based mixture—PG are shown in Figures 6 and 7, respectively. The formation of secondary oxidation products expressed as TBARS for the same two sets of samples are shown in Figures 8 and 9. The results indicate that surfactants played a decisive role in nanoemulsion oxidative stability and that the addition of antioxidants may be necessary to optimize it at different storage temperatures.

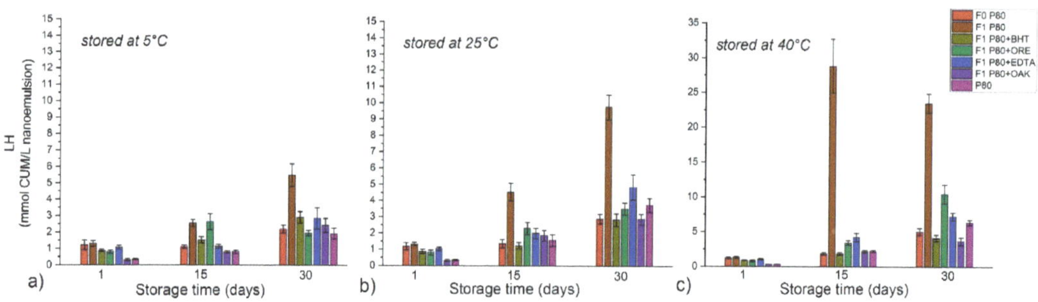


**Figure 6.** Lipid hydroperoxides—LH of polysorbate 80 dispersion in water (P80) and P80-stabilized NEs during one month of storage at different temperatures: (**a**) 5 °C, (**b**) 25 °C and (**c**) 40 °C. Blank nanoemulsion was prepared without raspberry seed oil—RO (F0 P80), while RO loaded NEs are marked as F1 P80; antioxidants tested in nanoemulsions were marked as: butylated hydroxytoluene—BHT, oregano essential oil—ORE, ethylenediaminetetraacetic acid disodium salt dihydrate—EDTA, or sessile oak extract—OAK.

P80 NE prepared without additional antioxidants (F1 P80) showed very high oxidative instability when stored at 40 °C, with a dramatic increase of LH and TBARS after only 15 days of storage (from 1.32 to 28.81 and from 0.79 to 4.79, for LH and TBARS, respectively) resulting in a distinctive rancid odor. This high instability can be attributed to synergistic interaction between RO and F1 P80 components, because the dispersion of pure surfactant (P80) and blank NE without RO both exhibited much slower formation of oxidation products. Such behavior was previously reported [49], and it was concluded that hydrophilic head groups in P80 molecules can generate hydroperoxides when in contact with prooxidants in water, resulting in faster oxidation of unsaturated fatty acids in oils. In this study it was shown that the addition of any of the tested antioxidants (especially BHT, OAK or EDTA) or storage in refrigerator significantly inhibited the nanoemulsion oxidation.

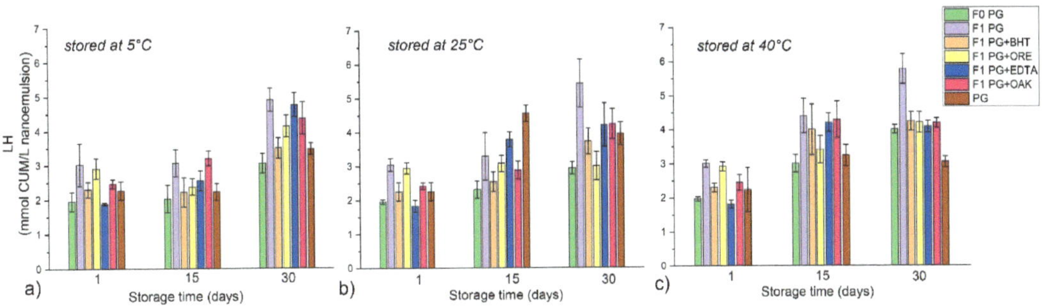


**Figure 7.** Lipid hydroperoxides—LH of polyglycerol ester-based mixture in water (PG) and PG-stabilized NEs during one month of storage at different temperatures: (**a**) 5 °C, (**b**) 25 °C, and (**c**) 40 °C. Blank nanoemulsion was prepared without raspberry seed oil—RO (F0 PG), while RO-loaded NEs are marked as F1 PG; antioxidants tested in nanoemulsions were marked as: butylated hydroxytoluene—BHT, oregano essential oil—ORE, ethylenediaminetetraacetic acid disodium salt dihydrate—EDTA, or sessile oak extract—OAK.

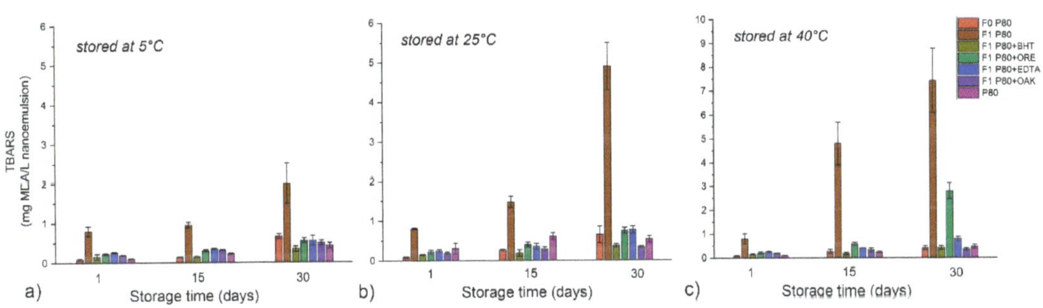


**Figure 8.** Thiobarbituric reactive substances—TBARS of polysorbate 80 dispersion in water (P80) and P80-stabilized NEs during one month of storage at different temperatures: (**a**) 5 °C, (**b**) 25 °C, and (**c**) 40 °C.

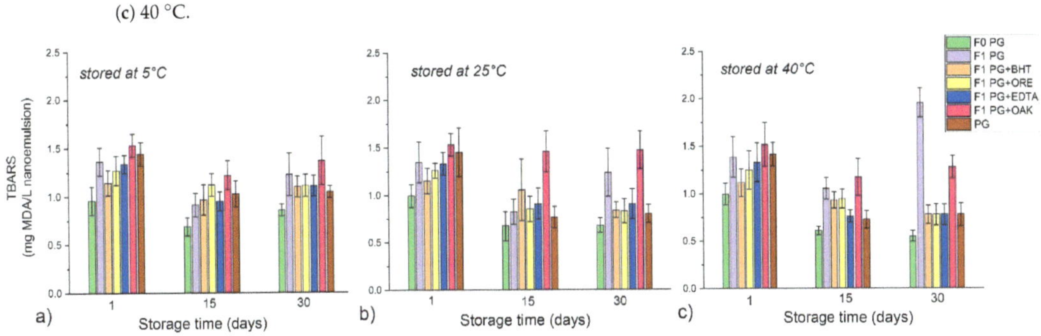


**Figure 9.** Thiobarbituric reactive substances—TBARS of polyglycerol ester-based mixture in water (PG) and PG-stabilized NEs during one month of storage at different temperatures: (**a**) 5 °C, (**b**) 25 °C, and (**c**) 40 °C.

Although in PG NEs LH values slightly increased with storage time, statistical analysis has revealed that after 30 days of storage there were no significant differences in LH values among the samples stored at three tested temperatures, and no significant differences in TBARS values among the samples stored at 25 °C and at 40 °C. Even in the case of the formulation F1 PG without additional antioxidants, LH and TBARS remained within acceptable limits, proving that the surfactant layer played an important role in the protection of RO within the NE core. As it was shown in Figure 2, at 40 °C PG-NEs transformed into a two-phase system, where RO was located in the top layer (yellowish colored concentrated nanoemulsion), while the remaining water phase was in the bottom. Therefore, this transformation could have played a role in RO stabilization, for example, by separating RO from prooxidants in the NE water phase or by thicker surfactant layer, although additional experiments should be performed to prove this hypothesis. As for the tested antioxidants, BHT, EDTA, and ORE showed similar positive effect on the oxidative stability of PG NEs, while OAK was the least effective. ORE could be used as antioxidant in both tested types of nanoemulsions, although it clearly exhibited prooxidative activity in combination with bulk RO. Therefore, its antioxidant activity was successfully preserved in a nanoemulsion carrier, which is in line with our previous findings where it was revealed that antioxidant activity of essential oils can be significantly changed when incorporated into nanoemulsion carriers [30].

It should be noted that PG NEs had four-times lower main droplet size diameter compared to P80 (~45 nm compared to ~210 nm) but the increase in total oil-to-water surface area did not decrease NE stability. Other researchers also concluded that droplet sizes cannot be taken as a single factor in NE oxidative stability, due to the complexity of nanoemulsion composition and possible interactions among ingredients such as oil phase components and antioxidants [25,26,50]. PG esters are prepared by esterification of natural glycerol with fatty acid from natural oils with different chain length and unsaturation level [40,41]. In our study, a commercially available mixture comprised of polyglyceryl 4-laurate, polyglyceryl-6 caprylate, polyglyceryl-3 cocoate, polyglyceryl-4 caprate, polyglyceryl-6 ricinoleate. Therefore, it can be expected that this complex composition contributed to the more compact surfactant packing and improved barrier at the oil–water interface compared to P80 (polyoxyethylene-20 sorbitan monooleate) as a single surfactant with small molecular weight [51]. This PG surfactant layer probably slightly limited the action of antioxidants at the oil–water interface; therefore, the addition of the tested antioxidants was less pronounced than the protective role of surfactant itself [52]. It is worth mentioning that the majority of papers presented oxidative stability of NEs calculated as mmol or mg CUM and MDA/kg of oil [25,26,28,50]. Having in mind that nanoemulsions have complex composition, with multiple possible interactions among ingredients, we have expressed the results as mmol CUM/L and mg MDA/L of nanoemulsions, as reported previously [35,53]. However, as pointed out in publication involving fish oil in conventional vs. nanoemulsions NEs [54], the limits for LH and TBARS values for (nano)emulsions are yet to be determined. To the best of our knowledge, this study reports the oxidative stability of PG NEs for the first time. The results obtained are promising and could be applied to the products in the cosmetic, pharmaceutical, and food industries.

## 4. Conclusions

In this comprehensive study, we have investigated the simultaneous effect of the compositional factors (type of surfactant, oil phase composition, and addition of antioxidants) and temperature (5, 25, or 40 °C) on the physicochemical stability of low energy nanoemulsions, aiming to improve oxidative stability of red raspberry seed oil with high PUFA content. Pure red raspberry seed oil exhibited satisfactory oxidative stability during the first 15 days at all tested temperatures, which can be attributed to its innate antioxidant properties, while slight deterioration was evidenced after 30 days of storage at 40 °C. RO was compatible with BHT, which showed moderate protection against lipid oxidation, whereas ORE acted as prooxidant. Our results have shown that the nanoemulsification of RO with traditional polyethoxylated surfactant polysorbate 80 enhances the oxidative degradation of the oil, necessitating the use of antioxidants. This is in contrast with the effect of polyglycerol ester mixture, which stabilizes RO. It was found that polyglycerol ester-based surfactants, a new generation of biodegradable and biocompatible surfactants suitable for applications in food, medicine, and cosmetics, could be used to obtain nanoemulsions with very fine droplet sizes of < 50 nm and narrow droplet size distribution of PDI < 0.11. Moreover, PG NEs showed improved physicochemical stability at room temperature, and especially at 40 °C, compared to the P80 NE. It was also shown that natural antioxidants ORE and OAK are compatible with nanoemulsion carriers used in this study. Therefore, their use is proposed as an effective natural alternative to the synthetic molecules such as BHT and EDTA in natural formulations.

**Supplementary Materials:** The following supporting information can be downloaded at: https://www.mdpi.com/article/10.3390/antiox11101898/s1. Table S1. Z-average droplet sizes (Z-ave) and PDI values polysorbate 80 (P80) and polyglycerol ester-based (PG)-based nanoemulsions during one month of storage at different temperatures. Table S2. pH values and electrical conductivity of P80 and PG-based nanoemulsions during one month of storage at different temperatures.

**Author Contributions:** Conceptualization, A.G. and A.J.-L.; methodology, A.G., A.J.-L.; formal analysis, A.G., A.J.-L.; investigation, A.G., A.J.-L.; resources, S.S.; data curation, A.G.; writing—original draft preparation, A.G.; writing—review and editing, S.T., A.J.-L.; visualization, A.G.; supervision, S.S.; project administration, S.S.; funding acquisition, S.S. All authors have read and agreed to the published version of the manuscript.

**Funding:** This research was funded by the Ministry of Education, Science and Technological Development, Republic of Serbia (Grant Agreement with University of Belgrade—Faculty of Pharmacy No: 451-03-68/2022-14/200161). This funder did not have any additional role in the study design, data collection and analysis, decision to publish, or preparation of the manuscript.

**Institutional Review Board Statement:** Not applicable.

**Informed Consent Statement:** Not applicable.

**Data Availability Statement:** Data is contained within the article and supplementary material.

**Acknowledgments:** The authors would like to express gratitude to Phoenix Chemical, USA and Evonik Nutrition and Care GmbH, Germany for kindly providing free samples of ethylhexyl pelargonate and polyglycerol esters.

**Conflicts of Interest:** The authors declare no conflict of interest.

## References

1. Faria-Silva, C.; Ascenso, A.; Costa, A.M.; Marto, J.; Carvalheiro, M.; Ribeiro, H.M.; Simões, S. Feeding the skin: A new trend in food and cosmetics convergence. *Trends Food Sci. Technol.* **2020**, *95*, 21–32. [CrossRef]
2. Bialek, A.; Bialek, M.; Jelinska, M.; Tokarz, A. Fatty acid profile of new promising unconventional plant oils for cosmetic use. *Int. J. Cosmet. Sci.* **2016**, *38*, 382–388. [CrossRef] [PubMed]
3. Krasodomska, O.; Jungnickel, C. Viability of fruit seed oil O/W emulsions in personal care products. *Colloids Surf. A Physicochem. Eng. Asp.* **2015**, *481*, 468–475. [CrossRef]
4. Bushman, B.S.; Phillips, B.; Isbell, T.; Ou, B.; Crane, J.M.; Knapp, S.J. Chemical composition of Caneberry (*Rubus* spp.) seeds and oils and their antioxidant potential. *J. Agric. Food Chem.* **2004**, *52*, 7982–7987. [CrossRef] [PubMed]
5. Van Hoed, V.; de Clerq, N.; Echim, C.; Andjelkovic, M.; Leber, E.; Dewettnick, K.; Verhe, R. Berry seeds: A source of specialty oils with high content of bioactives and nutritional value. *J. Food Lipids* **2009**, *16*, 33–49. [CrossRef]
6. Michalak, M.; Kiełtyka-Dadasiewicz, A. Oils from fruit seeds and their dietetic and cosmetic significance. *Herba Pol.* **2018**, *64*, 63–70. [CrossRef]
7. Pilkington, S.M.; Watson, R.E.B.; Nicolaou, A.; Rhodes, L.E. Omega-3 polyunsaturated fatty acids: Photoprotective macronutrients. *Exp. Dermatol.* **2011**, *20*, 537–543. [CrossRef]
8. Parry, J.; Su, L.; Luther, M.; Zhou, K.; Yurawecz, M.P.; Whittaker, P.; Yu, L. Fatty acid composition and antioxidant properties of cold-pressed marionberry, boysenberry, red raspberry, and blueberry seed oils. *J. Agric. Food Chem.* **2005**, *53*, 566–573. [CrossRef]
9. Yang, B.; Ahotupa, M.; Petri Määttä, P.; Kallio, H. Composition and antioxidative activities of supercritical CO2-extracted oils from seeds and soft parts of northern berries. *Food Res. Int.* **2011**, *44*, 2009–2017. [CrossRef]
10. Gledovic, A.; Janosevic Lezaic, A.; Krstonosic, V.; Djokovic, J.; Nikolic, I.; Bajuk-Bogdanovic, D.; Antic Stankovic, J.; Randjelovic, D.; Savic, S.M.; Filipovic, M.; et al. Low-energy nanoemulsions as carriers for red raspberry seed oil: Formulation approach based on Raman spectroscopy and textural analysis, physicochemical properties, stability and in vitro antioxidant/biological activity. *PLoS ONE* **2020**, *15*, e0230993. [CrossRef]

11. Oomah, B.D.; Ladet, S.; Godfrey, D.V.; Liang, J.; Girard, B. Characteristics of raspberry (*Rubus idaeus* L.) seed oil. *Food Chem.* **2000**, *69*, 187–193. [CrossRef]
12. Badea, G.; Lacatusu, I.; Badea, N.; Ott, C.; Meghea, A. Use of various vegetable oils in designing photoprotective nanostructured formulations for UV protection and antioxidant activity. *Ind. Crops Prod.* **2015**, *67*, 18–24. [CrossRef]
13. Prescha, A.; Grajzer, M.; Dedyk, M.; Grajeta, H. The antioxidant activity and oxidative stability of cold-pressed oils. *J. Am. Oil Chem. Soc.* **2014**, *91*, 1291–1301. [CrossRef] [PubMed]
14. Šućurović, A.; Vukelić, N.; Ignjatović, L.; Brčeski, I.; Jovanović, D. Physical-chemical characteristics and oxidative stability of oil obtained from lyophilized raspberry seed. *Eur. J. Lipid Sci. Technol.* **2009**, *111*, 1133–1141. [CrossRef]
15. Ahmadi, O.; Jafarizadeh-Malmiri, H. Green approach in food nanotechnology based on subcritical water: Effects of thyme oil and saponin on characteristics of the prepared oil in water nanoemulsions. *Food Sci. Biotechnol.* **2020**, *29*, 783–792. [CrossRef] [PubMed]
16. Demisli, S.; Theochari, I.; Christodoulo, P.; Zervou, M.; Xenakis, A.; Papadimitriou, V. Structure, activity and dynamics of extra virgin olive oil-in-water nanoemulsions loaded with vitamin D3 and calcium citrate. *J. Mol. Liq.* **2020**, *306*, 112908. [CrossRef]
17. Pavoni, L.; Perinelli, D.R.; Bonacucina, G.; Cespi, M.; Palmieri, G.F. An Overview of micro- and nanoemulsions as vehicles for essential oils: Formulation, preparation and stability. *Nanomaterials* **2020**, *10*, 135. [CrossRef]
18. Gupta, A.; Eral, H.B.; Hatton, T.A.; Doyle, P.S. Nanoemulsions: Formation, properties and applications. *Soft Matter* **2016**, *12*, 2826–2841. [CrossRef]
19. Komaiko, J.S.; McClements, D.J. Formation of food-grade Nanoemulsions using low-energy preparation methods: A review of available methods. *Compr. Rev. Food Sci. Food Saf.* **2016**, *15*, 331–351. [CrossRef]
20. Frankel, E.N.; Huang, S.-W.; Kanner, J.; German, J.B. Interfacial phenomena in the evaluation of antioxidants: Bulk oils vs emulsions. *J. Agric. Food Chem.* **1994**, *42*, 1054–1059. [CrossRef]
21. Horn, A.F.; Nielsen, N.S.; Andersen, U.; Søgaard, L.H.; Horsewell, A.; Jacobsen, C. Oxidative stability of 70% fish oil-in-water emulsions: Impact of emulsifiers and pH. *Eur. J. Lipid Sci. Technol.* **2011**, *113*, 1243–1257. [CrossRef]
22. Szumała, P.; Wysocka, I. Effect of gelation and storage conditions on the oxidative stability of microemulsion and nanoemulsion delivery systems. *Eur. J. Pharm. Sci.* **2018**, *124*, 17–25. [CrossRef] [PubMed]
23. Walker, R.; Decker, E.A.; McClements, D.J. Development of food-grade nanoemulsions and emulsions for delivery of omega-3 fatty acids: Opportunities and obstacles in the food industry. *Food Funct.* **2015**, *6*, 42–55. [CrossRef] [PubMed]
24. Sharma, S.; Cheng, S.-F.; Bhattacharya, B.; Chakkaravarthi, S. Efficacy of free and encapsulated natural antioxidants in oxidative stability of edible oil: Special emphasis on nanoemulsion-based encapsulation. *Trends Food Sci. Technol.* **2019**, *91*, 305–318. [CrossRef]
25. Rehman, A.; Jafari, S.M.; Tong, Q.; Karim, A.; Mahdi, A.A.; Iqbal, M.W.; Aadil, R.M.; Ali, A.; Manzoor, M.F. Role of peppermint oil in improving the oxidative stability and antioxidant capacity of borage seed oil-loaded nanoemulsions fabricated by modified starch. *Int. J. Biol. Macromol.* **2020**, *153*, 697–707. [CrossRef] [PubMed]
26. Kampa, J.; Frazier, R.; Rodriguez-Garcia, J. Physical and chemical characterisation of conventional and nano/emulsions: Influence of vegetable oils from different origins. *Foods* **2022**, *11*, 681. [CrossRef]
27. Costa, M.; Freiría-Gándara, J.; Losada-Barreiro, S.; Paiva-Martins, F.; Aliaga, C.; Bravo-Díaz, C. Interfacial kinetics in olive oil-in-water nanoemulsions: Relationships between rates of initiation of lipid peroxidation, induction times and effective interfacial antioxidant concentrations. *J. Colloid Interface Sci.* **2021**, *604*, 248–259. [CrossRef]
28. Ferreira da Silveira, T.F.; Laguerre, M.; Bourlieu-Lacanal, C.; Lecomte, J.; Durand, E.; Figueroa-Espinoza, M.C.; Baréa, B.; Barouh, N.; Alves Castro, I.; Villeneuve, P. Impact of surfactant concentration and antioxidant mode of incorporation on the oxidative stability of oil-in-water nanoemulsions. *LWT* **2021**, *141*, 110892. [CrossRef]
29. Gledovic, A.; Janosevic Lezaic, A.; Nikolic, I.; Tasic-Kostov, M.; Antic-Stankovic, J.; Krstonosic, V.; Randjelovic, D.; Bozic, D.; Ilic, D.; Tamburic, S.; et al. Polyglycerol ester-based low energy nanoemulsions with red raspberry reed oil and fruit extracts: Formulation development toward effective in vitro/in vivo bioperformance. *Nanomaterials* **2021**, *11*, 217. [CrossRef]
30. Gledovic, A.; Bajuk-Bogdanović, D.; Uskoković-Marković, S.; Pavun, L.; Savic, S.; Janošević Ležaić, A. Low energy nanoemulsions as carriers for essential oils in topical formulations for antioxidant skin protection. *Hem. Ind.* **2022**, *76*, 29–42. [CrossRef]
31. Drinić, Z.; Mudrić, J.; Zdunić, G.; Bigović, D.; Menković, N.; Šavikin, K. Effect of pomegranate peel extract on the oxidative stability of pomegranate seed oil. *Food Chem.* **2020**, *333*, 127501. [CrossRef] [PubMed]
32. Malinowska, P.; Gliszczynska-Swigło, A.; Szymusiak, H. Protective effect of commercial acerola, willow, and rose extracts against oxidation of cosmetic emulsions containing wheat germ oil. *Eur. J. Lipid Sci. Technol.* **2014**, *116*, 1553–1562. [CrossRef]
33. Shantha, N.; Decker, E.A. Rapid, sensitive, iron-based spectrophotometric methods for determination of peroxide values of food lipids. *R J. AOAC Int.* **1994**, *77*, 1994. [CrossRef]
34. AOCS. Available online: https://www.aocs.org/attain-lab-services/methods/methods/search-results?method=111529 (accessed on 24 September 2022).
35. Walker, R.M.; Gumus, C.E.; Decker, E.A.; McClements, D.J. Improvements in the formation and stability of fish oil-in-water nanoemulsions using carrier oils: MCT, thyme oil, & lemon oil. *J. Food Eng.* **2017**, *211*, 60–68.
36. Azmi, N.A.N.; Elgharbawy, A.A.M.; Motlagh, S.R.; Samsudin, N.; Salleh, H.M. Nanoemulsions: Factory for food, pharmaceutical and cosmetics. *Processes* **2019**, *7*, 617. [CrossRef]

37. Jurić, S.; Jurić, M.; Siddique, A.B.; Fathi, M. Vegetable oils rich in polyunsaturated fatty acids: Nanoencapsulation methods and stability enhancement. *Food Rev. Int.* **2022**, *38*, 32–69. [CrossRef]
38. Solè, I.; Pey, C.M.; Maestro, A.; González, C.; Porras, M.; Solans, C.; Gutiérrez, J.M. Nano-emulsions prepared by the phase inversion composition method: Preparation variables and scale up. *J. Colloid Interface Sci.* **2010**, *344*, 417–423. [CrossRef]
39. Chang, Y.Y.; McClements, D.J. Optimization of orange oil nanoemulsion formation by isothermal low-energy methods: Influence of the oil phase, surfactant, and temperature. *J. Agric. Food Chem.* **2014**, *62*, 2306–2312. [CrossRef]
40. Hanno, I.; Centini, M.; Anselmi, C.; Bibiani, C. Green cosmetic surfactant from rice: Characterization and application. *Cosmetics* **2015**, *2*, 322–341. [CrossRef]
41. Kato, T.; Nakamura, T.; Yamashita, M.; Kawaguchi, M.; Kato, T.; Itoh, T. Surfactants properties of purified polyglycerol monolaurates. *J. Surfact. Deterg.* **2003**, *6*, 331–337. [CrossRef]
42. Klang, V.; Matsko, N.B.; Valenta, C.; Hofer, F. Electron microscopy of nanoemulsions: An essential tool for characterisation and stability assessment. *Micron* **2012**, *43*, 85–103. [CrossRef]
43. Rebolleda, S.; Sanz, M.T.; Benito, J.M.; Beltran, S.; Escudero, I.; Gonzalez San-Jose, M.L. Formulation and characterisation of wheat bran oil-in-water nanoemulsions. *Food Chem.* **2015**, *167*, 16–23. [CrossRef] [PubMed]
44. Bernardi, D.S.; Pereira, T.A.; Macie, N.R.; Bortoloto, J.; Viera, G.S.; Oliveira, G.C.; Rocha-Filho, P.A. Formation and stability of oil-in-water nanoemulsions containing rice bran oil: In vitro and in vivo assessments. *J. Nanobiotechnol.* **2011**, *9*, 44–52. [CrossRef] [PubMed]
45. Neves, M.A.; Wang, Z.; Kobayashi, I.; Nakajima, M. Assessment of oxidative stability in fish oil-in-water emulsions: Effect of emulsification process, droplet size and storage temperature. *J. Food Process Eng.* **2015**, *40*, e12316. [CrossRef]
46. Kulisic, T.; Radonic, A.; Katalinic, V.; Milos, M. Use of different methods for testing antioxidative activity of orcgano essential oil. *Food Chem.* **2004**, *85*, 633–640. [CrossRef]
47. Dorman, H.J.D.; Surai, P.; Deans, S.G. In vitro antioxidant activity of a number of plant essential oils and phytoconstituents. *J. Essent. Oil Rex.* **2000**, *12*, 241–248. [CrossRef]
48. Lee, J.; Choi, S.-J. Influence of blending of nonionic emulsifiers having various hydrophilic head sizes on lipid oxidation: Investigation of antioxidant polarity—Interfacial characteristics relationship. *Antioxidants* **2021**, *10*, 886. [CrossRef]
49. Mc Clements, D.J.; Decker, E. Interfacial antioxidants: A review of natural and synthetic emulsifiers and coemulsifiers that can inhibit lipid oxidation. *J. Agric. Food Chem.* **2018**, *66*, 20–35. [CrossRef]
50. Guo, X.; Sun, X.-T.; Liang, L.; Shi, L.-K.; Liu, R.-J.; Chang, M.; Wang, X.-G. Physical stability, oxidative stability, and bioactivity of nanoemulsion delivery systems incorporating lipophilic ingredients: Impact of oil saturation degree. *J. Agric. Food Chem.* **2021**, *69*, 5405–5415. [CrossRef]
51. Marhamati, M.; Ranjbar, G.; Rezaie, M. Effects of emulsifiers on the physicochemical stability of oil-in-water nanoemulsions: A critical review. *J. Mol. Liq.* **2021**, *340*, 117218. [CrossRef]
52. Weigel, F.; Weiss, J.; Decker, E.A.; Mc Clements, D.J. Lutein-enriched emulsion-based delivery systems: Influence of emulsifiers and antioxidants on physical and chemical stability. *Food Chem.* **2018**, *242*, 395–403. [CrossRef] [PubMed]
53. Zhong, J.; Liu, X.; Wang, Y.; Qin, X.; Li, Z. γ-Oryzanol nanoemulsions produced by a low-energy emulsification method: An evaluation of process parameters and physicochemical stability. *Food Funct.* **2017**, *6*, 2202–2211. [CrossRef] [PubMed]
54. Nejadmansouri, M.; Hosseini, S.M.H.; Niakosari, M.; Yousefi, G.H.; Golmakani, M.T. Physicochemical properties and oxidative stability of fish oil nanoemulsions as affected by hydrophilic lipophilic balance, surfactant to oil ratio and storage temperature. *Colloids Surf. A Physicochem. Eng.* **2016**, *506*, 821–832. [CrossRef]

antioxidants

MDPI

*Article*

# Effect of Encapsulation Material on Lipid Bioaccessibility and Oxidation during In Vitro Digestion of Black Seed Oil

Jon Alberdi-Cedeño [1,2], Martha Aichner [1], Agnes Mistlberger-Reiner [1], Aimin Shi [3] and Marc Pignitter [1,*]

1 Department of Physiological Chemistry, Faculty of Chemistry, University of Vienna, 1090 Vienna, Austria
2 Food Technology, Faculty of Pharmacy, Lascaray Research Center, University of the Basque Country (UPV-EHU), 01006 Vitoria-Gasteiz, Spain
3 Institute of Food Science and Technology, Chinese Academy of Agricultural Sciences, Key Laboratory of Agro-Products Processing, Ministry of Agriculture and Rural Affairs, Beijing 100081, China
* Correspondence: marc.pignitter@univie.ac.at

**Abstract:** Different encapsulation materials might not only affect lipid hydrolysis but also lipid oxidation during in vitro digestion. Thus, this study aimed to investigate the effect of two commonly used shell materials, starch and gelatin, on the extent of lipolysis and bioaccessibility of the main and some minor lipid compounds, as well as on the oxidative status in encapsulated black seed oil (*Nigella sativa*) during in vitro digestion. The study was carried out using $^{1}H$ nuclear magnetic resonance spectroscopy, liquid chromatography-mass spectrometry and high-performance liquid chromatography-UV. It was shown that starch increased the level of lipid hydrolysis in black seed oil during gastric in vitro digestion, while no differences were observed in the intestinal digestates between starch-encapsulated oil and gelatin-encapsulated oil. Similarly, the bioaccessibility of minor compounds (tocopherols, sterols and thymoquinone) was not influenced by the shell materials. However, regarding lipid oxidation, a 20- and 10-fold rise of free oxylipins was obtained in oils encapsulated by starch and gelatin, respectively, after intestinal in vitro digestion. This study evidenced that gelatin rather than starch should be used for the encapsulation of oils to minimize the digestion-induced formation of bioactive oxylipins.

**Keywords:** black seed oil; encapsulation; free oxylipins; bioaccessibility; in vitro digestion

## 1. Introduction

Vegetable oils are popular for their use as supplements due to their bioactive ingredients. Their oxidation mechanism is a complex subject of great interest and has significant implications in terms of sensory and nutritional quality. As a result of the autoxidation reactions of lipids during storage, processing or even during digestion [1–8], oxidation products, which are related to several diseases, such as heart diseases, cancer, allergenic responses, atherosclerosis, obesity, Parkinson's disease, and Alzheimer disease [9–13], might be formed. Recently, oxylipins have been shown for the first time to be detected in oxidized corn oil, as well as in virgin flaxseed oil and corn oil subjected to in vitro digestion [4,5,7]. However, the fate of oxylipins during gastric and intestinal digestion is less well known.

In order to slow down the oxidation reactions, extend the shelf life of vegetable oils and reduce the formation of potentially toxic oxidation compounds, two main strategies could be used. On the one hand, natural and/or synthetic antioxidants can be added to the food matrix [5,14] and on the other hand, the lipids could be encapsulated to prevent their direct contact with oxygen, light, moisture and heat [15–17]. The latter could be a promising approach to protecting lipids from oxidation reactions, as well as masking off-flavors or colors [18].

Likewise, during in vitro gastrointestinal digestion processes, triacylglycerols contained in vegetable oils undergo hydrolysis. The extent of hydrolysis can be influenced by several factors, such as the proportions of oils/digestive fluids, the composition of digestive juices, the concentration, nature and activity of the enzymes in the juices and the unsaturation degree of the lipids [6,19,20]. In addition, encapsulation of vegetable oils could influence the extent, release and absorption of polyunsaturated fatty acids (PUFAs) and other minor compounds during digestion [21]. In this sense, it should be pointed out that most of the studies in the literature have generally focused on the release of PUFAs during the in vitro digestion of emulsions coated or encapsulated with different shell materials [22]. So far, only a few studies have described the release of PUFAs from encapsulated oils in different shell materials during their in vitro digestion [15,23]. However, these studies only focused on the release of free fatty acids, but as far as we know, none of these studies investigated the overall process of lipolysis in encapsulated oils, giving the molar % of the different acylglycerol structures formed, as well as their bioaccessibility after in vitro digestion. Moreover, there are no studies in the literature addressing the effect of shell material on the formation of potentially toxic compounds and the oxidative stability of edible oils exposed to in vitro digestion.

It can be hypothesized that capsule materials might influence the extent of lipid hydrolysis differently due to their different impacts on emulsification during digestion. As lipolytic products, such as free fatty acids, are more prone to lipid oxidation than triacylglycerols, shell materials might not only differently affect lipid hydrolysis but also oxidation.

In this context, the aim of this study was to analyze the effect of two different shell materials, starch and gelatin, on the extent of lipolysis, the bioaccessibility of main and some minor lipid compounds and on the oxidative status of encapsulated black seed oil (*Nigella sativa*) during in vitro digestion. The aim was achieved using $^1$H nuclear magnetic resonance ($^1$H NMR), liquid chromatography-mass spectrometry (LC-MS) and high-performance liquid chromatography-UV (HPLC-UV). It can be expected that free oxylipins might not only be detected in vegetable oils after in vitro digestion by NMR, as described previously [4,5], but also during digestion by LC-MS to investigate the fate of oxylipins after the gastric and intestinal phases. Black seed oil (*Nigella sativa*) was selected because of its high therapeutic value and widespread use as a food supplement. The results give a holistic view of the different reactions that take place during in vitro digestion, focusing not only on one compound but on several markers.

## 2. Materials and Methods

### 2.1. Samples

The study was carried out with commercially encapsulated black seed oil (*Nigella sativa*) provided by Goerlich Pharma, Edling, Germany. Each capsule contained 500 mg of Egyptian black seed oil encapsulated in either a soft starch capsule (BS1) or a soft gelatin capsule (BS2). The soft starch capsule consisted of modified corn starch (70.6 mg), glycerol (E422) (80.2 mg), carrageenan (E407) (28.2 mg), sodium carbonate (anhydrous) (0.1 mg) and water (14.9 mg), while the soft gelatin capsule was composed of gelatin (bovine) (129.3 mg), glycerol (E422) (66.0 mg) and water (14.7 mg). The composition of this oil in molar percentages of linoleic (L), oleic (O) and saturated (S) acyl groups was 55.3 ± 0.71%, 27.8 ± 0.75% and 16.9 ± 0.09%, respectively, as determined by $^1$H NMR and reported in previous studies [4–6].

### 2.2. Chemicals and Materials

Sodium hydroxide, potassium chloride, potassium dihydrogen phosphate, magnesium chloride hexahydrate, calcium chloride dihydrate, albumin, hydrochloric acid, dichloromethane, acetic acid and acetic acid ethyl ester were purchased from Carl Roth GmbH & CO., Karlsruhe, Germany. Sodium chloride, urea, uric acid, potassium thiocyanate, pancreatin from porcine pancreas, pepsin from porcine gastric mucosa, α-amylase from *Aspergillus oryzae*, sodium sulfate, sodium hydrogen carbonate, mucin, lipase from porcine pancreas, D-glucose, bile bovine, ammonium chloride, Amano lipase A from Aspergillus niger, α-tocopherol, γ-tocopherol and δ-tocopherol were obtained from Sigma Aldrich, St. Louis, MO, USA. Acetonitrile (LC-MS grade) and methanol (LC-MS grade) were acquired from Avantor/VWR International Inc. (Radnor, PA, USA). Chloroform D + 0.03% TMS was purchased from Eurisotop, Cambridge Isotope Laboratory Inc., Andover, USA. β-tocotrienol was acquired from Biomol GmbH, Hamburg, Germany. Sodium Dihydrogen Orthophosphate Dihydrate was obtained from Fisher Chemical Co., Fair lawn, NJ, USA and rac-Tocol was purchased from Abcam.

### 2.3. *In Vitro Digestion of the Capsules*

An in vitro digestion was employed based on a semi-static procedure that mimics digestive processes occurring in the mouth, stomach and duodenum by sequentially adding simulated digestive juices (saliva, gastric juice, duodenal juice and bile juice) to the sample, while incubated at 37 ± 2 °C and rotated head-over-heels throughout the duration of the experiment. This in vitro procedure was initially developed by Versantvoort et al. (2005) [24] and slightly modified by Nieva-Echervarría et al. (2016) [20] to reach a higher degree of lipolysis, similar to that occurring in vivo. The capsules containing 500 mg of oil were subjected in triplicate to in vitro gastrointestinal digestion. In short, 6 mL of simulated saliva juice were added to each sample and incubated on the analog rotator RS-RR 5 (Phoenix Instrument, Garbsen, Germany) in the oven at 37 ± 2 °C for 5 min. A total of 12 mL of gastric juice was added to the tubes afterwards. After one hour, the pH was measured and adjusted to 2–3 with NaOH (10 M). One hour later, in vitro gastric digestion was completed. The pH was measured and 2 mL $NaHCO_3$ (1 M), 12 mL simulated duodenal juice and 6 mL simulated bile juice were added. The pH was adjusted to 6–7 and the tubes were placed in the oven on the rotator for another 4 h. After the in vitro digestion and in order to stabilize the samples, the digestates were stored at −80 °C until extraction of the lipids. The composition of the juices is shown in the Supplementary Material. All samples were analyzed in triplicate after oral, gastric and intestinal in vitro digestion, as indicated in the results.

### 2.4. *Extraction of the Lipids from the Digestates*

After in vitro digestion, the lipids of the digestates were extracted with dichloromethane, as in previous studies [4–6]. Briefly, a six-fold extraction was performed, adding 15 mL dichloromethane to the samples, centrifuging them at 14,000 rpm for 7 min, taking the lower, non-polar phase, unifying the extracts and evaporating the solvent to obtain the digested lipids. The samples were stored at −40 °C under argon until further analysis.

### 2.5. *$^1$H NMR Analysis*

The $^1$H NMR spectra of the oils (BS1 and BS2) and of the lipid extracts of their digestates after gastric in vitro digestion (G-BS1, G-BS2) and intestinal in vitro digestion (I-BS1 and I-BS2) were acquired in triplicate using an Avance 400 spectrometer (Bruker, Billerica, MA, USA) operating at 400 MHz. For this purpose, 200 μL samples were dissolved in 400 μL of deuterated chloroform, which contained tetramethylsilane (TMS) as an internal reference. The acquisition conditions were the same as those used in previous studies [4]. The $^1$H NMR spectra were plotted at a fixed value of absolute intensity to be valid for comparative purposes using the MestreNova program (Mestrelab Research, Santiago de Compostela, Spain).

The identification of the acylglycerol structures, acyl groups, fatty acid composition, sterols and oxidation products was carried out on the basis of the bibliographic data [4–6,25,26]. The quantification of the molar percentage of each kind of acylglycerol structure was determined using the equations in previous studies [27]. The molar percentages of different acyl groups and fatty acids, in relation to the total moles of all kinds of fatty acids plus acyl groups (AG+FA), were estimated as in a previous study [6]. Likewise, the concentration of sterols and lipid oxidation products was analyzed based on their $^1$H NMR spectral signals using the general equation as previously published [6].

### 2.6. Nanoparticle Tracking Analysis

To characterize the oil droplets within the digestates, the particle concentration, particle size, and size distribution of all particles, including the oil droplets, were determined. Nanoparticle tracking analysis (NTA) was performed with ZetaView PMX 120-Z equipped with a 520 nm laser (Particle Metrix, Inning am Ammersee, Germany). Before the NTA measurements, the black seed oil was digested, as described above. Samples were analyzed after oral, gastric and intestinal in vitro digestion. The instrument settings were optimized to detect particles with a mean size of around 200 nm, and the following parameters were chosen: sensitivity: 60, shutter: 150, frame rate: 15; post-acquisition parameters: min. brightness: 20, min. area: 10, max. area: 10,000, tracelength: 30. All measurements were performed at a flow cell temperature of 22 °C, and the samples were diluted in water to reach about 100–200 particles per frame. For each replicate, 3 measurements were performed, each measuring 11 positions. At least 1000 particles per replicate were measured and included in the analysis for size determination. The data analysis was carried out using ZetaView software, version 8.05.14 SP7.

### 2.7. Sample Preparation for LC-MS Analysis

A solid phase extraction (SPE) was performed to extract the free oxylipins from the black seed oil or from the lipid extract of the digestates, based on a previous study [28], with some modifications. In short, the SPE cartridge, Strata™-x 33 µm Polymeric Reversed Phase, 30 mg/1 mL tube (Phenomenex, Torrance, CA, USA) was washed with 1 mL ethyl acetate and 2 × 1 mL methanol and conditioned with 2 × 1 mL buffer (5% methanol in water with 0.1% acetic acid). A total of 50 mg of oil or lipid extract of digestates was diluted in 2 mL ethyl acetate and loaded onto the cartridge. Then, the cartridge was washed twice with 1 mL buffer. After drying the cartridge 20 min under vacuum, the free oxylipins were eluted from the cartridge with 1 mL methanol and 3 × 1 mL ethyl acetate. The samples were dried under nitrogen and reconstituted with 300 µL of methanol. After filtration with syringe filters (Phenomenex, Phenex™-NY 15 mm, 0.2 u), the samples were analyzed using LC-MS.

### 2.8. LC-MS Analysis of Free Oxylipins

To determine the free oxylipins present in the oil and digestates, the samples (20 µL) were injected onto an LC-20 system with an LCMS-8040 detector (Shimadzu, Korneuburg, Austria). The free oxylipins were separated by the LC-20 using a C12 column (Synergi 4 µm Max-RP 80 Å, 150 × 2 mm, Phenomenex, Torrance, CA, USA) at 25 °C. The mobile phase was $H_2O$ with 0.1% acetic acid (solvent A) and acetonitrile/methanol (80/15) (solvent B). The LC gradient was 25% solvent B for the first minute and increased to 30% until 1.5 min; from 30% to 53% until 10 min; from 53% to 68% until 19.5 min; 68% to 95% until 24.5 min; hold it until 27 min; 95% to 25% until 27.1 min; and hold it until 30.5 min. The flow rate was 0.3 mL/min. The following ESI ion source settings were applied: nebulizing flow 3 L/min, drying gas flow 10 L/min, desolvation line temperature 150 °C and heat block temperature 350 °C. Argon with different collision energies was used for MS/MS (Table S2). For the identification, the mass spectrometric analysis for each free oxylipin was performed in the multiple reaction monitoring (MRM) mode on the ESI triple quadrupole mass spectrometer in negative mode, according to a previous study [28]. For semi-quantification, the abundance of free oxylipins was determined based on the area under the curve of the corresponding MRM transition. The free oxylipins with the corresponding transitions, collision energies and retention times are listed in Table S2. Data analysis was performed with Labsolution software version 5.99 SP2.

### 2.9. Determination and Quantification of Tocols by HPLC-UV

Tocols were analyzed as described previously [8]. Briefly, a 100 mg sample (oil or lipid extract of the digestates) was dissolved in 300 µL 2-propanol. A total of 1 µL internal standard was added to the sample before the sample was filtered with syringe filters (Phenomenex, Phenex™-NY 15 mm, 0.2 u). The samples were analyzed by the LC-20

(Shimadzu, Korneuburg, Austria) equipped with a C18 column (C18 Kinetex, 5 μm, EVO C18, 150 mm × 2.6 mm) at 10 °C and a photodiode array detector (SPD-M30A, Shimadzu, Korneuburg, Austria). The flow rate was 0.5 mL/min. The mobile phase was water (solvent A) and methanol (solvent B). The gradient was: 95–100% from 0 min to 4 min, 100% from 4 min to 14 min, 100–95% from 14 min to 16 min and 95% from 16 min to 18 min. The tocols were measured at a wavelength of 294 nm. Standard curves of α-tocopherol and γ-tocopherol with the linear range concentrations from 0.98–62.5 μg/mL and from 7.81–500 μg/mL for β-tocotrienol were used to quantify the tocols in the samples. The limit of quantification was determined by a signal to noise ratio of 10. The recovery of tocols for fresh black seed oil, digestate after gastric in vitro digestion and digestate after intestinal in vitro digestion were 110 ± 11%, 71 ± 9% and 79 ± 12%, respectively.

### 2.10. Statistical Analysis

Statistically significant differences were determined by one-way ANOVA or two-way ANOVA followed by Tukey's post-hoc test or by Student's *t*-test with GraphPad Prism 9.1.2, as indicated in the results section. The results were considered to be significant with a *p*-value lower than 0.05 ($p < 0.05$). For each experiment, at least 3 independent samples were measured.

## 3. Results and Discussion

### 3.1. The Extent of Lipid Hydrolysis during In Vitro Digestion

In order to evaluate the impact of in vitro digestion and capsule materials on the degree of lipid hydrolysis in black seed oil, the lipolytic products, diacylglycerols (1,2-DG and 1,3-DG) and monoacylglycerols (1-MG and 2-MG), derived from the hydrolysis of triacylglycerols (TG), were quantified by $^1$H NMR (Figure S1). Moreover, lipid bioaccessibility, described by the ratio between the concentration of these absorbable species and all fatty acids (FA) plus acyl groups (AG) present in the corresponding digestate, was calculated as in previous studies [4–6] (see Table 1).

**Table 1.** Molar percentages of triacyglycerols (TG%), diacylglycerols (1,2-DG% and 1,3-DG%), monoacylglycerols (1-MG% and 2-MG%) and glycerol (Gol%) in relation to the total number of glyceryl structures and lipid bioaccessibility (LB %) in black seed oil samples (BS1 and BS2), encapsulated black seed oil samples after gastric (G-BS1, G-BS2) and intestinal (I-BS1, I-BS2) in vitro digestion.

| | Acylglycerol Species (Molar %) | | | | | | LB (%) |
|---|---|---|---|---|---|---|---|
| **Samples** | **TG%** | **1,2-DG%** | **1,3%-DG%** | **1-MG%** | **2-MG%** | **Gol%** | |
| BS1 | 76.9 ± 0.19 a | 5.29 ± 0.11 a | - | 1.50 ± 0.04 a | 0.17 ± 0.01 a | 16.2 ± 0.34 a | 19.6 ± 0.26 a |
| G-BS1 | 42.6 ± 1.10 b | 27.9 ± 0.36 b | 2.40 ± 0.29 a | 0.84 ± 0.04 b | 6.34 ± 0.39 b | 20.0 ± 0.64 b | 37.2 ± 1.03 b |
| I-BS1 | 24.0 ± 5.41 c | 14.2 ± 1.41 c | 4.83 ± 1.14 b | 6.82 ± 2.37 c | 17.1 ± 4.10 c | 33.0 ± 2.80 c | 63.3 ± 7.05 c |
| BS2 | 82.7 ± 0.22 d | 4.09 ± 0.10 d | - | 1.39 ± 0.11 a | 0.12 ± 0.00 d | 11.8 ± 0.23 d | 14.6 ± 0.15 d |
| G-BS2 | 64.2 ± 4.01 e | 17.1 ± 2.41 c | 4.57 ± 0.41 b | 0.87 ± 0.09 b | 2.12 ± 0.48 e | 11.1 ± 1.45 e | 21.3 ± 2.65 e |
| I-BS2 | 25.8 ± 3.93 c | 16.7 ± 1.04 c | 4.21 ± 0.47 b | 5.26 ± 0.44 c | 18.8 ± 1.75 c | 29.2 ± 2.37 c | 60.2 ± 3.70 c |

Different letters within each column indicate a significant difference among the samples analyzed by one-way ANOVA followed by Tukey's post-hoc test ($p < 0.05$). BS1 and BS2: black seed oils before digestion; G-BS1 and B-BS2: encapsulated black seed oil samples after gastric in vitro digestion; I-BS1 and I-BS2: encapsulated black seed oil samples after intestinal in vitro digestion; -: not detected.

Vegetable oils are complex mixtures represented mainly by triacylglycerols, with around 99% [4–6]. However, as can be observed in Table 1, BS1 and BS2 oils were already slightly hydrolyzed, showing a lower percentage of TG and the presence of some other acylglycerol species, such as 1,2-DG, 1-MG and 2-MG. Furthermore, as can be observed in Table 1, the molar percentage of the acylglycerol species varied slightly between the BS1 and BS2 oils. In more detail, BS1 presents lower TG% but higher amounts of 1,2-DG, 2-MG and Gol, resulting in statistically significantly higher lipid bioaccessibility (LB (%), $p > 0.05$) when compared with BS2.

The triacylglycerol hydrolysis pattern of encapsulated black seed oils after intestinal in vitro digestion (I-BS1 and I-BS2) was very similar, with the presence and the same concentration of diacylglycerols, monoacylglycerols and glycerol in both digestates (Table 1). No significant ($p > 0.05$) differences were found among the molar percentages of the different acylglycerols present in the lipid extracts of encapsulated (I-BS1 and I-BS2) oil digestates, nor regarding the lipid bioaccessibility (LB%) after intestinal in vitro digestion. Furthermore, the percentages of TG decreased from 76.9% (BS1) and 82.7% (BS2), near to 24–25%, reaching triacylglycerols' transformation of around 70%. The molar percentages of the lipolytic products of TG, 1,2-DG, 1,3-DG, 1-MG and 2-MG, were 14–16%, 4%, 5–6% and 17–18%, respectively (Table 1). Likewise, the lipid bioaccessibility parameter showed values of 60–63%. These results are similar to those reported in previous studies on sunflower and corn oils rich in omega-6 acyl groups after in vitro digestion [4,29]. These findings showed that the encapsulation of black seed oil with either starch or gelatin did not have any effect on the extent of lipolysis after intestinal in vitro digestion.

However, significant differences ($p < 0.05$) related to the capsule material were identified in the digestates after gastric in vitro digestion. Black seed oil encapsulated in starch material showed higher hydrolysis, with a decrease in the molar percentage of TG of 34.3%, compared with black seed oil encapsulated in gelatin, revealing a decline of only 18.5%. Moreover, 2-MG and Gol were 3-fold and almost 2-fold higher, respectively, in G-BS1 than in G-BS2. Therefore, lipid bioaccessibility after gastric in vitro digestion was statistically significantly higher for the oil encapsulated in starch material (37%) than for the oil encapsulated in gelatin (21%), which means that during gastric in vitro digestion, the shell material had an effect on the lipolysis of black seed oil. This effect could be due to the presence of α-amylase in the oral juices, which can start hydrolyzing starch material enhancing, therefore, the access of the lipase added with the gastric juices. These results are in agreement with previous studies [15], which showed that the release of FFA during oil digestion was higher when the oil was encapsulated in gum material than in protein. This effect could be explained by the finding that the shell materials affected the interfacial behavior and might interfere with the diffusion and adsorption of lipase to the oil droplets, as explained by Timilsena and coworkers (2017) [15]. Furthermore, Mun and coworkers (2007) [22] reported that the extent and rate of lipid hydrolysis depends on the interfacial composition. Therefore, shell materials or surfactants with much higher surface activity than lipase could create an interfacial layer around the oil droplets that would prevent the enzyme from approaching the triacylglycerols [22]. Considering this hypothesis, it could be suggested that gelatin might have a higher surface activity than starch and lipase, so the lipolytic enzyme cannot easily access the triacylglycerols in the oil.

Another explanation for the difference in lipid hydrolysis might be based on the oil droplet size within the oil-in-water-emulsion that is formed during digestion. When the oil droplets are smaller, the total surface area available for the adsorption of lipase becomes larger; therefore, enzymatic digestion becomes faster [15]. Since the capsule material could influence the oil droplet size or emulsion stability during digestion [15], this could affect lipid hydrolysis. To test this hypothesis, digestates of black seed oil encapsulated in starch and gelatin after oral (O-BS1 and O-BS2), gastric (G-BS1 and G-BS2), and intestinal (I-BS1 and I-BS2) in vitro digestion were analyzed using nanoparticle-tracking analysis. The concentration and size distribution of particles, including the oil droplets, in the size range of about 50 to 1000 nm were assessed.

After oral in vitro digestion, the particle concentration in the digestate of the starch-encapsulated black seed oil (O-BS1) was almost twice as high as that in the gelatin-encapsulated oil (O-BS2) (Figure 1A). Additionally, the particles from O-BS1 were smaller (Figure 1B–D) compared to those from O-BS2, suggesting that after oral in vitro digestion, there were more and smaller oil droplets present in O-BS1, which could provide a greater surface area for lipase adsorption and lipid hydrolysis during the following gastric in vitro digestion. After gastric in vitro digestion, the particle concentration was reduced and the median particle size was increased in both samples compared to each sample before this digestion step, as visible in Figure 1A,B. This could mean that during gastric in vitro digestion, oil droplets merged and formed bigger and fewer droplets. When comparing G-BS1 to G-BS2, the median particle size did not differ significantly (Figure 1B), but the size distribution profile clearly showed that much bigger particles were present in G-BS2 (see Figure 1C,D).

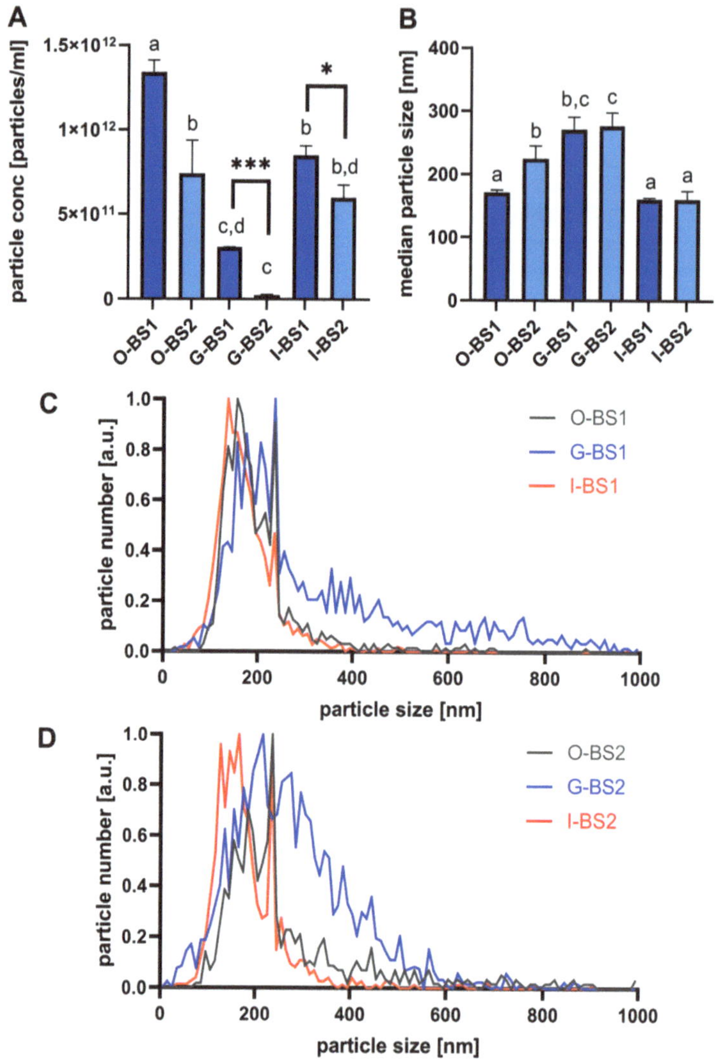


**Figure 1.** Particle concentration (particles/mL) (**A**) and median particle size (nm) (**B**) in the digestates of starch-encapsulated black seed oil (BS1) or gelatin-encapsulated black seed oil (BS2) after oral (O-BS), gastric (G-BS), or intestinal (I-BS) in vitro digestion. Particle size distribution in the different digestates of BS1 (**C**) and BS2 (**D**), respectively. Different letters indicate statistically significant differences among the samples analyzed by one-way ANOVA ($p < 0.05$). Asterisks indicate statistically significant differences between the indicated samples analyzed by student's *t*-test (*** = $p < 0.001$, * = $p < 0.05$).

In addition, the particle concentration was still higher in G-BS1 when directly compared to G-BS2 (Student's *t*-test, $p < 0.001$). Gelatin might not favor emulsification of the oil at the gastric stage, leading to reduced hydrolysis due to limited access of lipase to the triacylglycerols. The degree of emulsification might be the determining factor for the slower digestion of gelatin-encapsulated oil. Starch seemed to show suitable emulsifier properties during simulated gastric digestion. Therefore, the difference in the particle concentration, as well as the different size distribution profiles, suggests that the starch-encapsulated oil contained more and smaller particles compared to the gelatin-encapsulated oil after gastric in vitro digestion. Finally, after intestinal in vitro digestion, the median particle size decreased in both samples to a similar level (Figure 1B), which could also be observed for the size distribution profiles (Figure 1C,D). In addition, the particle concentration increased with decreasing size (Figure 1A–D), meaning that the bigger particles formed during gastric digestion were digested into smaller and more homogenous particles during intestinal in vitro digestion. Nevertheless, I-BS1 still contained more particles directly compared to I-BS2 (student's *t*-test, $p < 0.05$, Figure 1A).

These data support the hypothesis that a higher surface area of oil droplets was available during oral and gastric in vitro digestion of starch-encapsulated oil compared to gelatin-encapsulated oil. This might explain the increased levels of lipid hydrolysis for starch-encapsulated oil after gastric in vitro digestion (G-BS1). Therefore, it seems plausible that the encapsulation material influenced the digestive processes via the oil droplet size. It must be highlighted that this is the first time that the effect of oil encapsulation on the progress of lipid hydrolysis during in vitro digestion has been addressed.

### *3.2. Assessment of Lipid Oxidation of Encapsulated Black Seed Oil during In Vitro Digestion by $^{1}H$ NMR*

#### 3.2.1. Changes in the Molar Percentages (%) of Black Seed Oil Acyl Groups and Fatty Acids

During in vitro digestion, the changes in the molar percentages of acyl groups might be more or less marked, depending on their unsaturation degree and the presence or absence of prooxidant or antioxidant compounds [4,30,31]. Figure 2 depicts the molar percentages (%) of different acyl groups in relation to the total moles of AG+FA present in black seed oils before in vitro digestion (BS1 and BS2) and in encapsulated black seed oil samples after gastric and intestinal in vitro digestion (G-BS1, I-BS1, G-BS2 and I-BS2). No significant differences ($p > 0.05$) were found in the molar percentages of linoleic (L), oleic (O) and saturated (S) acyl groups and fatty acids after gastric and intestinal in vitro digestion compared to the oils before digestion. Moreover, no significant changes were observed due to capsule material. The lack of differences indicates that if oxidation has taken place, this has not produced variations in the concentration of the different kinds of acyl groups + fatty acids at a level detectable by means of $^{1}H$ NMR spectroscopy.

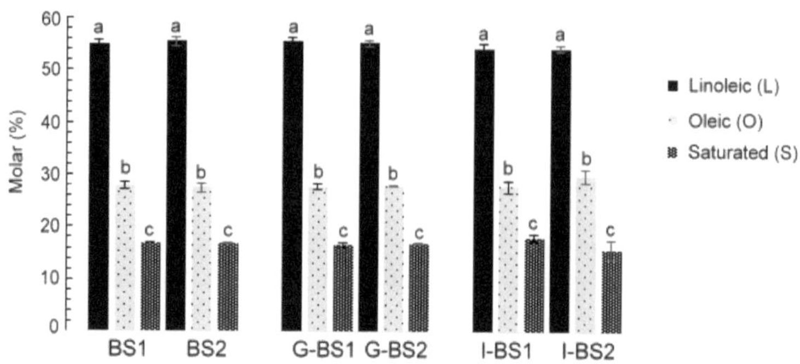


**Figure 2.** Molar percentage of linoleic, oleic and saturated fatty acids (FA) plus acyl groups (AG) (FA+AG) in relation to the total moles of all kinds of AG and FA (FA+AG) in black seed oils 1 and 2 (BS1 and BS2) and in the digestates of these oils after gastric (G-BS1 and G-BS2) and intestinal (I-BS1 and I-BS2) digestion. Different letters indicate statistically significant differences among the samples analyzed by one-way ANOVA ($p < 0.05$).

3.2.2. Formation of n-Alkanals

As has been described in several studies before, under in vitro digestion conditions, vegetable oil might undergo degradation and therefore some oxidation compounds might be generated [4–6,8]. It is well known that during the lipid oxidation process, lipid hydroperoxides are formed. These primary oxidation compounds are intermediate compounds in the oxidation process and can evolve to form secondary oxidation compounds, such as aldehydes. During the in vitro digestion of black seed oil, n-alkanals were detected, indicating that oxidation took place (Figure 3). The main increase in the concentration of n-alkanals occurred after gastric in vitro digestion. Moreover, the lipid extract of the digestates of all samples after intestinal in vitro digestion presented a higher oxidation degree than those after gastric in vitro digestion, showing a statistically significant ($p < 0.05$) increase in the concentration of n-alkanals.

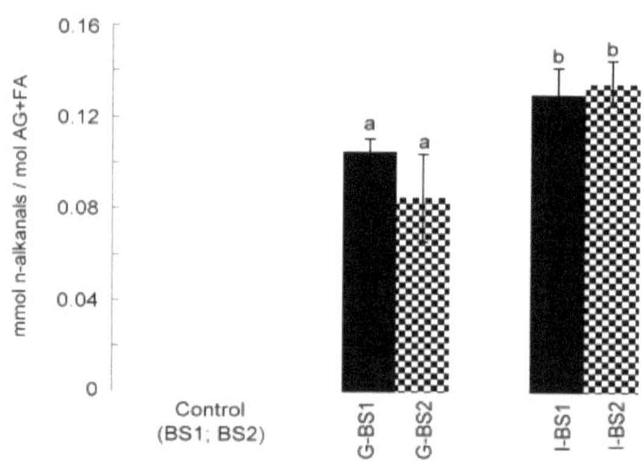


**Figure 3.** Concentration of n-alkanals expressed as mmol per mol of AG+FA in black seed oils 1 and 2 (BS1 and BS2) and in the digestates of these oils after gastric (G-BS1 and G-BS2) and intestinal (I-BS1 and I-BS2) in vitro digestion. Different letters indicate statistically significant differences among the samples analyzed by one-way ANOVA ($p < 0.05$).

### *3.3. Assessment of Lipid Oxidation of Encapsulated Black Seed Oil during In Vitro Digestion by LC-MS: Formation and Detection of Free Oxylipins*

With the purpose of corroborating and reinforcing the results obtained by $^{1}$H NMR and to obtain an overall view of the oxidation process during in vitro digestion, LC-MS was used, paying attention to the formation of several free oxylipins derived from the oxidation of omega-6 acyl fatty acids, since these are the main ones in black seed oil. These oxidation compounds can be defined as a set of structurally very diverse molecules derived from the oxidation of polyunsaturated fatty acids (PUFAs) and possessing at least one oxygenated group in their structure, in addition to the carboxylic group. It should be noted that these free oxylipins can be absorbed after digestion and play a key role in diseases such as cancer, Alzheimer's, Parkinson's, acute respiratory distress syndrome (ARDS), circulatory shock, disseminated intravascular coagulation and multiple organ failure, among others [12].

Both black seed oils, BS1 and BS2, presented the same pattern and amount of oxylipins, which indicates that these oils were already slightly oxidized. Among the oxylipins detected in the fresh oils and in the digestates, there were 9-hydroperoxy and 13-hydroperoxy octadecadienoic acids (9-HpODE and 13-HpODE), 9-hydroxy and 13-hydroxy octadecadienoic acids (9-HODE and 13-HODE), 9-oxo and 13-oxo octadecadienoic acids (9-oxoODE and 13-oxoODE), 9,10-epoxy and 12,13-epoxy octadecenoic acids, (9(10)-EpOME and 12(13)-EpOME), 12,13-dihydroxy and 9,10-dihyroxy octadecenoic acids, (9,10-DiHOME and 12,13-DiHOME), and 9,10,13-trihydroxy and 9,12,13-trihydroxy octadecenoic acids (9,10,13-TriHOME and 9,12,13-TriHOME). Table 2 shows the abundances of different classes of oxylipins, such as monohydroperoxydienes (HpODE), monohydroxydienes (HODE), monoketodienes (oxoODE), monoepoxymonoenes (EpOME), dihydroxymonoenes (DiHOME) and trihydroxymonoenes (TriHOME), in the different black seed oils and their digestates.

**Table 2.** Abundances of different classes of oxylipins in black seed oil (BS1 and BS2) and lipid extracts of the digestates after gastric (G-BS1 and G-BS2) and intestinal (I-BS1 and I-BS2) in vitro digestion, expressed as the area under the peak of the corresponding MRM transition.

| Oxylipins | BS1 | G-BS1 | I-BS1 | BS2 | G-BS2 | I-BS2 |
|---|---|---|---|---|---|---|
| HpODE | 3.00 ± 0.57 a | 61.6 ± 9.70 b | 147 ± 28.8 c | 4.47 ± 0.72 a | 24.3 ± 8.02 d | 85.6 ± 6.11 e |
| HODE | 2.25 ± 0.26 a | 27.8 ± 1.14 b | 24.5 ± 1.98 c | 2.32 ± 0.17 a | 6.19 ± 0.83 d | 5.82 ± 0.44 d |
| oxoODE | 1.14 ± 0.10 a | 10.6 ± 0.73 b | 8.72 ± 0.77 c | 1.06 ± 0.06 a | 2.92 ± 0.39 d | 2.71 ± 0.24 d |
| EpOME | 0.42 ± 0.07 a | 5.21 ± 0.22 b | 4.48 ± 0.41 c | 0.46 ± 0.05 a | 1.21 ± 0.20 d | 1.13 ± 0.09 d |
| DiHOME | 2.67 ± 0.40 a | 13.1 ± 1.47 b | 9.23 ± 0.93 c | 1.10 ± 0.10 d | 1.83 ± 0.09 ad | 1.19 ± 0.09 d |
| TriHOME | 0.28 ± 0.09 a | 5.89 ± 0.66 b | 1.98 ± 0.36 c | 0.25 ± 0.03 a | 1.28 ± 0.14 d | 0.63 ± 0.06 e |
| Total | 9.76 ± 1.50 a | 124 ± 13.9 b | 196 ± 33.2 c | 9.68 ± 1.13 a | 37.7 ± 9.69 d | 97.0 ± 7.04 e |

Different letters within each row indicate a significant difference among the samples analyzed by one-way ANOVA followed by Tukey's post-hoc test ($p < 0.05$). BS1 and BS2: black seed oils before digestion; G-BS1 and B-BS2: encapsulated black seed oil samples after gastric in vitro digestion; I-BS1 and I-BS2: encapsulated black seed oil samples after intestinal in vitro digestion.

As mentioned above, both fresh black seed oils (BS1 and BS2) already showed a basal oxidation level (Table 2). Although no significant changes were observed in the acyl groups + fatty acid concentrations before digestion and after gastric and intestinal in vitro digestions, an increase in the oxidation level with respect to their control was observed in the lipid extract of all samples. However, the abundance of the oxidation compounds formed varied depending on the digestion phase and the shell material. During gastric in vitro digestion, the highest increase of oxylipins was observed in G-BS1, with HpODE and HODE forming at the highest level. G-BS1 showed a 4-fold higher level of oxylipins than G-BS2, suggesting that the oxidation was more pronounced in the oil encapsulated in starch than in gelatin. This could be explained, on the one hand, by the fact that BS1 contained an already higher concentration of FFA than BS2 (Table 1). Moreover, as commented before, G-BS1 digestate presented higher hydrolysis levels, therefore showing a higher concentration of FFA than G-BS2. This higher concentration in FFA could make the digestates more prone to oxidation, since FFAs are more susceptible to oxidation than triacylglycerols [32]. On the other hand, G-BS1 presented a higher surface area for oil droplets than G-BS2. This finding might favor the presence of more FFA at the surface where oxidation is considered to occur.

Likewise, the formation of oxylipins increased after intestinal digestion in both I-BS1 and I-BS2 digestates. However, differences in the level of oxidation were found between the digestates. Thus, I-BS1 presented a 2-fold higher oxidation level than I-BS2, corroborating that the lipids of oil encapsulated in starch capsules are more susceptible to oxidation. The presence of some of these oxylipins, such as HpODE, HODE, oxoODE and EpOME, have been described previously in the digestates of corn oil enriched with α-tocopherol by $^1$H NMR [4]. However, as far as we know, this is the first time that LC-MS has been used for the identification and semi-quantification of free oxylipins in oils subjected to in vitro digestion. Moreover, no studies have demonstrated so far the effect of capsule materials on the oxidative stability of edible oils during in vitro digestion, nor that gelatin could be a superior capsule material in comparison with starch to limit digestion-induced oxidative deterioration of lipids.

Overall, it should be pointed out that gelatin capsules showed better protection against oxidation, which is of great importance not only for food safety and quality but also for human health. The enhanced formation of free oxylipins during the in vitro digestion of oils encapsulated in starch material might lead to higher bioaccessibility and availability for absorption, promoting pathophysiological conditions.

### 3.4. *In Vitro Assessment of the Bioaccessibility of Some Minor Compounds Present in Black Seed Oil*

As is well known, during in vitro digestion, not only the main components present in the oils are degraded but also the minor compounds (tocopherols, sterols, terpenes) might be degraded or transformed [5,6,31,32]. Black seed oils contain minor compounds such as tocopherols, tocotrienols, sterols, terpenes and thymoquinone [26,33,34]. This study focused on the degradation and bioaccessibility of tocols (γ-tocopherol, α-tocopherol, β-tocotrienol), sterols (Δ7-avenasterol, cholestanol and esters of cycloartenol), and thymoquinone, since they have important biological activities [35–39]. Some of them give signals in the $^1$H NMR spectra, which do not overlap with any others, enabling their quantification with this method [6,25,26], and others, such as tocols, can be quantified by means of HPLC-UV [8], allowing the estimation of their evolution and bioaccessibility during in vitro digestion. The bioaccessibility of these compounds was calculated as the ratio between their concentration in the digestate ($C_D$) and their concentration in the oil before digestion ($C_O$); bioaccessibility = $[C_D]/[C_O]$.

#### 3.4.1. In Vitro Bioaccessibility of Tocols Present in Black Seed Oil

As presented in Table 3, the main tocols in BS1 were γ-tocopherol and β-tocotrienol followed by α-tocopherol, while BS2 showed lower concentrations of both γ-tocopherol and α-tocopherol. During in vitro digestion, these compounds can be subjected to degradation [4,5,30–32]. Their bioaccessibility depends mainly on its initial concentration in foods, the type of fluids used and/or the conditions during digestion, among other factors [4,5,30,31].

**Table 3.** Concentration of tocols expressed in μg/g oil and of some sterols and thymoquinone given in molar percentage in relation to the total moles of AG+FA present in black seed oils (BS1 and BS2) and in the digestates after gastric (BS1-G and BS2-G) and intestinal (BS1-I and BS2-I) in vitro digestion.

| Minor Compounds | BS1 | G-BS1 | I-BS1 | BS2 | G-BS2 | I-BS2 |
|---|---|---|---|---|---|---|
| α-Tocopherol (μg/g oil) | 92.3 ± 3.19 a | 64.7 ± 12.8 b | - | 54.2 ± 9.74 b | 18.8 ± 3.22 c | - |
| γ-Tocopherol (μg/g oil) | 388 ± 77.4 a | 318 ± 19.0 a | 243 ± 26.0 b | 112 ± 14.7 c | 117 ± 8.72 c | 37.2 ± 8.50 d |
| β-Tocotrienol (μg/g oil) | 363 ± 12.8 a | 326 ± 12.3 a | 243 ± 29.2 b | 394 ± 35.8 a | 334 ± 17.6 a | 226 ± 13.7 b |
| Δ7-Avenasterol (mmol/mol AG + FA) | 0.10 ± 0.00 a | 0.10 ± 0.01 a | 0.07 ± 0.01 b | 0.14 ± 0.01 c | 0.14 ± 0.00 c | 0.09 ± 0.01 d |
| Esters of cycloartenol (mmol/mol AG + FA) | 0.29 ± 0.01 a | 0.28 ± 0.01 a | 0.27 ± 0.02 a | 0.32 ± 0.02 a | 0.28 ± 0.01 a | 0.25 ± 0.02 a |
| Sitostanol (mmol/mol AG + FA) | 0.05 ± 0.00 a | 0.04 ± 0.01 a | - | 0.05 ± 0.00 a | 0.05 ± 0.00 a | - |
| Thymoquinone (mmol/mol AG + FA) | 1.75 ± 0.01 a | 1.26 ± 0.06 b | 1.14 ± 0.06 c | 2.77 ± 0.04 d | 1.88 ± 0.11 e | 1.66 ± 0.18 f |

Different letters within each row indicate a significant difference among the samples analyzed by one-way ANOVA followed by Tukey's post-hoc test ($p < 0.05$). BS1 and BS2: black seed oils before digestion; G-BS1 and B-BS2: encapsulated black seed oil samples after gastric in vitro digestion; I-BS1 and I-BS2: encapsulated black seed oil samples after intestinal in vitro digestion; -: not detected.

The bioaccessibility of α-tocopherol after gastric in vitro digestion was more pronounced the higher the concentration in the fresh oil, showing values around 70% for G-BS1 and 35% for BS2. However, this tocol was not detectable in either intestinal digestate (I-BS1 and I-BS2), suggesting that it would not be able to be absorbed. Regarding the γ-tocopherol after intestinal digestion, a bioaccessibility of 63% and 33% was observed for I-BS1 and I-BS2, respectively, again indicating that a higher concentration of tocols in the fresh oil led to a lower degradation and a higher bioaccessibility of tocols. This tendency is in agreement with a recently published study [4]. Likewise, the β-tocotrienol concentration remained almost stable during the gastric in vitro digestion, although it was partially degraded after intestinal in vitro digestion, indicating that up to around 67% (I-BS1) and 57% (I-BS2) of it remained in the digestates without degrading and being available for absorption. Overall, tocols present in the black seed oil encapsulated in gelatin (BS2) showed a higher degradation rate than BS1, although as above-commented, BS2 was less oxidized during in vitro digestion. This could be explained by the fact that mainly during gastric digestion, BS2 suffered from a low degree of hydrolysis with the release of low amounts of FFA. The low abundance of FFA at the surface might facilitate the passage of oxygen. Thus, the oxygen could have more access to the oil droplets where the lipophilic tocols might have been located, thereby promoting their oxidation.

3.4.2. In Vitro Bioaccessibility of Other Minor Compounds Present in Black Seed Oil

Using $^1$H NMR Δ7-avenasterol, sitostanol, and the esters of cycloartenol and thymoquinone were quantified in BS1 and BS2 and in their corresponding digestates. Sterols were quantified using the area of the signals at 0.54 ppm (Δ7-avenasterol), at 0.58 ppm (esters of cyloartenol) and at 0.65 ppm (sitostanol) and thymoquinone at 6.58 ppm based on the assignment given in previous studies [7,25,26]. As can be observed in Table 3, the concentrations of some sterols and thymoquinone were slightly higher in BS2 than in BS1, showing statistically significant differences for Δ7-avenasterol and thymoquinone. The concentration of the esters of cycloartenol remained unchanged during in vitro digestion, which means that this sterol is fully bioaccessible after both gastric and intestinal in vitro digestion. These results are in accordance with those shown in a previous study during the in vitro digestion of olive oil [6]. However, both Δ7-avenasterol and sitostanol, as well as thymoquinone, were degraded during the in vitro digestion of BS1 and BS2. As Table 3 shows, the concentration of Δ7-avenasterol remained stable during gastric digestion in both the BS1 and BS2 digestates. However, a significant degradation was observed for this compound after intestinal digestion, I-BS1 and I-BS2, resulting in a bioaccessibility of around 70%. Likewise, significant changes in the thymoquinone concentration were observed after gastric and intestinal digestion, with around 60% remaining bioaccessible after in vitro digestion. This fact is of great importance since these compounds have been related to several health benefits [35,36]. However, as can be observed in Table 3, sitostanol is totally degraded after intestinal digestion. Thus, overall, no significant differences were observed in the degradation or bioaccessibility of the minor compounds due to the capsule material. This section may be divided into subheadings. It should provide a concise and precise description of the experimental results and their interpretation, as well as the experimental conclusions that can be drawn.

## 4. Conclusions

This study evidenced for the first time that capsule material has an impact on the extent of lipolysis and oxidative status of encapsulated black seed oil during in vitro digestion processes. An increase in the level of lipid hydrolysis for the starch-encapsulated oil after gastric in vitro digestion compared to the gelatin-encapsulated oil was observed, although no significant differences were found after intestinal in vitro digestion. On the one hand, this effect could be explained by the fact that gelatin might have a higher surface activity than starch and lipase during gastric in vitro digestion. Therefore, the lipolytic enzyme cannot easily access the triacylglycerols in the oil since gelatin could create an interfacial layer around the oil droplets, preventing the enzyme from approaching the triacylglycerols. On the other hand, the difference in lipid hydrolysis might result from the influence of the encapsulation material on the oil droplet size and, therefore, the surface area accessible to digestive enzymes. The latter hypothesis was supported by the analysis of the lipid droplets, which proved to be smaller and had a higher concentration in the digestates of the starch-encapsulated oil. Likewise, it has been shown for the first time that during in vitro digestion of encapsulated black seed oil, gelatin capsules showed better protection against oxidation, which is of great importance not only for food safety and quality but also for human health. The enhanced formation of free oxylipins during in vitro digestion of oils encapsulated in starch material might lead to higher bioaccessibility and absorption, promoting pathophysiological conditions. Overall, no significant differences were observed in the degradation or bioaccessibility of minor compounds naturally present in the oil due to capsule material. Therefore, this study evidenced that gelatin could be a better option than starch for the encapsulation and stabilization of edible oils. It not only provides novel mechanistic results regarding the formation of potentially toxic oxylipins in food supplements during in vitro digestion, thereby advancing science in this field but also clear recommendations for the food supplement industry regarding the use of capsule materials to limit the extent of lipid oxidation.

**Supplementary Materials:** The following supporting information can be downloaded at: https://www.mdpi.com/article/10.3390/antiox12010191/s1, composition of the juices employed in the in vitro digestion model, Table S1: Chemical shift assignments of $^{1}$H NMR signals of protons of acyl groups (AG), fatty acids (FA), and acylglycerols (MG: monoacylglycerols. DG: diacylglycerols. TG: triacylglycerols), alkanals and some sterols in CDCl3, Table S2: Free oxylipins detected by LC-MS based on previous publications [28,40], Figure S1: $^{1}$H NMR spectra of black seed oils 1 and 2 (BS1 and BS2) before the digestion, and the of the lipid extract from the digestates after gastric (G) and intestinal (I) in vitro digestion (G-BS1, I-BS1, G-BS2 and I-BS2), for the region between 3.6 and 4.4 ppm The signals are in agreement with those of Table S1.

**Author Contributions:** J.A.-C.: Conceptualization, Methodology, Formal analysis, Investigation, Data curation, Writing—original draft & editing, Visualization. M.A.: Methodology, Formal analysis, Investigation, Data curation, Writing—original draft. A.M.-R.: Methodology, Formal analysis, Data curation, Writing—original draft & review. A.S.: Writing—review & editing. M.P.: Conceptualization, Resources, Writing—review & editing, Methodology, Supervision, Project administration, Funding acquisition. All authors have read and agreed to the published version of the manuscript.

**Funding:** This work was kindly funded by the Austrian Science Fund (FWF): P 33636. J. Alberdi-Cedeño thanks the Eusko Jaurlaritza—Gobierno Vasco (EJ-GV) for a postdoctoral grant (POS_2020_1_0040).

**Institutional Review Board Statement:** Not applicable.

**Informed Consent Statement:** Not applicable.

**Data Availability Statement:** Data are contained within the article or Supplementary Material.

**Acknowledgments:** The authors want to thank the NMR core facility of the Division of Pharmaceutical Chemistry at the University of Vienna for providing access to the NMR instruments. Open Access Funding by the Austrian Science Fund (FWF).

**Conflicts of Interest:** The authors declare no conflict of interest.

## References


1. Guillen, M.D.; Goicoechea, E. Formation of oxygenated $\alpha$, $\beta$-unsaturated aldehydes and other toxic compounds in sunflower oil oxidation at room temperature in closed receptacles. *Food Chem.* **2008**, *111*, 157–164. [CrossRef]
2. Martínez-Yusta, A.; Guillén, M.D. Monitoring compositional changes in sunflower oil-derived deep-frying media by $^1$H Nuclear Magnetic Resonance. *Eur. J. Lipid Sci. Technol.* **2016**, *118*, 984–996. [CrossRef]
3. Nieva-Echevarría, B.; Goicoechea, E.; Guillén, M.D. Food lipid oxidation under gastrointestinal digestion conditions: A review. *Crit. Rev. Food Sci. Nutr.* **2020**, *60*, 461–478. [CrossRef] [PubMed]
4. Alberdi-Cedeño, J.; Ibargoitia, M.L.; Guillén, M.D. Effect of the enrichment of corn oil with *alpha*- or *gamma*-tocopherol on its *in vitro* digestion studied by $^1$H NMR and SPME-GC/MS; formation of hydroperoxy-, hydroxy-, keto-dienes and keto-*E*-epoxy-*E*-monoenes in the more *alpha*-tocopherol enriched samples. *Antioxidants* **2020**, *9*, 246. [CrossRef] [PubMed]
5. Alberdi-Cedeño, J.; Ibargoitia, M.L.; Guillén, M.D. A global study by $^1$H NMR spectroscopy and SPME-GC/MS of the in vitro digestion of virgin flaxseed oil enriched or not with mono-, di-or tri-phenolic derivatives. Antioxidant efficiency of these compounds. *Antioxidants* **2020**, *9*, 312. [CrossRef]
6. Alberdi-Cedeño, J.; Ibargoitia, M.L.; Guillén, M.D. Study of the in vitro digestion of olive oil enriched or not with antioxidant phenolic compounds. Relationships between bioaccessibility of main components of different oils and their composition. *Antioxidants* **2020**, *9*, 543. [CrossRef]
7. Alberdi-Cedeño, J.; Ibargoitia, M.L.; Guillén, M.D. Oxylipins associated to current diseases detected for the first time in the oxidation of corn oil as a model system of oils rich in omega-6 polyunsaturated groups. A global, broad and in-depth study by $^1$H NMR spectroscopy. *Antioxidants* **2020**, *9*, 544. [CrossRef]
8. Fruehwirth, S.; Zehentner, S.; Salim, M.; Sterneder, S.; Tiroch, J.; Lieder, B.; Pignitter, M. In vitro digestion of grape seed oil inhibits phospholipid-regulating effects of oxidized lipids. *Biomolecules* **2020**, *10*, 708. [CrossRef]
9. Esterbauer, H.; Schaur, R.J.; Zollner, H. Chemistry and biochemistry of 4-hydroxynonenal, malonaldehyde and related aldehydes. *Free Radic. Biol. Med.* **1991**, *11*, 81–128. [CrossRef]
10. Guillén, M.D.; Goicoechea, E. Toxic oxygenated $\alpha$, $\beta$-unsaturated aldehydes and their study in foods: A review. *Crit. Rev. Food Sci. Nutr.* **2008**, *48*, 119–136. [CrossRef]
11. Deol, P.; Fahrmann, J.; Yang, J.; Evans, J.R.; Rizo, A.; Grapov, D.; Salemi, M.; Wanichthanarak, K.; Fiehn, O.; Phinney, B.; et al. Omega-6 and omega-3 oxylipins are implicated in soybean oil-induced obesity in mice. *Sci. Rep.* **2017**, *7*, 12488. [CrossRef] [PubMed]
12. Ramsden, C.E.; Domenichiello, A.F.; Yuan, Z.-X.; Sapio, M.R.; Keyes, G.S.; Mishra, S.K.; Gross, J.R.; Majchrzak-Hong, S.; Zamora, D.; Horowitz, M.S.; et al. A systems approach for discovering linoleic acid derivatives that potentially mediate pain and itch. *Sci. Sig.* **2017**, *10*, eaal5241. [CrossRef]
13. Schulze, M.B.; Minihane, A.M.; Saleh, R.N.M.; Risérus, U. Intake and metabolism of omega-3 and omega-6 polyunsaturated fatty acids: Nutritional implications for cardiometabolic diseases. *Lancet Diabetes Endocrinol.* **2020**, *8*, 915–930. [CrossRef] [PubMed]
14. da Silva, M.V.; Santos, M.R.C.; Silva, I.R.A.; Viana, E.B.M.; Dos Anjos, D.A.; Santos, I.A.; de Lima, N.G.B.; Wobeto, C.; Jorge, N.; Lannes, S.C.D.S. Synthetic and natural antioxidants used in the oxidative stability of edible oils: An overview. *Food Rev. Int.* **2021**, *38*, 349–372. [CrossRef]
15. Timilsena, Y.P.; Adhikari, R.; Barrow, C.J.; Adhikari, B. Digestion behaviour of chia seed oil encapsulated in chia seed protein-gum complex coacervates. *Food Hydrocoll.* **2017**, *66*, 71–81. [CrossRef]
16. Fu, J.; Song, L.; Liu, Y.; Bai, C.; Zhou, D.; Zhu, B.; Wang, T. Improving oxidative stability and release behavior of docosahexaenoic acid algae oil by microencapsulation. *J. Sci. Food Agric.* **2020**, *100*, 2774–2781. [CrossRef]
17. Mu, H.; Song, Z.; Wang, X.; Wang, D.; Zheng, X.; Li, X. Microencapsulation of algae oil by complex coacervation of chitosan and modified starch: Characterization and oxidative stability. *Int. J. Biol. Macromol.* **2022**, *194*, 66–73. [CrossRef]
18. Abang, S.; Chan, E.S.; Poncelet, D. Effects of process variables on the encapsulation of oil in ca-alginate capsules using an inverse gelation technique. *J. Microencapsul.* **2012**, *29*, 417–428. [CrossRef]
19. Hur, S.J.; Lim, B.O.; Decker, E.A.; McClements, D.J. In vitro human digestion models for food applications. *Food Chem.* **2011**, *125*, 1–12. [CrossRef]

20. Nieva-Echevarría, B.; Goicoechea, E.; Manzanos, M.J.; Guillén, M.D. A study by $^1$H NMR on the influence of some factors affecting lipid in vitro digestion. *Food Chem.* **2016**, *211*, 17–26. [CrossRef]
21. Augustin, M.A.; Sanguansri, L. Challenges in developing delivery systems for food additives, nutraceuticals and dietary supplements. In *Encapsulation Technologies and Delivery Systems for Food Ingredients and Nutraceuticals*; Woodhead Publishing: Sawston, UK, 2012; pp. 19–48.
22. Mun, S.; Decker, E.A.; McClements, D.J. Influence of emulsifier type on *in vitro* digestibility of lipid droplets by pancreatic lipase. *Food Res. Int.* **2007**, *40*, 770–781. [CrossRef]
23. Sathasivam, T.; Muniyandy, S.; Chuah, L.H.; Janarthanan, P. Encapsulation of red palm oil in carboxymethyl sago cellulose beads by emulsification and vibration technology: Physicochemical characterization and in vitro digestion. *J. Food Eng.* **2018**, *231*, 10–21. [CrossRef]
24. Versantvoort, C.H.; Oomen, A.G.; Van de Kamp, E.; Rompelberg, C.J.; Sips, A.J. Applicability of an *in vitro* digestion model in assessing the bioaccessibility of mycotoxins from food. *Food Chem. Toxicol.* **2005**, *43*, 31–40. [CrossRef] [PubMed]
25. Ruiz-Aracama, A.; Goicoechea, E.; Guillén, M.D. Direct study of minor extra-virgin olive oil components without any sample modification. 1H NMR multisupression experiment: A powerful tool. *Food Chem.* **2017**, *228*, 301–314. [CrossRef] [PubMed]
26. Goryainov, S.V.; Khromov, A.V.; Bakureza, G.; Cesar, E.; Ivlev, V.A.; Vorobyev, A.N.; Abramovich, R.A.; Potanina, O.G.; Novikov, O.O. Results of a comparative study of Nigella Sativa L. seeds oils composition. *Sci. Pract. J.* **2020**, *8*, 29–39. [CrossRef]
27. Nieva-Echevarría, B.; Goicoechea, E.; Manzanos, M.J.; Guillén, M.D. A method based on $^1$H NMR spectral data useful to evaluate the hydrolysis level in complex lipid mixtures. *Food Res. Int.* **2014**, *66*, 379–387. [CrossRef]
28. Emami, S.; Zhang, Z.; Taha, A.Y. Quantitation of oxylipins in fish and algae oil supplements using optimized hydrolysis procedures and ultra-high performance liquid chromatography coupled to tandem mass-spectrometry. *J. Agric. Food Chem.* **2020**, *68*, 9329–9344. [CrossRef]
29. Nieva-Echevarría, B.; Goicoechea, E.; Manzanos, M.J.; Guillén, M.D. $^1$H NMR and SPME-GC/MS study of hydrolysis, oxidation and other reactions occurring during in vitro digestion of non-oxidized and oxidized sunflower oil. Formation of hydroxy-octadecadienoates. *Food Res. Int.* **2017**, *91*, 171–182. [CrossRef]
30. Martin-Rubio, A.S.; Sopelana, P.; Guillén, M.D. The key role of ovalbumin in lipid bioaccessibility and oxidation product profile during the in vitro digestion of slightly oxidized soybean oil. *Food Funct.* **2019**, *10*, 4440–4451. [CrossRef]
31. Nieva-Echevarría, B.; Goicoechea, E.; Guillén, M.D. Effect of adding *alpha*-tocopherol on the oxidation advance during *in vitro* gastrointestinal digestion of sunflower and flaxseed oils. *Food Res. Int.* **2019**, *125*, 108558. [CrossRef]
32. Choe, E.; Min, D.B. Mechanisms and factors for edible oil oxidation. *Compr. Rev. Food Sci. Food Saf.* **2006**, *5*, 169–186. [CrossRef]
33. Matthaus, B.; Özcan, M.M. Fatty acids, tocopherol, and sterol contents of some Nigella species seed oil. *Czech J. Food Sci.* **2011**, *29*, 145–150. [CrossRef]
34. Mazaheri, Y.; Torbati, M.; Azadmard-Damirchi, S.; Savage, G.P. A comprehensive review of the physicochemical, quality and nutritional properties of Nigella sativa oil. *Food Rev. Int.* **2019**, *35*, 342–362. [CrossRef]
35. Ermumcu, M.Ş.K.; Şanlıer, N. Black cumin (*Nigella sativa*) and its active component of thymoquinone: Effects on health. *Food Health* **2017**, *3*, 170–183. [CrossRef]
36. Hannan, M.A.; Rahman, M.A.; Sohag, A.A.M.; Uddin, M.J.; Dash, R.; Sikder, M.H.; Rahman, M.S.; Timalsina, B.; Munni, Y.A.; Sarker, P.P.; et al. Black cumin (*Nigella sativa L.*): A comprehensive review on phytochemistry, health benefits, molecular pharmacology, and safety. *Nutrients* **2021**, *13*, 1784. [CrossRef] [PubMed]
37. Imran, M.; Rauf, A.; Khan, I.A.; Shahbaz, M.; Qaisrani, T.B.; Fatmawati, S.; Abu-Izneid, T.; Imran, A.; Rahman, K.U.; Gondal, T.A. Thymoquinone: A novel strategy to combat cancer: A review. *Biomed. Pharmacother.* **2018**, *106*, 390–402. [CrossRef]
38. Abraham, A.; Kattoor, A.J.; Saldeen, T.; Mehta, J.L. Vitamin E and its anticancer effects. *Crit. Rev. Food Sci. Nutr.* **2019**, *59*, 2831–2838. [CrossRef]
39. Zhang, T.; Liu, R.; Chang, M.; Jin, Q.; Zhang, H.; Wang, X. Health benefits of 4,4-dimethyl phytosterols: An exploration beyond 4-desmethyl phytosterols. *Food Funct.* **2020**, *11*, 93–110. [CrossRef]
40. Dufour, C.; Loonis, M. Regio- and stereoselective oxidation of linoleic acid bound to serum albumin: Identification by ESI–mass spectrometry and NMR of the oxidation products. *Chem. Phys. Lipids* **2005**, *138*, 60–68. [CrossRef]

*antioxidants*

MDPI

*Article*

# Green Carbon Dots as Additives of Biopolymer Films for Preserving from Oxidation of Oil-Based Products

**Sandra Rodríguez-Varillas †, Clarissa Murru †, Marta Elena Díaz-García and Rosana Badía-Laíño ***

Department of Physical and Analytical Chemistry, Faculty of Chemistry, University of Oviedo, 33006 Asturias, Spain

* Correspondence: rbadia@uniovi.es

† These authors contributed equally to this work.

**Abstract:** The deterioration of oil-based products during processing, distribution and storage has a major negative impact on the industry from an economic point of view. The spoilage of oil is mainly due to its oxidation which can be triggered by various factors, such as UV light, heating or the presence of impurities that result in the formation of radical species. In this context, several packaging alternatives have recently been developed with the aim to protect and extend the shelf life of oil-based products. This work aimed to study the antioxidant properties of bio-polymer-based films (BPFs) obtained from high methoxylated pectin (HMP) and sodium caseinate (CAS) and enriched with different concentrations of green carbon dots (gCDs), 0.25%, 0.50 and 1% *w*/*w*, obtained from apple pomace (APCDs) and rosemary powder (RCDs). The resulting films (gCDs-BPFs) have shown that the presence of gCDs not only modified the surface roughness of the films, but also positively affected their antioxidant properties. The addition of gCDs enhanced the radical inhibiting capacity of the raw BPFs by 42 and 62% for the films containing 1% RCDs and 1% APCDs, respectively. As a proof of the concept, two oil samples (edible and cosmetic) were treated with the obtained antioxidant films, and the results demonstrated that in both types of samples the oxidation process was minimized during the five days of the experiment. These results are promising and suggest that the antioxidant bio-polymer-based films could be excellent candidates for further production of active packaging.

**Keywords:** active packaging; composite films; nano-additives; green carbon dots; antioxidant properties

## 1. Introduction

Oxidation is a process common to all types of lipids and involves a series of chemical reactions that degrade their quality and intrinsic value. The deteriorative changes caused by oxidation include unpleasant flavors, reduction in nutritional value and health properties of edible oils and lipids, but also a decrease in the physical and chemical properties for those oils with technical and industrial machine applications (industrial lubricants, hydraulic oils, etc.), thus shortening, in all cases, the useful life of oils and lipids [1,2].

Oxidation of edible oils (as well as many cosmetic products containing vegetable essential oils) can occur through three mechanisms: autoxidation, photo-oxidation, and enzyme-catalyzed oxidation. Autoxidation is a chain process that involves, during the initiation step, the formation of alkyl radicals mediated by the activation of oxygen by an energy source (light or heat) and/or by a catalytic compound such as transition metals (e.g., iron). These alkyl radicals initiate lipid oxidation and, simultaneously, a propagation phase is started with the formation of highly reactive hydroperoxide type radicals, $LOO^{\bullet}$. The process continues until a final termination step is reached where stable non-radical compounds are formed. Photooxidation requires the presence of photosensitizers (e.g., chlorophyll, riboflavin) and light for the generation of singlet oxygen responsible for reaction with unsaturated fatty acids to produce hydroperoxides. Finally, enzymatic oxidation initiates by cleavage of unsaturated fatty acids by lipoxygenase, the main enzyme involved in the process. The active site of lipoxygenase contains iron in ferrous form, responsible to trigger the oxidation process forming alkyl radicals. The oxidation of oils and lipids is a complex one, and the main reactions and factors involved in each have been documented and reviewed in recent papers [3,4].

It is known that prolonged exposure to UV radiation can promote the oxidation of oil-based products and shorten their useful life [5]. Therefore, to address this problem, researchers are trying to design sustainable materials capable of UV blocking as well as biodegradable packaging. With this aim, there are numerous works focused on the development of new biodegradable biopolymers for packages which incorporate natural antioxidant extracts or bioactive compounds obtained mainly from agro-industrial wastes [6–9]. On the other hand, nanotechnology is a burgeoning field of study in the manufacture of nano-based active packages; consequently, edible coatings enriched with nanoparticles have been demonstrated to be more advantageous than conventional packaging materials to preserve food products [10,11]. Nanomaterials like ZnO, $TiO_2$, $CeO_2$, $SiO_2$, and Cu nanoparticles are widely used in UV blocking applications [12–14]; however, the addition of carbon nanoparticles, such as carbon dots (CDs) as antioxidant agents, to such films and coatings is a more innovative approach to the manufacture of active films with antioxidant properties to increase not only products' shelf life but also their quality, freshness and safety. Although most papers dealing with the synthesis of CDs claim their biocompatibility and non-toxic properties, their impact on human health and on bioma is still in its infancy [15]. Nanoparticles, as metal-based as carbon-based, can modify the physical and mechanical properties of packaging polymers by improving their strength, durability, flexibility, and/or gas-barrier properties [16–20]. Proteins and polysaccharides have been widely used in the food industry as starting materials for the fabrication of films and coatings, [21–23]. Among natural materials, protein-polysaccharide based biopolymers reinforced with carbon nanomaterials have demonstrated to be promising for developing new biodegradable and edible wraps depending on their formulations [24–28].

Herein, high methoxylated pectin (HMP) and sodium caseinate (CAS) biopolymer films (BPFs) were produced based on a previous study by M. Jahromi et al. [29]. In this study, we combined the production of the films with the addition of green carbon dots (gCDs) obtained from rosemary (RCDs) and apple pomace (APCDs) to potentiate the valorization and use of industrial waste by-products. To this end, the synthesis, and the antioxidant activity of both types of gCDs and the corresponding gCD-BPFs, were investigated by the peroxide value method and the radical DPPH assay. Moreover, the optical properties of the materials were evaluated by UV and fluorescence spectrophotometry. As a proof of the concept, the gCD-BPFs were assayed as protective materials to prevent the oxidation of two model products: olive oil and cosmetic oil. The results obtained were promising as the oxidation of samples treated with gCD-BPFs was clearly minimized, thus paving the way for the use of antioxidant gCDs as active additives in the fabrication of biofilms and/or coatings that can increase the useful life of the oil-based products during their storage.

## 2. Materials and Methods

### 2.1. Chemicals

Olive Oil of the brand "Hacendado" was obtained from a Spanish local supermarket (Mercadona, Oviedo, Spain). Cosmetic Oil "Johnson&Johnson" was obtained from the local market Alimerka (Oviedo, Spain). Rosemary powder was purchased from a local supermarket (Netto, Regensburg, Germany) and depectinized apple pomace, Herbavital F12, was supplied by Herbstreith & Fox GmbH & Co. KG Pektin-Fabriken (Neuenbürg, Germany). Sodium Caseinate powder (CAS), HPLC grade methanol, ethanol, and chloroform were supplied by Merck Co. (Darmstadt, Germany). High-methoxylated pectin powder (HMP) (GENU® Pectin, 50–75% esterification) was obtained from CPKelco (Lille Skensved, Denmark) and Glycerol 99% (PlusOne Glycerol) was from GE Healthcare (Chicago, IL, USA). Glacial acetic acid (95%), potassium iodate and sodium thiosulfate were obtained from Merck Co. (Darmstadt, Germany). DPPH reactive was obtained from Sigma Aldrich (Madrid, Spain). All solutions were prepared using distilled water.

### 2.2. Instrumentation

Carbon dots were characterized using UV–visible spectroscopy with a Lambda 900 UV/VIS/NIR spectrophotometer (Perkin Elmer Spain, Madrid, Spain), while photoluminescence emission spectra were recorded on a Varian Cary Eclipse spectrofluorometer from Agilent Technologies (Las Rozas, Madrid, Spain), equipped with a xenon flash lamp. The absorbance and fluorescent spectra of the films were collected by an absolute measurement system on an FS5 fluorometer (Edinburgh Instruments, Livingston, UK) equipped with an integrating sphere module and FluOracle® software (Livingston, UK). The redox potential measurements were performed using a pH/ion-meter GLP 22 (Hach Lange Spain, S.L.U., Barcelona, Spain) with an accuracy of $\pm$ 0.2 mV. Surface characterization was carried out using High Resolution Scanning Probe Microscopy from Agilent Technologies (Las Rozas, Madrid, Spain) and WSxM software (Madrid, Spain) for data treatment. The zeta potential and the particle size distribution were measured at room temperature using a Zeta Potential Analyser (Zetasizer ZS 3600, Malvern Instruments, Malvern, UK) and Zetasizer software (Malvern, UK). The Zeta potential of water-diluted gCDs solution was evaluated after 10 min of sonication in an ultrasonic bath. All experiments were carried out in triplicate.

### 2.3. Preparation of Bio-Films Enriched with Antioxidant gCDs

The synthesis of the gCDs by a hydrothermal method and the fabrication of enriched bio-films were carried out following the general procedure described by Jing et al. [30] without introducing significant modifications: 20 g of Rosemary (or Apple Pomace) were placed in a Pyrex glass beaker, heated in an oven at 200 °C (5 h along which 5 mL of distilled water were added to prevent the product from scorching every 30 min), and finally cooled to room temperature. Then, 20 mL of distilled water were added to the obtained black powders and centrifuged at 12,000 rpm for 50 min. Any remnant solid was removed by filtration through a 0.40 µm pore paper filter and the gCDs suspension was purified using a continuous nanofiltration system (MWCO 5 kDa) for 1 h. Finally, the suspension was lyophilized and the obtained gCDs were stored in the dark in a fridge until further use. For the preparation gCDs-BPFs, 25 mL of a sodium caseinate suspension (2% $v/v$ CAS) and 75 mL of a high-methoxylated pectin suspension (2% $v/v$ HMP) were mixed adding different concentrations of gCDs as active agents (0.25%, 0.5% and 1% $v/v$), using glycerol as plasticizer (0.75% $w/w$). Each mixture was poured into a plastic Petri dish, dried for 24 h at 40 °C and finally stored in a 53% relative humidity atmosphere (saturated magnesium nitrate solution) at room temperature.

### 2.4. AFM-Based Surface Characterization

Atomic Force Microscopy (AFM) analyses were performed using a 2 cm $\times$ 2 cm piece of gCD-BPFs using non-contact mode. Roughness parameters obtained included peak-to-peak intensity values (p-p), average roughness (Ra) and root mean square roughness (Rq). Peak-to-peak distance refers to the difference between the lowest and the highest values measured in the sample area studied. Ra is the mean of the absolute values between the detected surface and the peaks while Rq is the square root of the mean of all values after squaring them [31]. All experiments were carried out in triplicate.

### 2.5. DPPH Free Radical Scavenging Assay

The DPPH free radical scavenging assay for the determination of the antioxidant activity of gCDs was realized according to the following protocol: a methanol solution of DPPH (100 µM) was prepared, and 1 mL was added to a given volume of the different gCDs dispersions. The mixture was stirred, and the reaction was left to take place in darkness for 30 min. The absorbance at 517 nm was finally measured to determine the concentration of the remaining free DPPH• radical. For the study of the antioxidant activity of the gCDs-

BPFs with 1% RCD, 5 mg of the film (about 0.20 $cm^2$) were placed in Eppendorf tubes and 1.5 mL of methanol and 0.5 mL of DPPH reagent were added. The tubes were covered with foil to allow the reaction to take place in darkness. All tests were performed in triplicate and methanol was used as blank control. The scavenging efficiency was calculated using Equation (1):

$$DPPH\ inhibition\ \% = (1 - As/Ab) \times 100 \tag{1}$$

where $Ab$ is the absorbance of the control and $As$ is the absorbance of the sample.

### 2.6. Study of Olive Oil Oxidation in Presence of RCD-BPFs

A 2 cm × 2 cm piece of RCD-BPF was introduced into a 50 mL glass beaker with approximately 41 mm diameter containing 25 g of olive oil and was in contact with gentle stirring for 15 days at 25 °C (samples). Another beaker containing the same amount of oil in the absence of the biofilm sample was subjected to the same treatment and was used as a reference (blank). Every day, 0.5 g of the oils (blank and sample) were taken and dissolved in 5 mL of chloroform to measure the fluorescence signal at 348 nm, corresponding to the band emission of the RCDs, and at 322 nm, corresponding to the band emission of oil. All experiments were carried out in triplicate.

### 2.7. Olive and Cosmetic Oils Peroxide Value Determination

Peroxide Value method (PV) allowed to determine the iodine released by the oxidation of peroxides in the presence of potassium iodide. The iodine was volumetrically analyzed using a standard solution of sodium thiosulfate and the final point was potentiometrically determined using a pH/ion-meter equipped with a combined Pt/Ag/AgCl indicator electrode and a AgCl as the reference electrode. The thiosulfate solution was added using a burette. The PV of oil samples treated with gCD-BPFs was performed according to the following procedure: 36 glass tubes were filled with olive oil and placed in an oven, half of them containing a sample of a film with a size of 2 cm × 2 cm and a weight of 0.2 g. The remaining tubes were used as blank. The samples were heated for 5 days at 40 °C to allow a slow formation of reactive radical species. To evaluate the PV, 10 g of each oil sample was weighed and dissolved in 50 mL of a mixture of $Cl_3CH$/glacial acetic acid (3:2 $v/v$). Then, 0.5 mL of a saturated solution of KI aqueous solution was added. The mixture was gently stirred up to the finish of the analysis. Measurements were performed after each addition of 0.0050 M sodium thiosulfate. Similar operations were performed for the blank solution. The PV was calculated using Equation (2):

$$PV = \frac{(V - V_0) \times C_t \times 1000}{w} \tag{2}$$

where $V$ and $V_0$ were the volumes of the sodium thiosulfate standard solution added at the final point to the analyzed oil sample and blank solution, respectively. $C_t$ is the molar concentration of the standard solution of sodium thiosulfate and $w$ is the weight (g) of the oil sample. A parallel study was carried out using cosmetic oil and following the same procedure explained above but heating the samples at 70 °C.

## 3. Results and Discussion

Optical, morphological and antioxidant properties of the gCD-BPFs were studied to evaluate their potential as active protection of oil base products. The different BPFs obtained were shown to have a homogeneous structure (See the photographs inserted in Figure 1 and manageability.

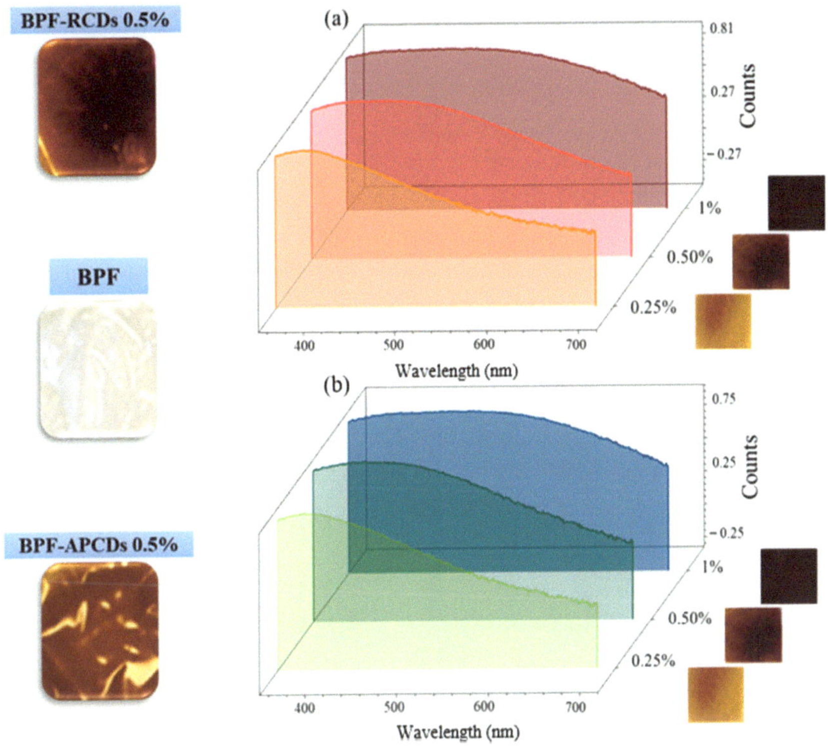


**Figure 1.** Absorption spectra and photographs of BPF and gCDs-BPFs: (**a**) APCDs, (**b**) RCDs.

### *3.1. Optical Characterization of gCDs and gCD-BPFs*

In a previous work [17] we could determine that the RCDs and APCDs were approximately spherical shape with an average diameter of 3.8 ± 0.75 nm for RCDs and 4.2 ± 0.94 nm for APCDs. The UV absorption spectra of RCDs and APCDs (Figure 2a) showed similar spectral characteristics, with a band at 280 nm corresponding to $\pi \rightarrow \pi^*$ transitions due to $sp^2$ hybridized groups (C-C double bonds) and another band at 330 nm assigned to $n \rightarrow \pi^*$ transitions due to C=O groups, C-N or C-OH in $sp^3$ hybridized bonds. There are no significant differences in the photoluminescence spectra of both types of gCDs: an emission band with a maximum intensity at 440 nm and an excitation band with a maximum intensity at 350 nm (Figure 2b).

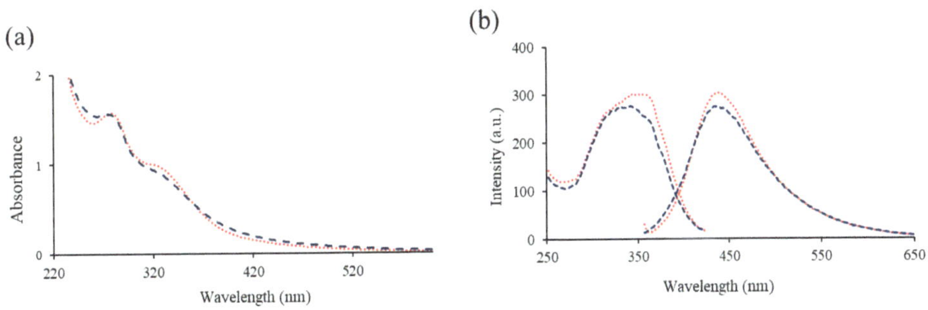


**Figure 2.** (**a**) Absorption and (**b**) excitation and emission spectra of gCDs: APCDs (...); RCDs (---).

Starting from the absorption spectra it was also possible to estimate the band gap ($E_g$) for the gCDs using Tauc's law for direct transitions (Equation (3)) [32]:

$$(\alpha h\nu)^2 = k(h\nu - E_g) \quad (3)$$

where $\alpha$ is the absorption coefficient, $h\nu$ is the energy of the incident light and $k$ is a constant. $E_g$ was estimated at the intersection point of the extrapolated straight line with the abscissa axis from the graphical representation of $(\alpha h\nu)^2$ vs. $h\nu$ (Figure 3). The band gap of gCDs was found to be 3.34 and 3.30 eV for RCDs and APCDs, respectively. These values are consistent with those in works already published [33–36]. The meaning of these data concerning the reactivity of the gCDs [37] in the context of the present paper has been discussed in detail in the Supplementary Information.

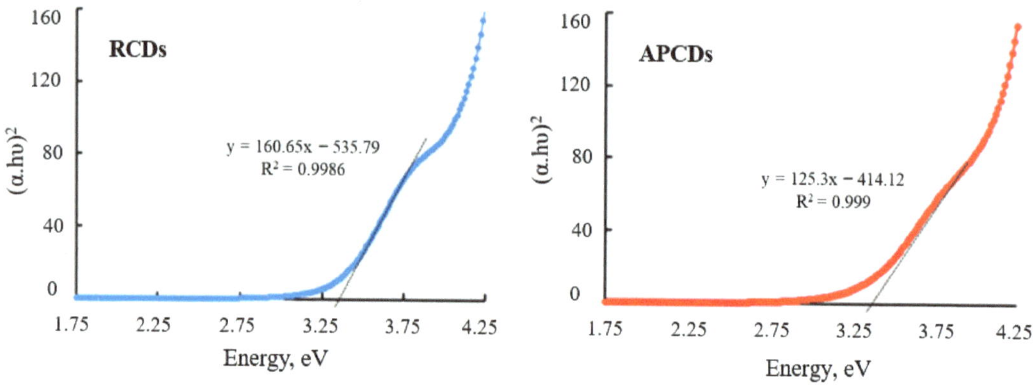


**Figure 3.** Tauc plots for estimation of the gCDs bandgap.

As can be observed in Figure 1, the RCD-BPFs with a low RCDs content (0.25%) showed a well-defined absorption band with a maximum at 420 nm, while those with a higher concentration presented broad and poorly defined absorption bands, probably due to internal filter effect. This phenomenon could be a factor to take into account when considering the possible practical applications of this type of BPF since those with the highest concentration of gCDs (0.5 and 1.0%) could act as a screen for external radiation if they were used to wrap light sensitive products (e.g., oils that oxidize and produce free radicals).

To determine the photoluminescent properties of the films, several samples of the same size (1 cm × 1 cm) were prepared and placed on top of the front-face support to record the excitation and fluorescence spectra. BPFs in the absence of gCDs were used as reference blanks. The fluorescence spectra of the films were also recorded, as seen in Figure 4. Using an excitation wavelength at 260 nm, the BPF without gCDs presented an intense fluorescence band with a maximum intensity at 325 nm, ascribed to the intrinsic fluorescence of casein [38]. BPF excitation at 348 nm, the maximum of the excitation band of gCDs, produced only a weak emission band at 425 nm. It was observed also that gCD-BPFs, when excited at 348 nm, also presented very weak emission bands at 438 nm despite the presence of gCDs. This fact could be explained by the presence of two fluorescent systems in gCD-BPFs that interact through a deactivation redox process: photoinduced electron transfer (PET). The possible mechanism for PET is proposed in the scheme inserted in Figure 4: when the gCDs were excited at their wavelength of excitation maximum, an electron was promoted from the valence band to the conduction band. In the nano-environment surrounding the gCDs, the electron may be transferred to casein and/or pectin, both materials with antioxidant capacity [39,40], thus preventing the electron its return to the valence band and, therefore, the fluorescence emission of the gCDs-BPFs seriously decreased. This process has been explained in terms of HOMO-LUMO energies [20,41,42] (Supplementary Information).

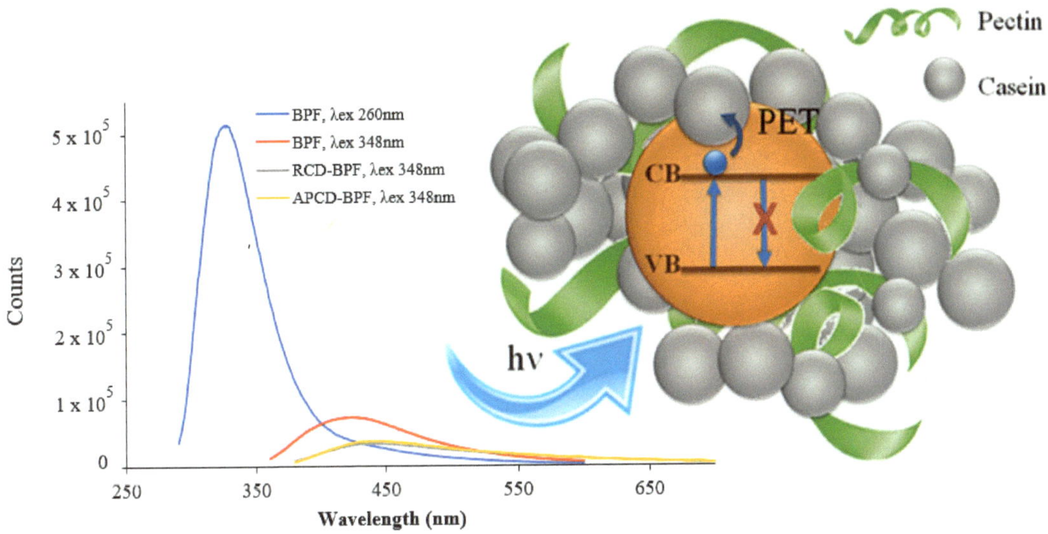


**Figure 4.** Fluorescence emission spectra of BPF (blue λex = 260 nm, red, λex = 348 nm) and gCD-BPFs containing 0.25% RCDs (gray, λex = 348 nm) and 0.25% APCDs (orange, λex = 348 nm). Inset: gCD-BPFs fluorescence deactivation through a PET process.

Functional Characterization of gCDs and gCD-BPFs

In order to obtain a characterization of functional groups in the nanomaterials, ATR-FTIR spectra were obtained. As can be seen in Figure 5, the region with the most information on the FTIR spectra BPFs and gCD-BPFs is the one that ranges between 900 and 1800 $cm^{-1}$. This region is important for defining the degree of esterification of pectin: the bands at 1735 $cm^{-1}$ and 1612 $cm^{-1}$ can be assigned to the stretching of the C=O in the methyl-ester group and to the asymmetric stretching of the carbonyl group in the unesterified carboxylate of pectin [43], respectively. Other detected bands related to the structural conformation of glycosidic bonds in pectin were those observed in the 1100–920 $cm^{-1}$ region: the band at 1140 $cm^{-1}$ is characteristic of O-C-O asymmetric stretching, while the band at 1020 $cm^{-1}$ is assigned C–O stretching and C–C stretching [44]. The FTIR spectrum of RCDs presented a broad band at 3200–3600 $cm^{-1}$ attributed to stretching vibrations of -O-H bonds and N-H bonds. The bands at 1590 $cm^{-1}$, 1014 $cm^{-1}$ and 1700 $cm^{-1}$ were assigned to the absorption of superficial carboxylic groups, due to vibrations of C-O bonds present in polysaccharides and to simple ketones and carboxylic groups, respectively. The band at 1520 $cm^{-1}$ observed in BPFs was assigned to amide II vibrations (stretching vibrations of CN bonds in combination with bending vibrations of NH bonds) of casein [39]. This band has practically disappeared in the gCD-BPFs spectrum. On the other hand, the band observed at 1612 $cm^{-1}$ BPFs has shifted to 1590 $cm^{-1}$ in the gCD-BPFs spectrum, probably by interaction with the RCDs of casein and pectin, respectively. The relative intensity of the film bands did not change with the presence of the RCDs. The ATR-FTIR spectra of the APCDs and 1% APCD-BPFs are shown in Figure S1 (Supplementary Materials), exhibiting a similar pattern to the RCD-based systems.

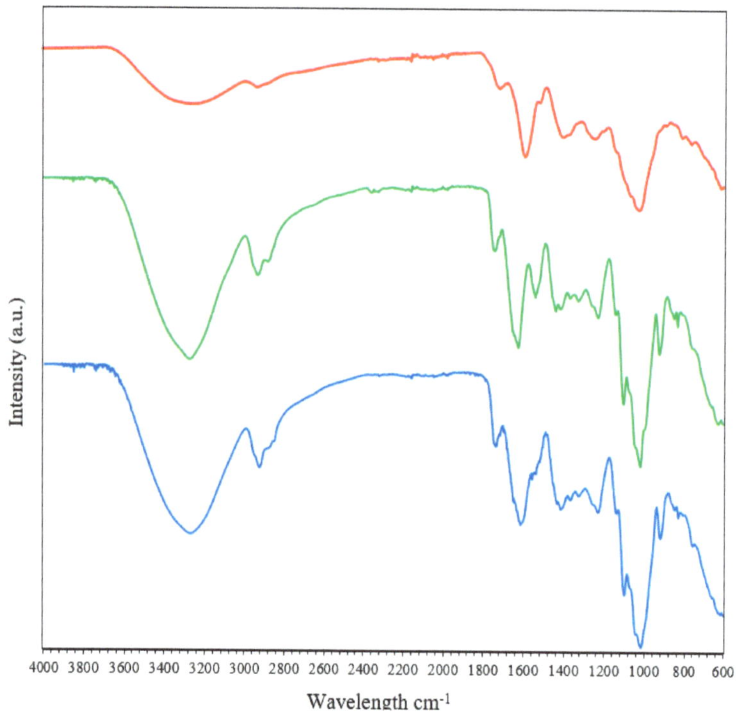


**Figure 5.** ATR-FTIR spectra of RCDs (—); BPF (—) and 1% RCD-BPF (—).

### 3.2. *Surface Characterization and Colloidal Stability of CDs*

In order to obtain a morphological characterization of the surface, studies using AFM and DLS were carried out. Based on the AFM images (Figure 6), the roughness parameters of BPFs changed when the gCDs were added and these variations depended on the type of gCDs used and their concentration. The presence of gCDs into the BPFs matrix seemed to modify their structure, filling the pores initially observed in BPFs (Figure 6d). The results obtained for average roughness (Ra), root mean square roughness (Rq) and peak-to-peak intensity values (p-p) for each sample are collected in Table 1. In comparison with the neat BPF, the roughness Ra values were lower in RCD-BPFs, independently of their concentration. These results could be explained by the RCDs filling up the BPF pores, thus resulting in a smoother BPF surface.

**Table 1.** Roughness parameters of BPFs with and without CDs reinforcement.

| Sample | Ra (nm) | Rq (nm) | p-p (nm) |
|---|---|---|---|
| BPF | 12.4 | 16.9 | 120.1 |
| RCD-BPF 0.25% | 10.6 | 13.0 | 62.2 |
| RCD-BPF 0.5% | 10.0 | 12.5 | 61.1 |
| RCD-BPF 1% | 8.1 | 10.2 | 50.0 |
| APCD-BPF 0.25% | 10.0 | 12.2 | 55.0 |
| APCD-BPF 0.5% | 14.0 | 17.2 | 77.0 |
| APCD-BPF 1% | 93.9 | 114.0 | 531.1 |

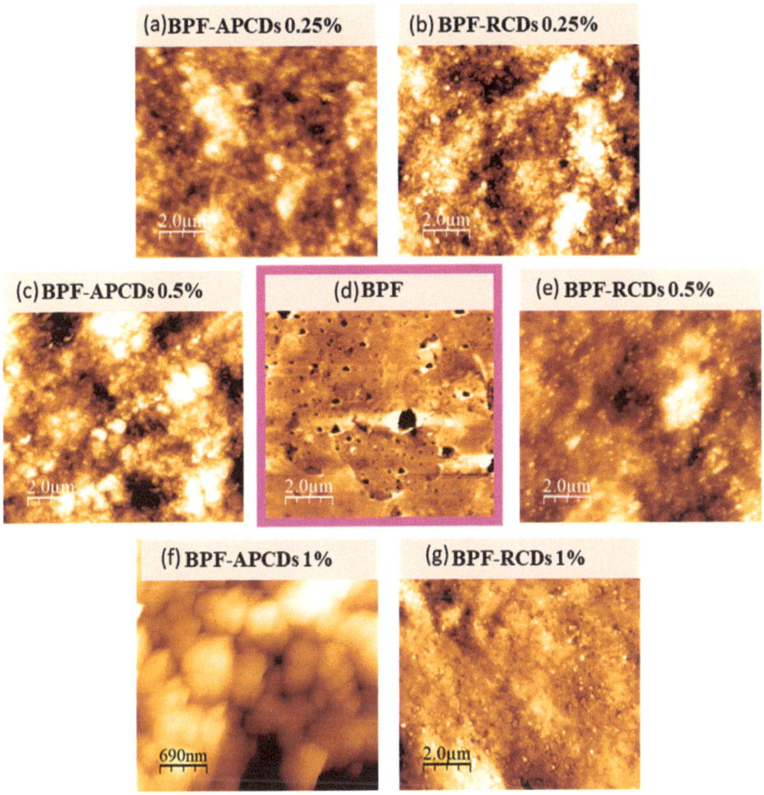


**Figure 6.** Surface images of: biofilms reinforced with (**a**) 0.25 %, (**c**) 0.5 % and (**f**) 1 % of APCDs; (**d**) biofilm without gCDs, and biofilms reinforced with (**b**) 0.25 %, (**e**) 0.5 % and (**g**) 1 % of RCDs.

However, for APCD-BPFs the value of Ra only decreased for an APCDs concentration of 0.25%, while higher concentrations raised the Ra, particularly for the 1% APCD-BPF. This effect could be ascribed to the formation of agglomerates (see magnification Figure 6f).

ELS measurements provide information about the surface charge of nanoparticles. The zeta potential (ζ) is the potential established between medium and charged nanoparticles in an aqueous solution, and it is related to colloidal stability [45]: particles with large ζ-potential (>±30 mV) means strong repulsive forces that avoid the particles to aggregate (stable dispersion), while low ζ-potential values (<±10 mV) indicate that dispersions are highly unstable, showing flocculation processes. In this study, the experimental values of ζ-potential for APCDs and RCDs were $-4 \pm 1$ mV and $-13 \pm 2$ mV, respectively. These results suggested that the RCDs suspensions were more stable than those of APCDs which, in turn, support the formation of aggregates observed in AFM images when 0.5 and 1% APCDs were added to the BPFs. Consequently, the addition of RCDs (even at high contents as 1%) to BPFs resulted in homogeneous and less rough films because coagulation and flocculation phenomena did not manifest during the formation of the RCD-BPFs. Furthermore, zeta potential values provided information about the electrical charge on the nanoparticle surface. Negative values obtained in both, RCDs and APCDs, demonstrate that their surface is negatively charged due to the presence of carboxylic acid and hydroxyl groups [46]. The more positive ζ-potential of APCDs could be explained by the probable presence of residual amino components (e.g., proteins, vitamins, etc.) [47] that may impart some positive charges to the synthesized APCDs.

### 3.3. *Antioxidant Activity of gCDs and gCD-BPFs*

Antioxidants interfere with the action of free radicals by intervening at any of the three major steps of the free radical-mediated oxidative process, viz., initiation, propagation and/or termination. DPPH assay allows for evaluation of the antioxidant activity due to the presence of an odd electron on the nitrogen atom in the DPPH molecule, forming the DPPH•. This radical can be reduced to DPPH-H by attaching an electron or a hydrogen atom. The purple colour of an ethanol DPPH• solution (absorbance at 517 nm) changes to yellow as the electron pairs off. The colour change at 517 nm is used to calculate the radical scavenging activity (% of inhibition) using Equation (1).

The results of the DPPH assay to determine the antioxidant activity of the gCDs in solution are shown in Figure 7. It can be observed that the DPPH inhibition activity of gCDs depended on their concentration. For RCDs, it was observed that the percentage of inhibition of the DPPH• increases up to a concentration of 0.2 mg/mL, over which a slight decrease was observed. In the case of APCDs, a steady growth in the percentage of inhibition of DPPH• was also observed up to a concentration of 0.5% mg/mL. For the same concentration of gCDs used, the RCDs present a higher activity; so, at a concentration of 0.2 mg/mL, the percentage of inhibition DPPH• by RCDs and APCDs was found to be 45% and 20%, respectively. Figure 8 shows the results obtained for the inhibition assays of gCD-BPFs as a function of the concentration of CDs used in their preparation. It is important to note that neat BPFs exhibit a relatively high antioxidant activity itself, with a 55% inhibition of DPPH• under the experimental conditions used. This fact could be attributed to the antioxidant activity of casein [39] and pectin [48]. It has been demonstrated that the presence of aromatic amino acids such as Trp, Tyr and Phe favour the antioxidant activity of peptides and that the indole and pyrrolidine rings in Trp and Pro, respectively, can act as hydrogen donors (via hydroxyl groups) in the deactivation of hydroxyl radicals [40,49]. It has also been studied that pectin has antioxidant properties due to contaminants (e.g., phenols) that are co-extracted with pectin [50]. The addition of gCDs notably increases the radical inhibiting capacity of the films: 62% for 1% RCD-BPFs and 42% for 1% APCD-BPFs.

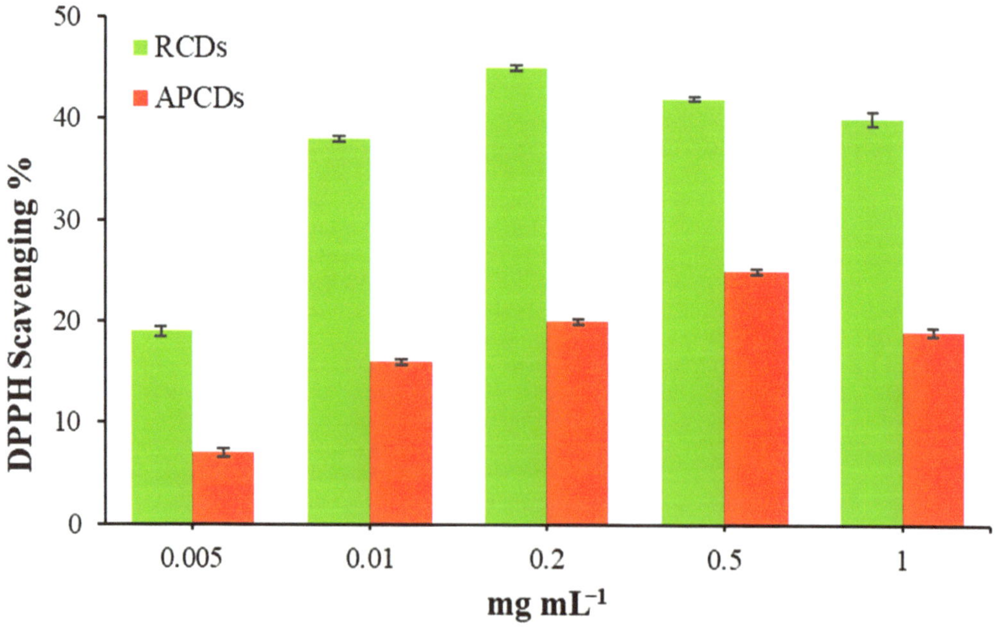


**Figure 7.** Antioxidant activity of RCDs and APCDs using the DPPH assay.

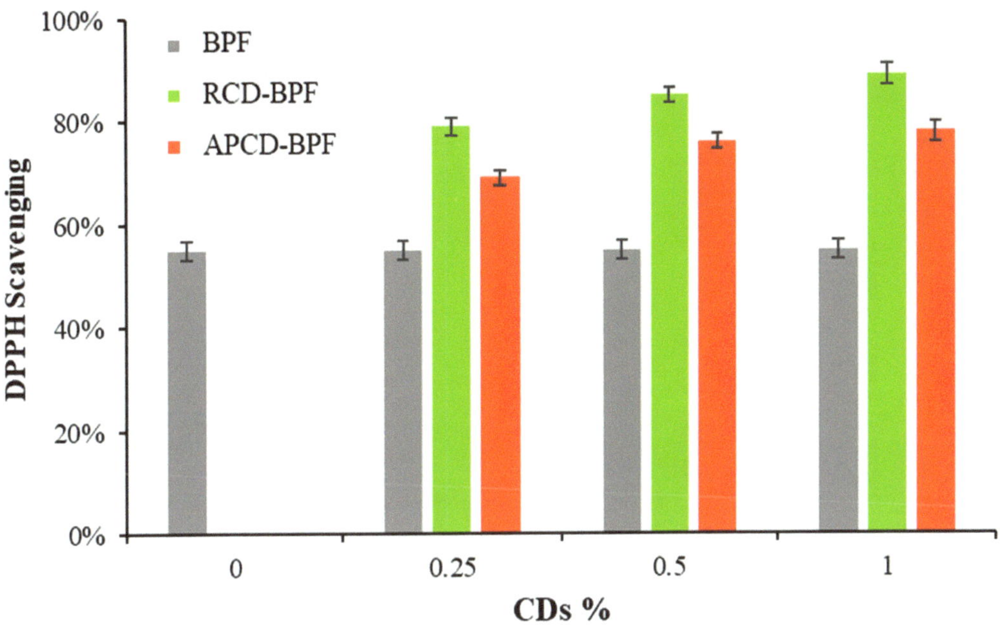


**Figure 8.** Antioxidant activity of gCDs-BPFs.

*3.4. Study of the Stability of gCD-BPFs in Oily Medium*

Since the objective of developing gCDs-BPFs was their final use as a material to produce active packaging, it was necessary to investigate their stability with respect to the leaching of the gCDs from the films towards the oily sample. Given the intrinsic fluorescence presented by gCDs dispersed in oils, the study was carried out by immersing a 2 cm × 2 cm piece of the 1% RCD-BPFs in a sample of 25 mL of commercial olive oil, with constant gently stirring during 15 days. The fluorescence intensity of the oil at the maximum excitation wavelength of the RCDs was measured every three days. Figure 9 shows the variation in the fluorescence signal along the time assayed. It was possible to observe that the emission intensity of the oil, with an emission maximum at 322 nm, was slightly higher than the oil containing the RCD-BPFs, and it remained statistically constant over time. However, the emission of the oil containing the 1% RCD-BPFs slightly, which seemed to indicate that the RCDs were not released from the film due to the hydrophobic nature of the medium and the low dispersibility of the nanoparticles in it. In Figure 9, it can also be observed that at 348 nm (maximum emission wavelength of RCDs), the trend followed a similar pattern although the difference in emission intensities between the neat oil and the oil containing the 1% RCD-BPFs was smaller. A plausible explanation for the decrease in fluorescence intensity would be related to the presence of water/humidity in the oil containing the 1% RCD-BPFs: the films were stored under a relative humidity of 57% and it was plausible that during the test, low traces of water were released forming reverse micelles modifying the polarity of the medium [4]. A parallel study was carried out using cosmetic oil and the results, similar to those described for olive oil, are collected in Figure S2 (Supplementary Materials).

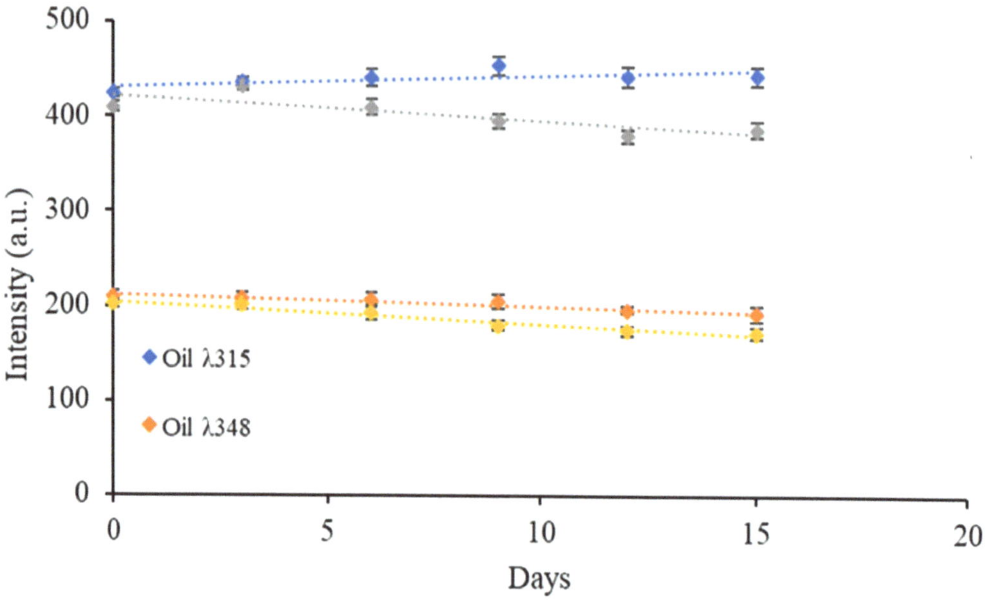


**Figure 9.** Study of the stability of RCDs contained in RCD-BPFs in contact with olive oil.

### 3.5. *Proof of Concept*

#### 3.5.1. Influence of CD-BPFs on the Peroxide Value of Edible and Cosmetic Oils

Over the years, different methods have been developed to determine the degree of oxidation of oils (and fats), both for industrial and human or animal consumption. Among these methods, the Peroxide Value (PV) is one of the main indicators used to evaluate the formation of hydroperoxides. PV is an indicator used to evaluate the amount of hydroperoxides formed in terms of milliequivalents per g of oil, which oxidize potassium iodide under the standard conditions of the text [51]. In order to verify the efficacy of the gCDs-BPFs to decrease and/or to stop the initial oxidation process, the PV of the oils (edible and cosmetic) was determined in the presence and absence of gCDs-BPFs using the iodometric titration by potentiometric determination of the endpoint by signing the protocol described in the Materials and Methods section. To carry out this experiment, only the films that contained a higher percentage of gCDs (1% gCD-BPFs) were used, as they turned out to have the highest antioxidant potential as demonstrated by the DPPH test. The results are collected in Figure 10 where it can be observed that the most significant changes in PV took place during the first days of the experiment. In the absence of films, there was an increase in PV during the first 2 days due to the oxidation of the oil, followed by a rapid decrease in PV, probably due to the termination of the free radical chain reactions. Similar behavior was observed in the measurements of the cosmetic oil samples, although the initial PV was significantly lower than in olive oil. Still, it rose continuously during the time tested. Samples containing 1% gCD-BPFs have shown that the PV in olive oil began to decrease from day one, suggesting that the films deactivated the peroxides generated in these early stages of the oxidation process. In cosmetic oil, the PV remained at a low value throughout the study time, indicating that the 1% gCD-BPFs stabilized the oil due to the consumption of the peroxides generated.

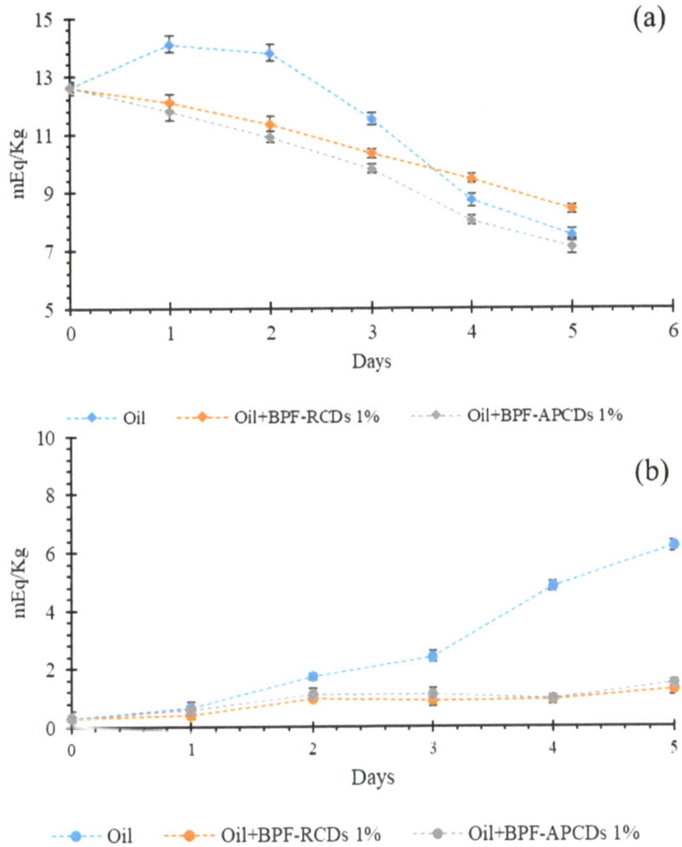


**Figure 10.** Study of the formation of radical species in samples of olive oil (**a**) and cosmetic oil (**b**) treated with antioxidant 1% gCD-BPFs.

3.5.2. Potential Applications of Bio-Polymer Films in Wet Food Samples

In order to check whether the gCD-BPFs developed offered practical potential use as antioxidant wrappers in products with a high humidity, apples were selected as they oxidize very quickly. Four bio-polymer films were tested: BPF, gCD-BPFs 0.25%, gCD-BPFs 0.5%, and gCD-BPFs 1%. The films were cut into similarly sized circular sections and placed on the surface of half apples, as shown in Figure 11. The samples were kept at room temperature of 25 °C for 90 min. The films were then gently removed after 45 min, and it was observed that those areas where they were placed were paler than the rest of the fruit surface. Removal of the gCD-BPFs from the apple surface was difficult due to their hydrophilicity. Once the films were removed, the apples were allowed to continue to oxidize for another 45 min and it could be observed that those zones previously covered with the gCD-BPFs continued being oxidized, indicating that the gCDs were not released on the surface of the wet fruit. At first glance, it was not possible to assess the contribution that the presence and/or nature of gCDs had on this behaviour. Likewise, the problem of the stability of neat BPFs films in aqueous media is still to be solved [9,52]. However, the results allow foreseeing that the system could have a good performance in products with a high degree of humidity, once the stability is achieved.

**Figure 11.** Oxidation of an apple treated with BPF and gCD-BPFs.

## 4. Conclusions

In this work, the optical and antioxidant properties of gCDs obtained from rosemary leaves (RCDs) and apple pomace leaves (APCDs) have been investigated. Some gCDs have been used as additives for the fabrication of active biopolymer-based films, gCD-BPFs. Both types of gCDs have fluorescent and antioxidant properties, being the RCDs the ones with the highest antioxidant activity against the radical DPPH•. BPFs also showed antioxidant activity (due to casein and pectin), which increased with the addition of gCDs. Through AFM surface characterization it was observed that BPFs reinforced with RCDs were less rough than biofilms containing APCDs, as the latter showed a greater tendency to form aggregates at high concentrations as demonstrated by DLS measurements. It was possible to demonstrate, by determining the peroxide value, that gCD-BPFs decreased the oxidation process of edible and cosmetic oils. However, films in contact with edible products with high water content were destabilized due to their hydrophilicity. In order to expand the potential of BPFs and gCD-BPFs as active antioxidant materials as wrappers for products with high water content, it will be mandatory in the future to delve into their stabilization and optimization of physical properties.

**Supplementary Materials:** The following supporting information can be downloaded at: https://www.mdpi.com/article/10.3390/antiox11112193/s1, Figure S1: ATR-FTIR spectra of gCDs, Figure S2: Study of the stability of RCDs contained in RCD-BPFs in contact with edible oil, Tauc plots: Implications of the RCDs band-gap value and HOMO-LUMO analysis.

**Author Contributions:** Conceptualization, R.B.-L. and M.E.D.-G.; methodology, C.M. and S.R.-V.; formal analysis C.M., S.R.-V. and R.B.-L.; investigation, C.M.and S.R.-V.; writing—original draft preparation, C.M. and S.R.-V.; writing—review and editing, R.B.-L. and M.E.D.-G.; supervision, R.B.-L. and M.E.D.-G.; project administration, R.B.-L.; funding acquisition, R.B.-L. All authors have read and agreed to the published version of the manuscript.

**Funding:** This research was funded by the Ministerio de Ciencia, Innovación y Universidades (MCIU), Agencia Estatal de Investigación (AEI) and European Regional Development Fund (FEDER), Project # RTI2018-099756-B-I00 (MCIU/AEI/FEDER, UE), and Foundation for the Promotion in Asturias of Applied Scientific Research and Technology (FICYT), Project # AYUD/2021/57506. Sandra Rodríguez Varillas was financially supported by The Principality of Asturias Administration under the Program Severo Ochoa for the research and education formation in the Principality of Asturias.

**Institutional Review Board Statement:** Not applicable.

**Informed Consent Statement:** Not applicable.

**Data Availability Statement:** Data is contained within the article.

**Acknowledgments:** Sandra Rodríguez Varillas wants to acknowledge the Administration of the Principality of Asturias for the grant of a Severo Ochoa fellowship. The authors acknowledge the Scientific technical services of the University of Oviedo for the measurements.

**Conflicts of Interest:** The authors declare no conflict of interest. The funders had no role in the design of the study; in the collection, analyses, or interpretation of data; in the writing of the manuscript, or in the decision to publish the results.

## References


1. Grootveld, M.; Percival, B.C.; Leenders, J.; Wilson, P.B. Potential Adverse Public Health Effects Afforded by the Ingestion of Dietary Lipid Oxidation Product Toxins: Significance of Fried Food Sources. *Nutrients* **2020**, *12*, 974. [CrossRef] [PubMed]
2. Cen, H.; Morina, A.; Neville, A. Effect of lubricant ageing on lubricants' physical and chemical properties and tribological performance; Part I: Effect of lubricant chemistry. *Ind. Lubr. Tribol.* **2018**, *7*, 385–392. [CrossRef]
3. Domínguez, R.; Pateiro, M.; Gagaoua, M.; Barba, F.J.; Zhang, W.; Lorenzo, J.M. A Comprehensive Review on Lipid Oxidation in Meat and Meat Products. *Antioxidants* **2019**, *8*, 429. [CrossRef]
4. Murru, C.; Badía-Laiño, R.; Díaz García, M.E. Oxidative Stability of Vegetal Oil-Based Lubricants. *ACS Sustain. Chem. Eng.* **2021**, *9*, 1459–1476. [CrossRef] [PubMed]
5. Gore, A.H.; Prajapat, A.L. Biopolymer Nanocomposites for Sustainable UV Protective Packaging. *Front. Mater.* **2022**, *9*, 855727. [CrossRef]
6. Lombo Vidal, O.; Barros Santos, M.C.; Batista, A.P.; Franceschi Andrigo, F.; Baréa, B.; Lecomte, J.; Figueroa-Espinoza, M.C.; Gontard, N.; Villeneuve, P.; Guillard, V.; et al. Active packaging films containing antioxidant extracts from green coffee oil by-products to prevent lipid oxidation. *J. Food Eng.* **2022**, *312*, 110744. [CrossRef]
7. de Moraes Crizel, T.; de Oliveira Rios, A.; Alves, V.D.; Bandarra, N.; Moldão-Martins, M.; Flôres, S.H. Active food packaging prepared with chitosan and olive pomace. *Food Hydrocoll.* **2018**, *74*, 139–150. [CrossRef]
8. Serrano-León, J.S.; Bergamaschi, K.B.; Yoshida, C.M.P.; Saldaña, E.; Selani, M.M.; Rios-Mera, J.D.; Alencar, S.M.; Contreras-Castillo, C.J. Chitosan active films containing agro-industrial residue extracts for shelf life extension of chicken restructured product. *Food Res. Int.* **2018**, *108*, 93–100. [CrossRef]
9. Moccia, F.; Agustin-Salazar, S.; Berg, A.-L.; Setaro, B.; Micillo, R.; Pizzo, E.; Weber, F.; Gamez-Meza, N.; Schieber, A.; Cerruti, P.; et al. Pecan (Carya illinoinensis (Wagenh.) K. Koch) Nut Shell as an Accessible Polyphenol Source for Active Packaging and Food Colorant Stabilization. *ACS Sustain. Chem. Eng.* **2020**, *8*, 6700–6712. [CrossRef]
10. Ghosh, M.; Singh, A.K. Potential of engineered nanostructured biopolymer based coatings for perishable fruits with Coronavirus safety perspectives. *Prog. Org. Coat.* **2022**, *163*, 106632. [CrossRef]
11. Jafarzadeh, S.; Salehabadi, A.; Nafchi, A.M.; Oladzadabbasabadi, N.; Jafari, S.M. Cheese packaging by edible coatings and biodegradable nanocomposites; improvement in shelf life, physicochemical and sensory properties. *Trends Food Sci. Technol.* **2021**, *116*, 218–231. [CrossRef]
12. Li, M.; Li, G.; Jiang, J.; Zhang, Z.; Dai, X.; Mai, K. Ultraviolet Resistance and Antimicrobial Properties of ZnO in the Polypropylene Materials: A Review. *J. Mater. Sci. Technol.* **2015**, *31*, 331–339. [CrossRef]
13. Sani, M.A.; Maleki, M.; Eghbaljoo-Gharehgheshlaghi, H.; Khezerlou, A.; Esmaeil Mohammadian, E.; Liu, Q.; Jafari, S.M. Titanium dioxide nanoparticles as multifunctional surface active materials for smart/active nanocomposite packaging films. *Adv. Colloid Interface Sci.* **2022**, *300*, 102593. [CrossRef] [PubMed]
14. Ngang, H.P.; Ooi, B.S.; Ahmad, A.L.; Lai, S.O. Preparation of PVDF–$TiO_2$ mixed-matrix membrane and its evaluation on dye adsorption and UV-cleaning properties. *Chem. Eng. J.* **2012**, *197*, 359–367. [CrossRef]
15. Peralta Videa, J.; Sreenivasan, S.T.; Narayan, M. Influence of Carbon Quantum Dots on the Biome. *Processes* **2020**, *8*, 445. [CrossRef]
16. Dong, W.; Su, J.; Chen, Y.; Xu, D.; Cheng, L.; Mao, L.; Gao, Y.; Yuan, F. Characterization and antioxidant properties of chitosan film incorporated with modified silica nanoparticles as an active food packaging. *Food Chem.* **2022**, *373 Pt A*, 131414. [CrossRef]
17. Murru, C.; Mohammadifar, M.A.; Wagner, J.B.; Badía Laiño, R.; Díaz García, M.E. High Methoxyl Pectin and Sodium Caseinate Film Matrix Reinforced with Green Carbon Quantum Dots: Rheological and Mechanical Studies. *Membranes* **2022**, *12*, 695. [CrossRef]
18. Deshmukh, R.K.; Akhila, K.; Ramakanth, D.; Gaikwad, K.K. Guar gum/carboxymethyl cellulose based antioxidant film incorporated with halloysite nanotubes and litchi shell waste extract for active packaging. *Int. J. Biol. Macromol.* **2022**, *201*, 1–13. [CrossRef]
19. Wu, M.; Zhou, Z.; Yang, J.; Zhang, M.; Cai, F.; Lu, P. ZnO nanoparticles stabilized oregano essential oil Pickering emulsion for functional cellulose nanofibrils packaging films with antimicrobial and antioxidant activity. *Int. J. Biol. Macromol.* **2021**, *190*, 433–440. [CrossRef]
20. Murru, C.; Badía-Laíño, R.; Díaz-García, M.E. Synthesis and Characterization of Green Carbon Dots for Scavenging Radical Oxygen Species in Aqueous and Oil Samples. *Antioxidants* **2020**, *9*, 1147. [CrossRef]
21. Jamróz, E.; Kopel, P. Polysaccharide and Protein Films with Antimicrobial/Antioxidant Activity in the Food Industry: A Review. *Polymers* **2020**, *12*, 1289. [CrossRef] [PubMed]
22. Loke, X.; Chang, C.; Hou, C.; Cheng, K.; Hsieh, C. Plasma-treated polyethylene coated with polysaccharide and protein containing cinnamaldehyde for active packaging films and applications on tilapia (Orechromis niloticus) fillet preservation. *Food Control* **2021**, *125*, 108016. [CrossRef]

23. Roy, S.; Rhim, J. Preparation of pectin/agar-based functional films integrated with zinc sulfide nano petals for active packaging applications. *Colloids Surf. B* **2021**, *207*, 111999. [CrossRef] [PubMed]
24. Bealer, E.J.; Onissema-Karimu, S.; Rivera-Galletti, A.; Francis, M.; Wilkowski, J.; Salas de la Cruz, D.; Hu, X. Protein–Polysaccharide Composite Materials: Fabrication and Applications. *Polymers* **2020**, *12*, 464. [CrossRef] [PubMed]
25. Baranwal, J.; Barse, B.; Fais, A.; Delogu, G.L.; Kumar, A. Biopolymer: A Sustainable Material for Food and Medical Applications. *Polymers* **2022**, *14*, 983. [CrossRef] [PubMed]
26. Yan, F.; Xu, M.; Xu, J.; Zang, Y.; Sun, J.; Yi, C.; Wang, Y. Advances in Integrating Carbon Dots With Membranes and Their Applications. *ChemistrySelect* **2012**, *6*, 7443–7462. [CrossRef]
27. Jamróz, E.; Kulawik, P.; Kopel, P. The Effect of Nanofillers on the Functional Properties of Biopolymer-Based Films: A Review. *Polymers* **2019**, *11*, 675. [CrossRef]
28. Zhao, D.; Chung, T.S. Applications of Carbon Quantum Dots (CQDs) in Membrane Technologies: A Review. *Water Res.* **2018**, *147*, 43–49. [CrossRef]
29. Jahromi, M.; Niakousari, M.; Golmakani, M.T.; Mohammadifar, M.A. Physicochemical and structural characterization of sodium caseinate based film-forming solutions and edible films as affected by high methoxyl pectin. *Int. J. Biol. Macromol.* **2020**, *165 Pt B*, 1949–1959. [CrossRef]
30. Jing, S.; Zhao, Y.; Sun, R.; Zhong, L.; Peng, X. Facile and High-Yield Synthesis of Carbon Quantum Dots from Biomass-Derived Carbons at Mild Condition. *ACS Sustain. Chem. Eng.* **2019**, *7*, 7833–7843. [CrossRef]
31. Semnani, D. Geometrical characterization of electrospun nanofibers. In *Electrospun Nanofibers*; Woodhead Publishing: Sawston, UK, 2017; pp. 151–180. [CrossRef]
32. Zadeh, N.J.; Zarandi, M.B.; Nateghi, M.R. Optical properties of the perovskite films deposited on meso-porous $TiO_2$ by one step and hot casting techniques. *Thin Solid Film.* **2019**, *671*, 139–146. [CrossRef]
33. Cao, M.; Liu, Y.; Zhu, M.; Xia, J.; Xuan, T.; Jiang, D.; Zhou, G.; Li, H. A novel and highly stable dual-emission carbon dots-based phosphor. *J. Alloy. Compd.* **2021**, *873*, 159819. [CrossRef]
34. Sekar, A.; Yadav, R. Green fabrication of zinc oxide supported carbon dots for visible light-responsive photocatalytic decolourization of Malachite Green dye: Optimization and kinetic studies. *Optik* **2021**, *242*, 167311. [CrossRef]
35. Lakshmanan, A.; Surendran, P.; Manivannan, N.; Sathish, M.; Balalakshmi, C.; Suganthy, N.; Rameshkumar, P.; Kaviyarasu, K.; Ramalingam, G. Superficial preparation of biocompatible carbon quantum dots for antimicrobial applications. *Mater. Today Proc.* **2021**, *36 Pt 2*, 171–174. [CrossRef]
36. Rodríguez-Varillas, S.; Fontanil, T.; Obaya, Á.J.; Fernández-González, A.; Murru, C.; Badía-Laíño, R. Biocompatibility and Antioxidant Capabilities of Carbon Dots Obtained from Tomato (Solanum lycopersicum). *Appl. Sci.* **2022**, *12*, 773. [CrossRef]
37. Manolopoulos, D.E.; May, J.C.; Down, S.E. Theoretical studies of the fullerenes: C34 to C70. *Chem. Phys. Lett.* **1991**, *181*, 105–111. [CrossRef]
38. Ghisaidoobe, A.B.T.; Chung, S.J. Intrinsic Tryptophan Fluorescence in the Detection and Analysis of Proteins: A Focus on Förster Resonance Energy Transfer Techniques. *Int. J. Mol. Sci.* **2014**, *15*, 22518–22538. [CrossRef]
39. Kim, Y.-E.; Kim, J.W.; Cheon, S.; Nam, M.S.; Kim, K.K. Alpha-casein and beta-lactoglobulin from cow milk exhibit antioxidant activity: A plausible link to antiaging effects. *J. Food Sci.* **2019**, *84*, 3083–3090. [CrossRef]
40. Smirnov, V.V.; Golovchenko, V.V.; Vityazev, F.V.; Patova, O.A.; Selivanov, N.Y.; Selivanova, O.G.; Popov, S.V. The Antioxidant Properties of Pectin Fractions Isolated from Vegetables Using a Simulated Gastric Fluid. *J. Chem.* **2017**, *2017*, 5898594. [CrossRef]
41. Futera, Z. Amino-acid interactions with the Au (111) surface: Adsorption, band alignment, and interfacial electronic coupling, *Phys. Chem. Chem. Phys* **2021**, *23*, 10257–10266. [CrossRef]
42. Prabhu, D.; Sharma, S.; Prabhu, P.R.; Jomy, J.; Sadanand, R.V. Analysis of the inhibiting action of pectin on corrosion of AISI1040 dual-phase steel with ferrite–martensite and ferrite–bainite structure: A comparison in 0.5 M sulphuric acid. *Iranian Chem. Soc.* **2021**, *19*, 1109–1128. [CrossRef]
43. Szymanska-Chargot, M.; Zdunek, A. Use of FT-IR Spectra and PCA to the Bulk Characterization of Cell Wall Residues of Fruits and Vegetables Along a Fraction Process. *Food Biophys.* **2013**, *8*, 29–42. [CrossRef] [PubMed]
44. Sun, R.; Hughes, S. Extraction and physico-chemical characterization of pectins from sugar beet pulp. *Polym. J.* **1998**, *30*, 671–677. [CrossRef]
45. Karmakar, S. Chapter 5: Particle size distribution and zeta potential based on dynamic light scattering: Techniques to characterize stability and surface charge distribution of charged colloids. In *Recent Trends in Materials Physics and Chemistry*; Studium Press Pvt Ltd.: Delhi, India, 2019.
46. Hashemi, N.; Mousazadeh, M.H. Preparation of fluorescent nitrogen-doped carbon dots for a highly selective on-off detection of $Fe^{3+}$ ions in real samples. *Opt. Mater.* **2021**, *121*, 111515. [CrossRef]
47. Vendruscolo, F.; Albuquerque, P.M.; Streit, F.; Esposito, E.; Ninow, J.L. Apple Pomace: A Versatile Substrate for Biotechnological Application. *Crit. Rev. Biotechnol.* **2008**, *28*, 1–12. [CrossRef]
48. Millan-Linares, M.C.; Montserrat-de la Paz, S.; Martin, M.E. Pectins and Olive Pectins: From Biotechnology to Human Health. *Biology* **2021**, *10*, 860. [CrossRef]
49. Wang, J.; Hu, S.; Nie, S.; Yu, Q.; Xie, M. Reviews on mechanisms of in vitro antioxidant activity of polysaccharides. *Oxid. Med. Cell. Longev.* **2016**, *2016*, 1–13. [CrossRef]

50. Xu, F.; Zhang, S.; Waterhouse, G.I.; Zhou, T.; Du, Y.; Sun-Waterhouse, D.; Wu, P. Yeast fermentation of apple and grape pomaces affects subsequent aqueous pectin extraction: Composition, structure, functional and antioxidant properties of pectins. *Food Hydrocoll.* **2022**, *133*, 107945. [CrossRef]
51. AOCS Official Methods. *Peroxide Value Acetic Acid–Chloroform Method*; AOCS: Champaign, IL, USA, 1996; Volume II.
52. Lisitsyn, A.; Semenov, A.; Nasonova, V.; Polishchuk, E.; Revutskaya, N.; Kozyrev, I.; Kotenkova, E. Approaches in Animal Proteins and Natural Polysaccharides Application for Food Packaging: Edible Film Production and Quality Estimation. *Polymers* **2021**, *13*, 1592. [CrossRef]

MDPI AG
Grosspeteranlage 5
4052 Basel
Switzerland
Tel.: +41 61 683 77 34

*Antioxidants* Editorial Office
E-mail: antioxidants@mdpi.com
www.mdpi.com/journal/antioxidants

www.ingramcontent.com/pod-product-compliance
Lightning Source LLC
Chambersburg PA
CBHW041411210926
51850CB00013B/3239
* 9 7 8 3 7 2 5 8 4 1 9 8 1 *